The Molecular and Clinical Pathology of Neurodegenerative Disease

The Molecular and Clinical Pathology of Neurodegenerative Disease

Patrick A. Lewis

Jennifer E. Spillane

Academic Press is an imprint of Elsevier
125 London Wall, London EC2Y 5AS, United Kingdom
525 B Street, Suite 1650, San Diego, CA 92101, United States
50 Hampshire Street, 5th Floor, Cambridge, MA 02139, United States
The Boulevard, Langford Lane, Kidlington, Oxford OX5 1GB, United Kingdom

Library of Congress Cataloging-in-Publication Data
A catalog record for this book is available from the Library of Congress

British Library Cataloguing-in-Publication Data
A catalogue record for this book is available from the British Library

ISBN: 978-0-12-811069-0

For information on all Academic Press publications visit our website at https://www.elsevier.com/books-and-journals

Publisher: Nikki Levy
Acquisition Editor: Natalie Farra
Editorial Project Manager: Kristi Anderson
Production Project Manager: Sujatha Thirugnana Sambandam
Cover Designer: Mark Rogers

Typeset by TNQ Technologies

Contents

Preface

Neurodegenerative disorders such as Alzheimer disease, amyotrophic lateral sclerosis, and Parkinson disease, characterized by the death of neurons in the central nervous system, represent a grave challenge to healthy human aging and modern healthcare systems. Despite many decades of research, there are only a handful of therapies available for these disorders. In this textbook, the molecular basis and etiology of the most prominent forms of neurodegeneration are described and analyzed, bringing together the latest research into the causes of these disorders and the latest efforts to develop novel treatments that either slow down or halt the progress of these disorders. Looking across dementia, Parkinson disease, the prion disorders, the motor neuron diseases, Huntington chorea and multiple sclerosis, the clinical presentation, neuropathology, and molecular basis of these disorders are described—highlighting the similarities and differences in their disease pathogenesis. Although the recent history of drug development for neurodegenerative diseases has been characterized by a number of high-profile failures, recent advances in gene and immunotherapy offer hope that our increased understanding of how the brain degenerates will translate into treatments in the future.

Acknowledgments

The authors would like to thank their colleagues and students at the UCL Institute of Neurology, the Royal Free Hospital, the National Hospital for Neurology and Neurosurgery, and the University of Reading School of Pharmacy for helpful discussions and inspiration.

Particular thanks are due to Dr R. Kate Gordon and Mr James Clarke for their ongoing support and constructive criticism during the development of this textbook. Dr Lewis would like to acknowledge the contribution of Freyja Louis Eleanor Lewis and Mathilda Polly Esme Lewis, whose understanding and patience allowed him to devote time to writing. Dr Spillane would likewise like to thank Lucy Clarke for her intervention and contribution.

Acknowledgements

CHAPTER

AN INTRODUCTION TO NEURODEGENERATION

1

CHAPTER OUTLINE

1.1 WHAT IS NEURODEGENERATION?

The brain is the most complex organ in the body, the seat of cognition, and the control center for bodily function. It is made up of over 80 billion neurons and a supporting cast of tens of billions of other cells; it encapsulates much of what makes us human and is home to the knowledge, skills, and memories that define us as individuals. Small wonder then that diseases affecting this organ, and the consequences of these disorders upon human health, cognition, and function, are of great concern to society and health services across the world. Among the neurological disorders, neurodegenerative diseases, i.e., diseases that result in the death of neuronal cells in the central nervous system, have proved to be a particular challenge. Neurodegenerative diseases include a number of clinical entities, such as Alzheimer disease, Parkinson disease, Huntington chorea, and motor neuron disease. These disorders are characterized by the loss of cells in the central nervous system, coupled with the loss of specific brain function linked to the spatial distribution of cell death and the neuronal circuits impacted by degeneration. The impact that this has on health can be (and in the majority of cases, is) devastating, with a loss of both physical and mental ability resulting in decreased independence and incapacitation. This, in turn, leads to greatly increased mortality and huge costs on a personal level, for immediate family and friends, and more broadly at a societal level. This is most easily quantified in terms of healthcare costs, where some of the most common neurodegenerative disorders (Parkinson and Alzheimer diseases) have been estimated to afflict more than 7.5 million people in Europe alone, costing in excess of €110 billion [1,2] (Fig. 1.1). Worldwide, this represents a huge societal burden, excluding much of the unquantifiable damage and pain that these diseases bring to individuals and to families.

The Molecular and Clinical Pathology of Neurodegenerative Disease. https://doi.org/10.1016/B978-0-12-811069-0.00001-X

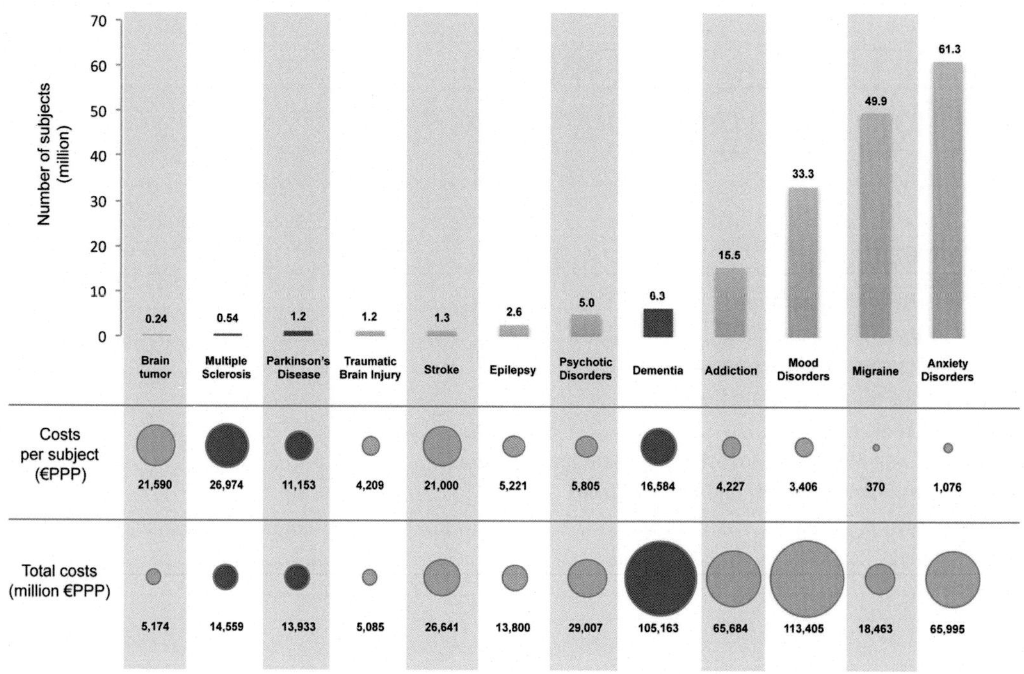

FIGURE 1.1

The estimated costs of brain diseases in Europe. The number of patients, costs per patient, and total costs are shown for a sample of the most common brain disorders. Highlighted in purple are the most common neurodegenerative disorders, indicating the high costs and disease burden inflicted by these disorders. *PPP*, purchasing power party.

Adapted from DiLuca M, Olesen J. The cost of brain diseases: a burden or a challenge? Neuron 2014;82:1205–8. Epub 2014/06/20.

The challenge presented to human health by neurodegenerative disorders is magnified by the fact that the majority of these disorders increase in incidence with age [3]. In a world where the global human population is aging, neurodegenerative disease is set to become an even bigger challenge to healthcare systems. Unfortunately, to date, this is a challenge that biomedical research has struggled to address. For symptomatic relief, there are drugs available for a subset of neurodegenerative diseases with variable clinical outcomes; for example, there are excellent drugs that combat the movement symptoms associated with Parkinson disease during the early stages of the disease (see Chapter 3 for further detail). For disease-modifying treatments, that is, a treatment that either delays or halts the progress of the disease, there are very few examples to hold up as evidence that we can change the clinical trajectory of these disorders. As such, the development of a detailed understanding of the molecular basis of these disorders—to know why cells are dying in the central nervous system—is of critical importance for drug development, as well as providing insight into the fundamental biology of the human brain. This molecular basis of neurodegenerative disease is the subject of this textbook.

1.2 HOW TO USE THIS TEXTBOOK

The goal of this textbook is to provide a detailed grounding in the molecular cause of neurodegenerative diseases. To achieve this, each chapter is divided into a number of relevant areas, including a brief historical introduction to the disorder/disorders, a description of the clinical presentation of the disease entity in question, an overview of the pathological condition observed in the brains of people with these disorders, a detailed description of our present understanding of the molecular basis of the disease, and finally an overview of the current state of therapies and drug development. It is important to emphasize at this point that this book is not intended to be a textbook of neurology or neuropathology, and for in-depth coverage of these areas the reader is directed to a number of excellent reference texts [4–6]. In order to understand the molecular basis of neurodegeneration, however, it is absolutely essential to have an overview of both the underlying changes in the brain (with regard to cellular pathology and spatial distribution) and the clinical consequences of these changes.

Although there is considerable variation in the presentation and the underlying pathological condition of neurodegenerative diseases, as well as in the molecular basis of these disorders, there are a number of key concepts and approaches that readers should be familiar with to gain most from the disease-specific chapters. This introduction aims to provide a brief overview of these concepts, and for a more detailed description of each of these topics there are excellent reviews and books available, some of which are indicated at the end of the chapter.

1.3 THE FUNDAMENTALS OF NEUROANATOMY

Central to our ability to explore how the brain goes wrong in neurodegeneration is a detailed understanding of how the brain functions as an organ and at a cellular level. This, in turn, is underpinned by insights into the organization of the brain—the study of neuroanatomy [7]. Our understanding of the structure and function of the human brain has undergone significant changes over the past several centuries (Box 1.1). Moving from early modern concepts of the brain being divided into discrete structures, we now have a very detailed understanding of the organization of the brain and how this connects outside the central nervous system to the rest of the body.

The macroscopic organization of the human central nervous system can be described in several different ways. It can be divided up into functionally and anatomically distinct macrostructures that are commonly noted as the cerebrum or cerebral hemispheres (the outer convoluted layers of the brain that are visible to the eye when a brain is removed from the skull combined with deeper structures), the diencephalon (made up of the thalamus and hypothalamus), the brain stem (consisting of the midbrain, pons, and medulla oblongata), the cerebellum at the base of the skull, and finally the spinal cord descending out of the brain and out to the rest of the body (Fig. 1.2A). Each of these structures can be further divided in several ways. The cerebral hemispheres, for example, can be functionally divided into several distinct structures: the amygdala, the hippocampus, the basal ganglia, and the cerebral cortex, which in turn can be subdivided into four lobes, the frontal, temporal, parietal, and occipital (Fig. 1.2B). These anatomical divisions relate to important functional distinctions. The hippocampus (the name of which derives from the Greek for seahorse, based on its distinctive shape), for example, plays a key role in spatial memory, which in turn allows us to navigate through our environment.

BOX 1.1 NEUROANATOMY THROUGH THE AGES

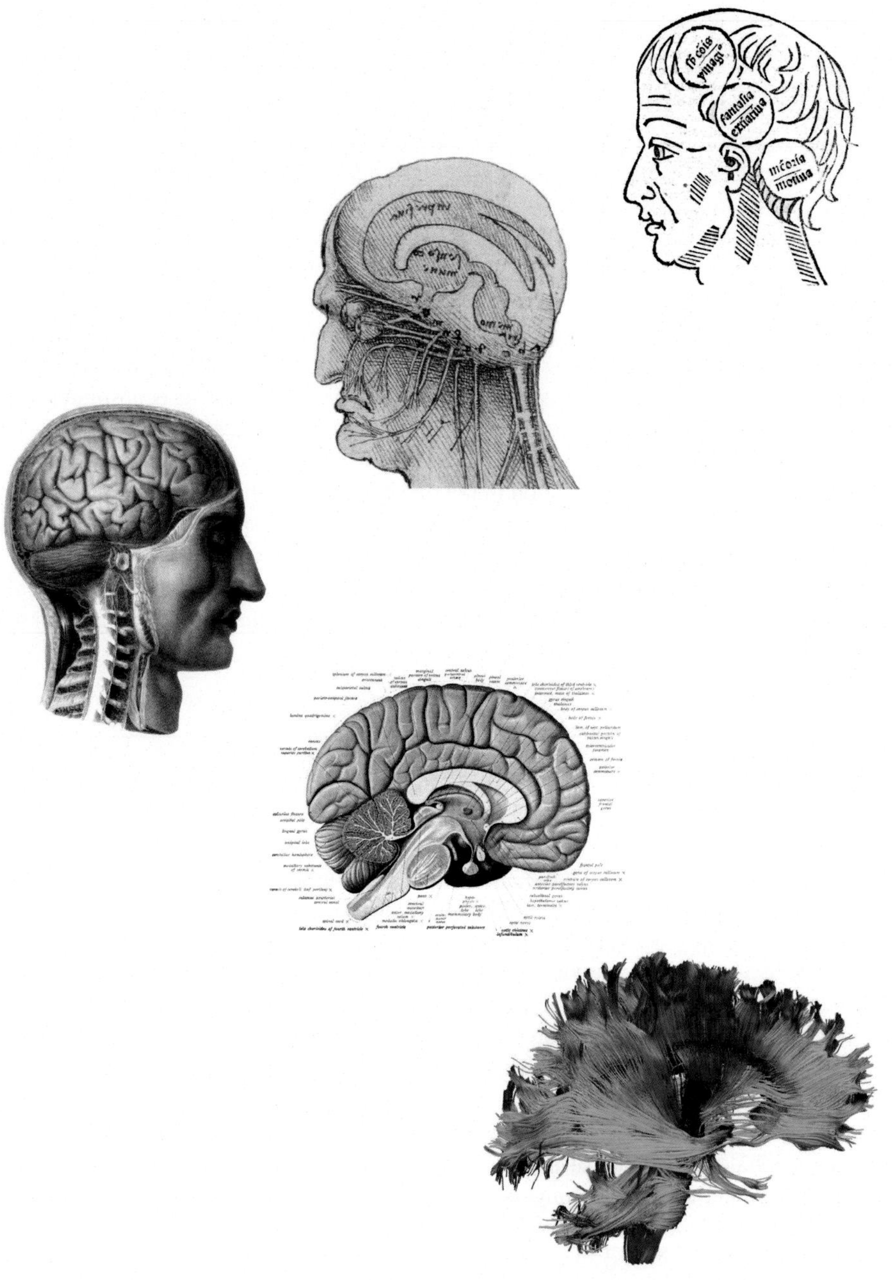

Our understanding of the brain has progressed from simple division of the brain into ventricles through more complex projections of the brain by Leonardo da Vinci and detailed anatomical descriptions to sophisticated computer-based imaging of the connections formed within the brain. Images from the top show a depiction of the brain dating from the 14th century, Leonardo da Vinci's diagram of the brain from the 16th century, a drawing of the brain by Jean-Baptiste Bourgery from the 19th century, a diagrammatic representation of some of the anatomical subdivisions of the brain from the early 20th century by Johannes Sobotta, and a connectomics map of the brain using advanced imaging techniques by Flavia Dell'Acqua. *Images re-used by permission of Creative Commons licence from the Wellcome Image library or in the public domain.*

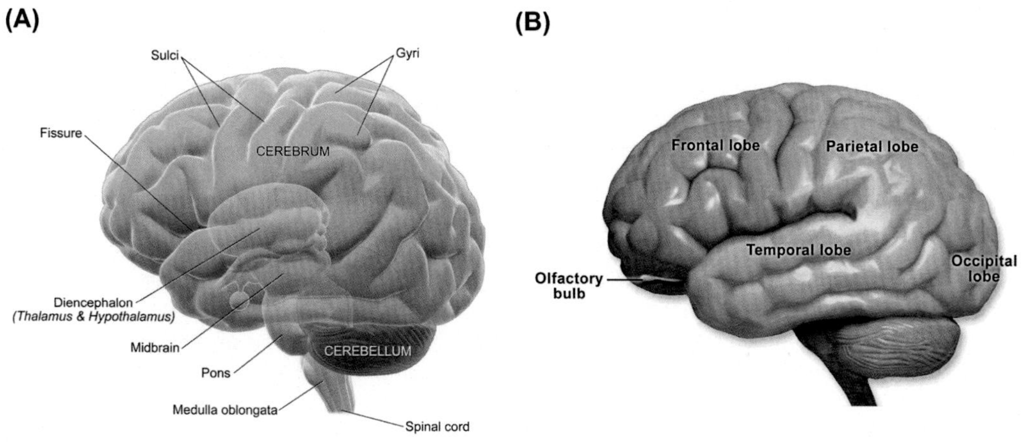

FIGURE 1.2

The neuroanatomy of the human brain. An overview of the macroscopic organization of the human brain, showing (A) the functionally and anatomically distinct major areas of the brain and (B) the division of the cerebrum into the frontal, parietal, temporal, and occipital lobes.

Images adapted from the Blausen collection using a Creative Commons Licence. Blausen.com staff. Medical gallery of Blausen Medical 2014. WikiJ Med 2014;1(2). https://doi.org/10.15347/wjm/2014.010.

A number of other terminologies and systems have been used to categorize and describe the organization of the brain at a macroscopic level. As a simple example, the brain can also be divided into three regions: the forebrain (incorporating the diencephalon and cerebrum), the midbrain, and the hindbrain (incorporating the medulla oblongata, pons, and cerebellum). Given the complexity of the central nervous system, it is perhaps unsurprising that microscopic examination of the structures of the brain reveals an intricate network of many smaller anatomically distinct units. These, again, have been the subject of a number of different systems of categorization. One of the earliest attempts to do this in a truly systematic manner was by the anatomist Korbinian Brodmann at the start of the 20th century [8]. Brodmann divided the cerebral cortex into 52 distinct areas based on the underlying cellular architecture as revealed by histochemical analysis of brain slices. This system, with some modification (e.g., subdivisions of certain Brodmann areas), is used to the present day [9].

With the advent of modern molecular analysis and imaging techniques, however, even more detailed characterization of the brain and its cellular make up has become possible. This is exemplified by efforts such as the Allen Brain Atlas initiative, which catalogues the structure, genetics, and gene expression profile of the human brain to a cellular level [10], and ongoing efforts to generate three-dimensional reconstructions of the human brain using electron microscopy to gain an exquisitely high resolution cell by cell of how the brain is constructed [11]. Increasingly, high-resolution imaging techniques can be used to look at the living brain. These include approaches such as structural and functional magnetic resonance imaging, as well as positron emission tomography. These techniques allow clinicians and researchers to peer inside the living brain and gain insights into both the structure and, in some cases, the real-time function of the central nervous system [12].

Another area where huge advances have been made is in terms of understanding the connections formed between the different anatomically distinct regions of the brain and how information flows between these [13]. Combined with computational neuroscience approaches, large-scale international efforts are ongoing to synthesize these into accurate in silico (computational) representations of the human brain [14,15]. As these efforts bear fruit, it is likely that they will provide important insights into brain function, cognition, and consciousness that are directly relevant to our understanding of neurodegenerative diseases.

A final critical area of brain biology and anatomy with great relevance to neurodegeneration is the immune status of the central nervous system. The immune system that acts to protect our body, organs, and cells from internal and external threats is a highly complex and multilayered defensive array. It comprises both innate immune mechanisms, existing within a wide range of cells and acting as a generic barrier to infection and damage, and an adaptive component that is shaped by and reacts to specific threats through the generation of specific antibodies. There is a distinction, however, between the peripheral immune system and immune function within the brain. Because of the presence of the blood–brain barrier, a selectively permeable cellular barricade between the cardiovascular system and the central nervous system, there is a physical impediment to both the components of the immune system and the potential external infectious threats such as bacteria [16]. The semipermeable nature of the blood–brain barrier also acts to control the passage of compounds into and out of the brain, a process that has important implications for drug penetrance into the brain and, because of this, drug development for neurodegeneration (see Section 1.7). Within the central nervous system, the guardian role played by circulating immune cells in the rest of the body is taken up by a specialized class of glial cells, the microglia [17]. These cells act to target and phagocytose, or engulf, unwanted invaders such as bacteria, as well as dysfunctional or damaged endogenous cells.

Understanding the structures that make up the brain and that protect the central nervous system from attack or damage is critical to comprehend the events and changes that lead to neurodegeneration and to the specific clinical presentation associated with discrete disease entities. For many of the diseases covered in this book, the symptoms with which they are associated can be linked back to localized regional degeneration in the brain—in the midbrain, impacting on movement control, for Parkinson disease or in the hippocampus, impacting on spatial memory, for Alzheimer disease.

1.4 A BEGINNER'S GUIDE TO BRAIN CELLS

As is to be expected for an organ of such exquisite complexity, the human brain consists of a large number of specialized cell types. Neurons, the key controllers of neuronal signaling within the brain, act to initiate and propagate the passage of information within the brain and to connect to the rest of the body, allowing the collection of information from the periphery and the environment, as well as the issuing of commands to the rest of the body. Neurons form a highly complex and intricate series of networks in the central nervous system and are integrated with (and supported by) a diverse range of cells collectively termed glia, derived from the Greek word for glue, the implication being that these cells act to glue the brain together [17]. The term glia is perhaps a disservice to the integral role that the different glial cell types play in the function of the brain. Their role ranges from insulating the action potentials that allow neurons to convey signals from one location to another (carried out by oligodendrocytes in the central nervous system and by Schwann cells in the periphery) to acting as guardians of

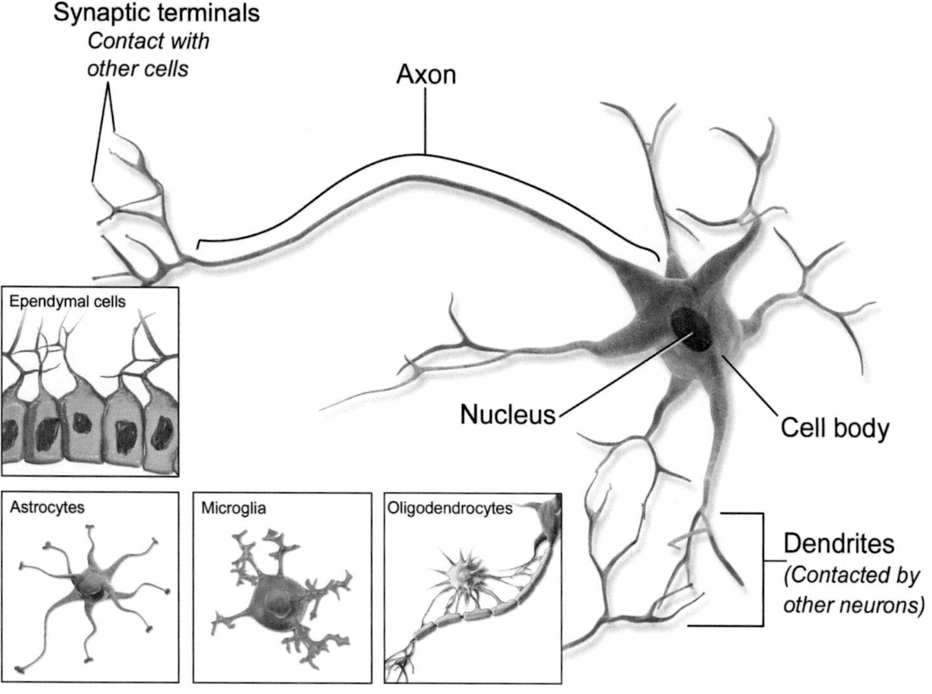

FIGURE 1.3

Brain cells: neurons and glia. Representations of an archetypal neuron, showing the cell body, dendrites, axon, and synaptic terminals and the four major classes of glial cells found within the human brain.

Images adapted from the Blausen collection using a Creative Commons Licence. Blausen.com staff. Medical gallery of Blausen Medical 2014. WikiJ Med 2014;1(2). https://doi.org/10.15347/wjm/2014.010.

the brain, fighting off invaders and infection (a role partly fulfilled by microglia). An overview of the different classes of human brain cell types is displayed in Fig. 1.3.

It has been estimated that the adult human brain is made up of in excess of 80 billion neurons, with a similar number of other cells types; however, given the sheer number of cells involved, there is, understandably, a range of estimates [18,19]. While possessing a stereotypical organization with a cell body and a variety of processes extending out from this body, including dendrites to receive signals from other cells and an axon projecting onward to pass on these signals to other cells, neurons display a quite startling diversity in the human central nervous system. The sheer range of neuronal cell types in the vertebrate nervous system became clear by the work of two titans of the neuroscience field: Camillo Golgi and Santiago Ramón y Cajal, working in the late 19th century Italy and Spain, respectively. Taking advantage of newly developing staining techniques, they were able to look at brain cells with a level of resolution, and with extraordinary results, that hitherto had not been possible (as shown in Box 1.2) [20]. Modern microscopic methods and staining techniques have yielded ever-increasing numbers of structurally and functionally distinct neurons, exemplified by the Allen Brain Atlas and the web resource NeuroMorpho [10,21]. Encompassing this diversity, there are a number of basic divisions

BOX 1.2 CAJAL AND GOLGI

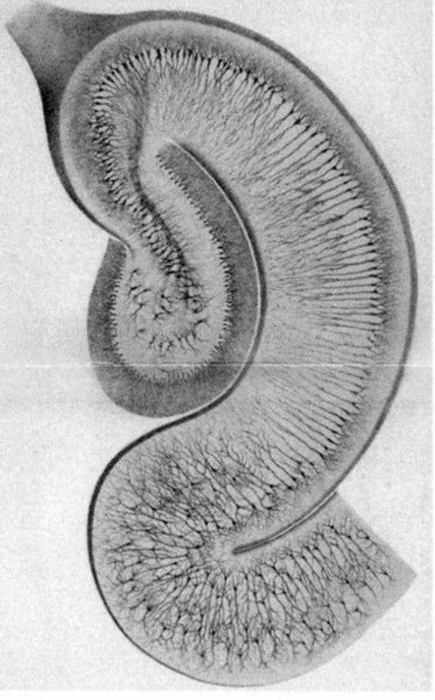

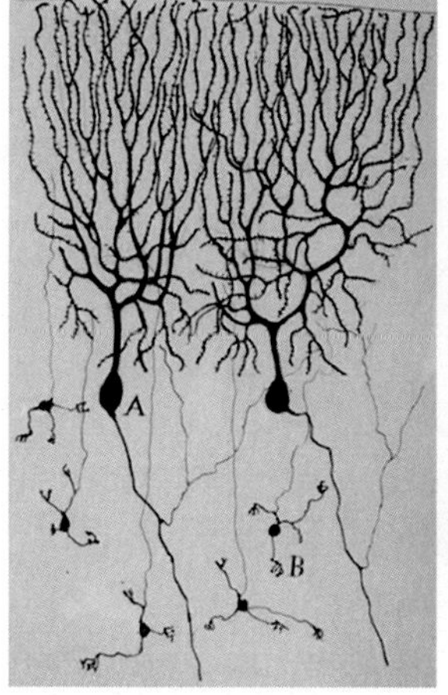

Santiago Ramón y Cajal (who was working in Spain, shown on the right), Camillo Golgi (who was working in Italy, shown on the left), and brain cells. The two standout figures of early cellular neuroscience are Santiago Ramón y Cajal and Camillo Golgi, whose elegant and detailed drawings of neurons provided the foundation for developing an understanding of the structure and function of these cells. Using novel staining techniques, Cajal and Golgi were able to provide a level of resolution to the microscopic structures in the brain that had hitherto been incredibly challenging. Cajal and Golgi shared the Nobel Prize in 1906 for their work in this area. The two images of cells in the brain show the hippocampus from the brain of newborn kitten (left image) and Purkinje cells in the cerebellum of a pigeon.

of neuronal cell types within which neurons fall based on their morphology, location within the nervous system, and favored chemical for neurotransmission [22]. Based on their morphology, neurons can be categorized as unipolar, bipolar, or multipolar, relating to the number of inputs/outputs that a neuron possesses (unipolar having one projection, bipolar having an input and an output, and multipolar having multiple inputs and an output). There is a fourth category of neurons named pseudounipolar, which possess one projection that divides into two. Purkinje cells, a specialized cell type of the cerebellum, illustrate the complexity that can be achieved by multipolar cells [23]. These cells, which are among the largest in the human brain, are characterized by an intricate and complex web of branched dendrites, providing multiple inputs to the processing activity undertaken by the cell (a Purkinje cell is shown in Box 1.2, as drawn by Ramón y Cajal). A second method of categorizing neurons is to divide them into classes based on their relative position in the nervous system. Neurons that bring information from the periphery into the brain are defined as sensory neurons; neurons that conduct signals between other neurons, interneurons; and those that relay instructions out to muscles in the rest of the body, motor neurons. Finally, neurons can be identified according to the neurotransmitters that they use to communicate with other cells. They include amino acids such as glutamate and gamma-aminobutyric acid (better known by its acronym GABA), monoamines such as dopamine and serotonin, gasotransmitters such as nitric oxide, and choline-based molecules such as acetylcholine.

Even taking into consideration all these different methods to categorize neuronal types does not begin to reveal the depth and breadth of diversity within the human nervous system, with distinct populations of cells in different regions of the brain distinguished by gene expression profiles, energetic requirements, and functional connections.

With regard to the last category, and focusing on the role of neurotransmission in neuronal function, it is important to have a basic understanding of how neurotransmitters work in the context of neurodegeneration. Many of the symptoms that characterize neurodegenerative diseases are due to the loss of specific neurotransmitter connections and, as will be seen in Chapters 2 and 3, some of the symptomatic therapies that exist for neurodegenerative disorders rely on supplementing or modulating neurotransmitters in the central nervous system. A simplified template for neurotransmission is displayed in Fig. 1.4. Neurons process and pass on information by means of propagating an action potential along an axonal projection, which is modulated by changes in the membrane potential and guided by the insulating presence of a myelin sheath (see discussion below with regard to the role of glial cells in

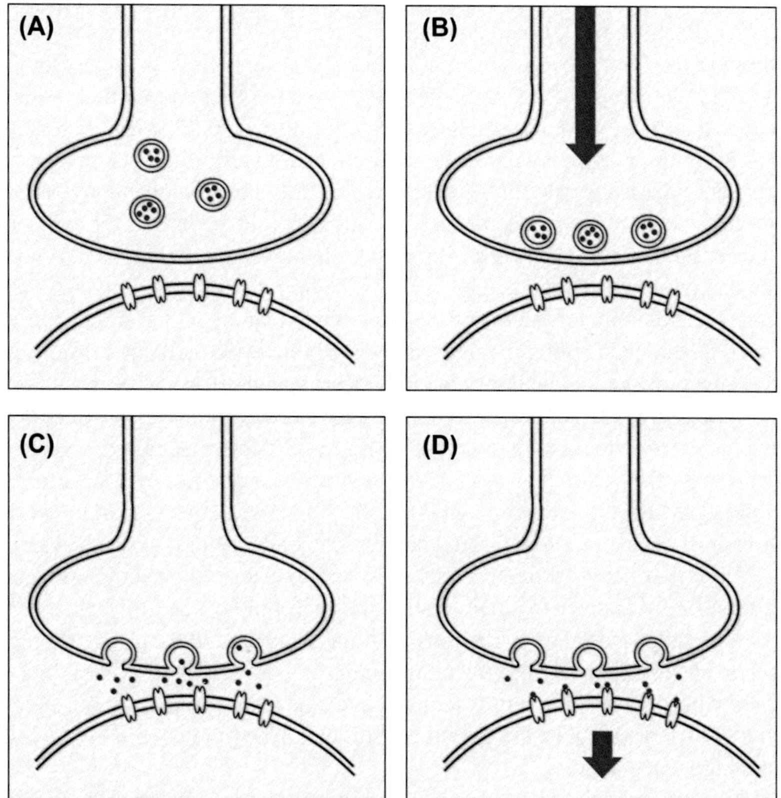

FIGURE 1.4

A simplified representation of a stereotypical neurotransmission event. (A) Vesicles containing neurotransmitters are localized at the synaptic terminal. (B) Upon receipt of an action potential, these vesicles move to the synaptic cell membrane. (C) Fusion of the vesicular membrane with the synaptic cell membrane occurs, releasing the neurotransmitters into the synaptic cleft. (D) The transmitter molecules bind to receptors on the receiving cell's dendrites, stimulating an intracellular cascade that passes on the message to the next cell.

achieving this). At the synaptic terminal, which is a specialized bulge at the end of an axon, vesicles full of neurotransmitters sit in the cytoplasm of the cell awaiting the signal to activate (Fig. 1.4A). When an action potential arrives at the synaptic terminal, these vesicles migrate to the cell membrane and undergo a complex process of membrane fusion with the cell membrane resulting in the release of their contents into the synaptic cleft (Fig. 1.4B and C). The transmitter molecules then diffuse across the synaptic cleft and bind to receptor proteins on the dendrites of the receiving cell, stimulating an intracellular signaling cascade that passes the signal on to the next cell (Fig. 1.4D). The remaining transmitter molecules are then taken up again into the synaptic terminal and repackaged ready for the next signal to come along, with glial cells also play a role in mopping up neurotransmitters at this point.

As noted earlier, the different classes of glial cell are integral and critical constituents of the brain, and without their support, neurons would not be able to carry out their function. The umbrella term glia applies to a wide range of cell types that can be broadly subdivided into four types: astrocytes, microglia, ependymal cells, and oligodendrocytes [17]. Although these cell classes contain diverse cell subtypes, the function of each type of glia can be simplified as follows. Astrocytes, named for their star-like morphology, provide a wide range of services within the brain. These include providing structural support, interweaving with neurons and their processes to supply a matrix within which neurons can nestle, providing energy in the form of nutrients, and controlling blood flow and regulating neurotransmitter and ion concentration at the synapse. In addition, astrocytes possess their own mechanisms to communicate, in particular using calcium signaling, and it is thought that they may play a direct role in the propagation and potentiation of neuron-centered signals within the brain. Microglia are the resident police within the central nervous system, fulfilling the role played by macrophages and other immune cells beyond the central nervous system. This is a particularly crucial function, as the presence of the blood–brain barrier and the immune status of the brain mean that the wider immune system has a limited role in protecting and maintaining the integrity of the brain. Microglia, therefore, act by defending the brain against infection, locating and eliminating bacteria, and removing dead or damaged cells. At a functional level, microglia also contribute to the formation and remodeling of synaptic connections in the brain, using phagocytosis to remove redundant synapses. Ependymal cells are a specialized type of epithelial cells that line the ventricular system of the brain and play a key role in the production of cerebrospinal fluid. They are critical for maintaining the homeostatic balance in the brain and for providing key nutrients and chemicals to neuronal populations in the central nervous system. Last, but by no means least, oligodendrocytes form a network of support cells for the axonal projections from neurons throughout the brain. Named for their branchlike morphology (dendro being derived from the Greek for branch), oligodendrocytes extend processes that wrap around the axon forming the lipid- and protein-rich myelin sheath that acts to protect and support the axon. They form an integral part of the structures that allow the passage of action potentials down the axon by insulating the axon and forming nodes of Ranvier, which are gaps in the myelin sheath that speed the conduct of changes in electrochemical potential along the axon.

Although neurodegeneration is characterized, and indeed defined, by the loss of neuronal cells in the brain, it is essential to recognize and consider the contribution of other cells within the brain to the processes that lead to cell death. Both astrocytes and microglia respond to initial damaging events within the brain, and there is increasing evidence that this reaction (with activated astrocytes and microglia crowding around damaged neurons) can act in a perverse way to amplify damage—a form of overreaction analogous to, and perhaps even mirroring, autoimmune dysfunction. Another example of the critical role of glia in neurodegeneration is the impact of demyelination leading to cell death, with loss of the myelin sheath extending from oligodendrocytes contributing to neuronal vulnerability and eventually cell death. In summary, although it is simpler to consider brain cells as discrete types functioning as independent actors, the complexity of the central nervous system and the pathways that lead to neurodegeneration mean that it is also critical to think about brain cell function/dysfunction in a holistic manner. To do this, the interplay of neurons, astrocytes, microglia, and oligodendrocytes across the brain needs to be considered.

1.5 CLINICAL TOOLS

The field of neurology, concerned with diseases afflicting the brain, was established through the later years of the 19th century. As modern concepts of medicine and nosology (the branch of medical science dealing with the classification of diseases) developed through the 1800s, clinicians began to carry out detailed observational studies of patients with disorders of cognition and movement by grouping patients with similar-presenting symptoms to establish distinct clinical entities. Two key players in this process were Jean-Martin Charcot, who was working at the Hôpital Salpêtrière in Paris, and William Gowers, who was working at what was then the National Hospital for the Paralysed and Epileptics (now the National Hospital for Neurology and Neurosurgery) in London (Box 1.3) [24,25].

The categorization of neurological, and hence neurodegenerative, diseases during the late 19th and early 20th centuries was based primarily on detailed clinical observation and evaluation of symptoms. This remains the foundation for much of the way in which we categorize, diagnose, and allocate treatments for neurodegenerative diseases today, but by taking advantage of the significant advances in medicine that have occurred in the intervening years our understanding has advanced. In particular, there is a growing awareness of how clinical symptoms, pathological conditions, and molecular causes (often driven by genetics) combine to create a complete picture of an individual disease, a point that will be returned to at the end of this section.

Clinical evaluation of neurological symptoms linked to neurodegenerative diseases can take a wide variety of guises and is covered in detail for each disorder in turn in the relevant chapter. As the central nervous system is involved in the processing of information from across the body, and controls function across all organs and senses, there is an equal breadth of bodily function assessment relevant to neurodegenerative diseases. Cognitive function, memory, and certain aspects of personality play a major part in many neurodegenerative diseases (particularly those with a dementia component), as do alterations in motor control (e.g., Parkinson and Huntington diseases). There are, however, a range of factors and causes that can alter these faculties. Equally as important, there is a great deal of normal variation in each of these functions in the human population—not everybody has the same level of fine motor control, or recall of past facts, people, and places. As such, and as part of the differential diagnosis process, a key consideration is the progressive degenerative nature of disease over time. This can be relatively rapid, over the course of months in disorders such as Creutzfeldt–Jakob disease, or much more insidious, lasting decades in the case of some of the inherited forms of parkinsonism.

A second, critical aspect of clinical evaluation is observing the pathological events underlying the disease. In years gone by, this occurred almost exclusively as a retrospective diagnostic effort by examining brain tissues post mortem to identify regional cell loss and to look for the presence of specific markers of disease. Neuropathological assessments of tissue include both macroscopic analysis of anatomy, for example, loss of total brain mass or obvious structural degeneration, and microscopic

BOX 1.3 CHARCOT AND GOWERS

The 19th century witnessed the birth of the field of modern neurology, the study of diseases of the brain. Two of the most important figures in the development of neurology were Jean-Martin Charcot (left), who was working at the Hôpital Salpêtrière in Paris, and William Gowers (right), who was working at the National Hospital for the Paralysed and Epileptics in London. Building on previous clinical descriptions across diseases of the brain and adding meticulous documentation of symptoms and disease course, first Charcot and then Gowers cataloged and ordered these disorders.

analysis of pathology using either generic chemical stains or antibodies that are specific for individual proteins associated with the disease. Although crucial for the confirmation of diagnosis and providing invaluable insights for research, postmortem neuropathological analysis does not aid in the diagnosis or treatment of a patient while alive. In recent years, however, advanced clinical imaging techniques have allowed neurologists to visualize degeneration in living patients. These are exemplified by the use of magnetic resonance imaging, computed tomography, and positron emission tomography (often referred to by their abbreviations, MRI, CT, and PET) [12]. These approaches use a variety of nuclear medicine approaches to generate detailed images of the brain and central nervous system. MRI, for example, uses strong magnetic fields coupled with the resonance properties of atoms within the brain to generate detailed structural and functional images of the brain. PET uses radioactive tracer probes specific for particular processes or proteins in the brain to pinpoint their activity and location. In neurodegeneration, the development of radioligands specific to the amyloid plaques associated with the disease has revolutionized clinicians' ability to monitor amyloid accumulation in this disease (see Chapter 2 for details). In addition to structural MRI technology that allows radiographers and neurologists to follow shrinkage of the brain as a proxy for cell death longitudinally over the course of many years, we now have an unprecedented ability to track the progress of neurodegeneration in the living brain. Although most of its benefits have been for research efforts, these approaches are increasingly being used in a diagnostic setting [26].

Beyond clinical analysis of symptoms and brain imaging, two further areas of advance have been in the fields of genetic discovery and biomarkers for disease. For the former, advances in sequencing technology now provide high-throughput, and often genome-wide, analysis of genetic risk for a disease. Over the past several decades, hundreds of mutations in the human genome have been identified as being linked either as strong risk factors or as the cause of neurodegeneration [27]. A more detailed description of genetic approaches is included in Section 1.6 and in each chapter discussing the diseases; however, genetic screening is now becoming more commonplace in a clinical setting for gaining greater insight into the nature of a disease. Genetic evaluation is even being used as a tool for presymptomatic diagnosis, i.e., identifying individuals at greater risk of a disease long before they present with the symptoms that define the disease.

Biomarkers, substances or alterations in biology that act as a proxy marker for a disease process, are an area where neurodegeneration as a field has struggled to identify changes that are specific to a disease and can be detected sensitively enough to be used in a clinical setting [28]. These include, for example, the presence of specific molecules in cerebral spinal fluid or blood, the levels of which correlate with disease progress in the central nervous system. There are important benefits in having sensitive and specific biomarkers for a disease with regard to both clinical diagnosis and research into a disorder; however, to date, there are few examples in the field of neurodegeneration that have moved beyond a research setting.

In summary, despite over a century of advances in medicine, much of the information used in a clinical setting for diagnosis of neurodegenerative diseases would be familiar to the 19th century neurologists who worked to define these disorders. This situation is, however, beginning to change as there is a growing realization that clinical appraisal, genetics, and neuropathology do not always point to the same underlying disease processes [29]. This is an issue that is particularly relevant to our understanding of motor neuron diseases and frontal temporal dementia (discussed in Chapter 5), but it has implications across neurodegenerative diseases as the boundaries between different disease designations blur based on our molecular understanding of the disease.

1.6 METHODS AND MODELS FOR INVESTIGATING NEURODEGENERATION

It is clear from the previous sections that the brain, and its cellular makeup, is highly complex. Identifying and deciphering the changes in the brain that result in neuronal cell death presents, therefore, a considerable challenge to biomedical research. Fortunately, there are a wide range of approaches and techniques that can be used to gain an insight into the cause of neurodegeneration, some examples of which are discussed in the following section.

As noted earlier, the revolution in our understanding of human disease genetics deriving from the development of DNA sequencing in the 1970s through the human genome project in the 1990s has had major implications for our understanding of neurodegenerative diseases. Genetic influence on the risk of developing a neurodegenerative disorder can be broadly divided into two categories [27]. First, there are familial forms of diseases where the presence of a deleterious mutation determines the development of the disease. These are inherited in a classical mendelian fashion, either autosomal dominant or autosomal recessive [30]; that is, requiring either one copy of a mutated gene or having both copies mutated, respectively (see Fig. 1.5). There are many examples of both across the spectrum of neurodegenerative diseases, including disorders that are purely genetically defined, such as Huntington disease (inherited in an autosomal dominant manner, see Chapter 6 for details), including single point mutations that change the amino acid sequence of proteins linked to neurodegeneration, nonsense mutations that act

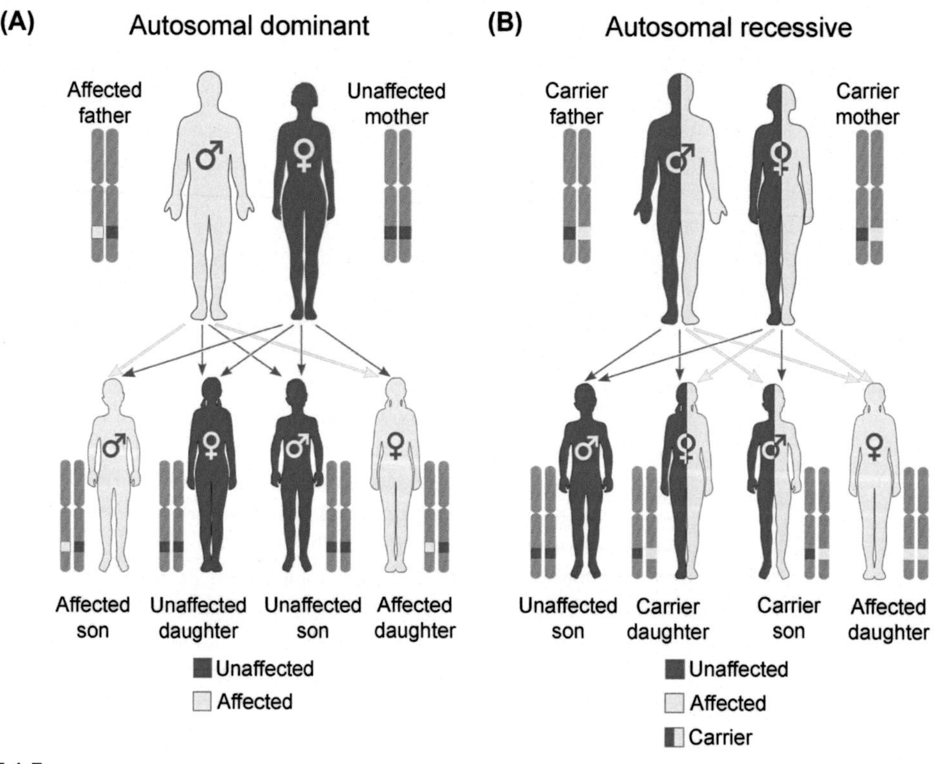

FIGURE 1.5

Modes of mendelian inheritance. A simplified representation of mendelian inheritance of disease in humans. (A) Autosomal dominant inheritance, where only one mutated version of a gene is required to cause disease. Here, an affected father passes on the mutated allele to one son and one daughter. (B) Autosomal recessive inheritance. In this case, both parents are heterozygous for a disease-related gene, each carrying one mutant allele. Two of their children also carry the mutated gene, but do not (or will not) present with disease. One of their children has two mutant alleles of the gene and will develop the disease.

Image derived from resources available from the National Library of Medicine.

as loss of function, and gene duplication or loss. There are also a small number of disorders that are inherited in an X-linked fashion, where mutations on the X chromosome segregate with disease. Mendelian forms of disease were historically identified by positional clonal approaches, mapping markers on chromosomes across the genome to narrow down the chromosomal location of a causative variant followed by exhaustive sequencing of genes in a candidate region. Once a variant was identified, this could then be assessed for segregation with disease in multiple individuals and (ideally) multiple independent families. The advent of next-generation sequencing, which has both sped up the process of genomic analysis and allowed this to expand across the genome relatively cheaply, has shifted sequencing approaches to assess either the human exome (all the transcribed sequences in the human genome) or the whole genome.

The second major component in genetic risk for neurodegenerative diseases comes from variants in the genome that do not cause the disease, but rather increase or decrease an individual's lifetime risk of developing a particular disorder. Until relatively recently, such variants had been challenging to identify, but the increasing use of DNA microarrays (microchips spotted with >1 million fragments of DNA probing for individual single-nucleotide polymorphisms in the genome) and genome sequencing has facilitated the identification of an ever-increasing number of genetic risk factors for a disease. This has been made possible by the use of genome-wide association studies, which are analyses that compare the frequency of genetic variants between a large number of individuals with and without a particular phenotype (in this case, neurodegenerative disease) (Fig. 1.6) [31]. Using statistical analyses, it is possible to determine whether a particular variant in the genome is significantly associated with increased or decreased risk of disease and to calculate an odds ratio for that variant—an odds ratio greater than one indicates increased association with disease, whereas an odds ratio less than one would indicate a protective variant. Genome-wide association studies have ushered in a step change in the way in which we comprehend genetic risk for a human disease, and this is certainly true of neurodegenerative diseases. The scale of these studies is also breathtaking because to achieve the statistical power needed to identify variants in the three billion or so base pairs of the human genome, tens of thousands of DNA samples from patients and controls are required.

Related to the underlying genetics, analysis of gene expression is a powerful tool to understand the changes that brain cells (including neurons and glial cells) and brain tissue undergo as part of the degenerative process, and it also provides potential insights into the cellular/tissue environment that immediately precedes degeneration and hence yield up clues as to what may be causing neuronal death [32]. The technology to examine gene expression at the RNA level has benefited from advances similar to those applied to DNA. The expression level of individual genes can be assessed by isolating messenger RNA (mRNA) and carrying out quantitative amplification using polymerase chain reaction–based investigations. By normalizing the observed levels of the gene of interest to housekeeping genes (normally, genes coding for critical cytoskeletal or essential metabolic proteins), an indication can be obtained as to whether the expression of this gene alters under experimental conditions. Of course, the nature of those experimental conditions is central to being able to derive meaningful data. This can include comparing the expression of genes in a postmortem healthy brain with that in a degenerated brain or using cellular or animal models of a disease and assessing genes that have been strongly implicated by genetic studies. As the technology allowing analysis of mRNA has improved, it has become possible to assess gene expression across the whole genome through expression microarrays and, most recently, by RNA sequencing. Both techniques provide coverage across the human genome and a quantitative readout for alterations in gene expression at the mRNA level. It is important to note that the

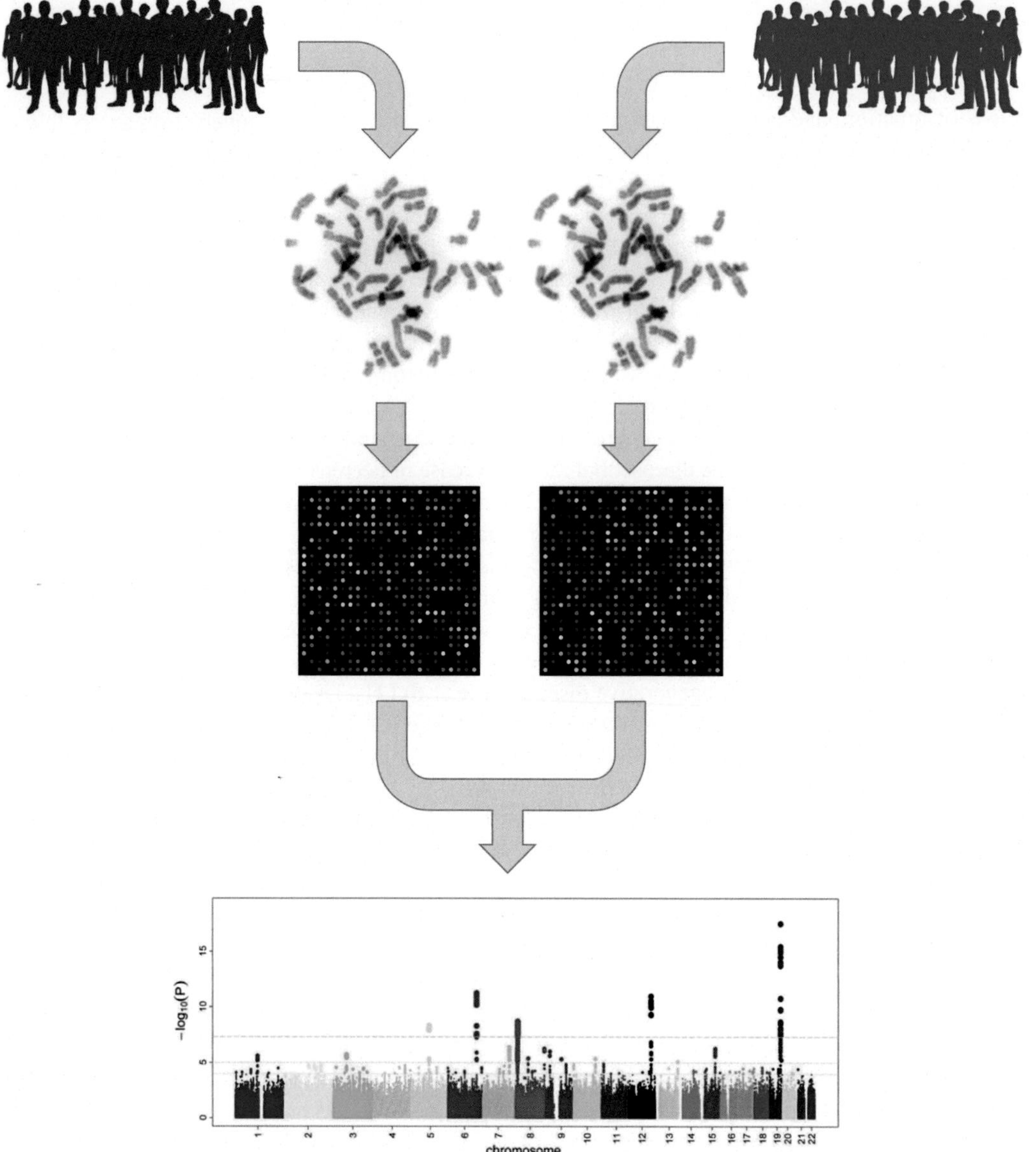

FIGURE 1.6

A diagrammatic representation of how a binary genome-wide association study works. Two populations of subjects are identified, one presenting with a particular trait (for example, Alzheimer disease) and one without it– (represented here in red and blue, respectively). After obtaining informed consent, DNA is isolated from both sets of subjects and is subjected to analysis either by DNA microarray (shown in the figure) or by exome/genome sequencing. These two datasets are analyzed to pick out DNA variants that are significantly over-represented in the disease compared with the control. The output of these analyses can be visualized using a Manhattan plot for the genome, indicating chromosomal regions associated with the trait of interest. Extensive functional work is then required to establish the nature of the change linked to these loci that causes the association with the trait of interest.

Manhattan plot image courtesy of Ikram MK, et al. Four novel loci (19q13, 6q24, 12q24, and 5q14) influence the microcirculation in vivo. PLoS Genet October 28, 2010;6(10):e1001184 via a creative commons licence. Other images modified from Wikimedia collection using a Creative Commons Licence.

expression levels of genes involved in neurodegeneration is clearly of great importance. This is best exemplified by the gene duplication and triplication events affecting the genes coding for the amyloid precursor protein and α-synuclein, key players in Alzheimer and Parkinson diseases, respectively [33]. In both cases, increasing the expression of these genes by 50% or 100% results in disease because of the increased accumulation of the proteins produced by these genes.

The ultimate product of the process of gene expression is, in the majority of cases, a protein that carries out a function in a cell. Understanding how protein function alters in neurodegeneration, both in isolated cells and in the diseased brain, has provided important insights into the processes that drive disease. Over the past three decades, in particular, this has been accelerated by focusing studies on the biology of proteins that have been implicated in neurodegeneration by genetic studies. This started with studies of huntingtin, the protein linked to the pathogenesis of Huntington disease (see Chapter 6), and has continued with intense scrutiny of genes linked to familial forms of neurodegeneration, as well as to those identified through genome-wide association studies. There is a wide range of tools to assess protein biology [34]. Protein expression levels in cells and tissue can be assessed by extracting and solubilizing proteins, followed by electrophoretic separation and analysis either using quantitative stains for protein (such as Coomassie blue) or specific antibodies by immunoblotting. High-throughput quantitative analysis of protein levels can be carried out by the enzyme-linked immunosorbent assay (frequently abbreviated as ELISA). Mass spectrometry, assessing molecules by the mass-to-charge ratio, allows the detailed characterization of individual proteins and is rapidly evolving to the point where advanced analysis is possible across the entire proteome (that is, the totality of proteins found in a sample, cell, or tissue).

Structural and biochemical characterization of individual proteins linked to a disease is an important area of research across biomedical science, and this is certainly true for neurodegeneration [35]. Proteins of interest can be isolated from a variety of expression systems, including bacterial, insect, and human cells, and purified to homogeneity. This allows structural biology approaches such as X-ray crystallography, nuclear magnetic resonance, and cryoelectron microscopy to be used to gain atomic resolution three-dimensional structures of these proteins. These can, in turn, provide important insights into the function of a protein and how a specific coding mutation can alter that function. Coupled to in vitro enzymatic and biophysical studies, a detailed picture can be constructed as to what a protein does and how this is altered in a disease, especially when mutations can be inserted into the sequence of the protein.

Moving beyond individual proteins, cellular investigative techniques allow a more complex profile of a protein of interest to be derived. Importantly, this includes the cellular consequences of manipulating the activities or function of that protein, whether this be using genetic approaches to alter the coding sequence, knocking out the gene, or using chemical modulators of function. For several neurodegenerative disorders, there are well-established toxin-based models for a disease. An standout example of this is the use of the mitochondrial respiratory chain inhibitor 1-methyl-4-phenyl-1,2,3,6-tetrahydropyridine (almost universally known by its abbreviation MPTP) that is used to model dopaminergic cell death in Parkinson disease [36].

Using a variety of cellular models, including neuronal cells, it is possible to observe and measure many aspects of cellular function. These range from mitochondrial function through waste disposal to synaptic function, all of which have been closely linked to neurodegeneration [37]. Microscopic imaging techniques, including light, confocal, and electron microscopy, as well as fluorescence-based techniques to label and image specific cellular functions or proteins, play a major part in allowing researchers to understand the processes that alter cellular function in neurodegeneration. For neuronal function, including synaptic biology, electrophysiological techniques using tiny electrodes to measure changes in membrane potential allow for the analysis of one of the key roles of neurons in the brain, i.e., the passing of information from cell to cell. More fundamentally, it is possible to assess cell viability and cell death pathways by directly examining neuronal survival and the contribution of other cells types to this process.

Clearly, the model system that is used to investigate a given neurodegenerative disease has a major impact on the nature of the data that will be gathered and how closely this translates to the human disorder. In the past century or more of research into the cause of neurodegeneration a large number of such systems have been used, ranging from simple cells grown in the laboratory to nonhuman primates that are among our closest relatives on Earth [38]. Different model systems, both cellular and animal (or organismal), have their own benefits and drawbacks. Those that are most simple, or reductionist, may be easy to investigate and manipulate but do not come close to the complexities of human brain function. Examples of these might include yeast cells or immortal mammalian cancer cell lines. Those that are most closely matched to humans, in particular the great apes such as chimpanzees, raise important ethical and moral issues around the use of sentient creatures in medical research and their well-being (as well as huge practical challenges centered around cost and tractability) [39]. In between these two ends of the spectrum lie a plethora of different cell models and organisms that have been used to investigate the neurodegenerative process. These range from human pluripotent stem cells (capable of giving rise to any cell type in the human body) that can be differentiated into specific neuronal populations, as well as glial cells, through simple nematode worms such as *Caenorhabditis elegans* and insect models such as the fruit fly *Drosophila melanogaster* to zebra fish, mice, rat, and nonhuman primate models (Fig. 1.7). Each of these systems has made a major contribution to our understanding of how the brain degenerates, with the balance between reductionism and complexity allowing complementary studies in different models to shed even greater light on the disease process.

1.7 DRUGS, DRUG DEVELOPMENT, AND CLINICAL TRIALS

Sadly, the current state of available treatments for neurodegenerative diseases does not provide much relief to patients suffering from these disorders [40]. Although efficacious drug therapies are available for specific symptoms of individual diseases (for example, Parkinson disease - see Chapter 3), there are very few neurodegenerative diseases for which disease-modifying treatments exist. This situation has begun to change, as huge investments in fundamental research have started to translate into clinical trials and novel treatments gaining approval for use in humans. To understand the challenges facing those seeking to develop new drugs, it is important to have an overview of the drug discovery process that operates across academic and pharmaceutical company research [41].

The starting point for this is the initial identification or characterization of a putative drug target. These can arise from a wide range of different sources, from human genetic studies pinpointing genes involved in the disease to animal models highlighting particular pathways or processes that can be manipulated with beneficial outcomes. This basic discovery phase leading to preclinical evaluation is the critical starting point for the development of drugs, but it can take a long time [42]. To take an example pertinent to the field of neurodegeneration, the *LRRK2* gene was first implicated in Parkinson disease in 2004 but it was not until 2017 that drugs targeting the activity of this gene began to be actively investigated in human clinical trials [43]. This period of target validation and initial candidate

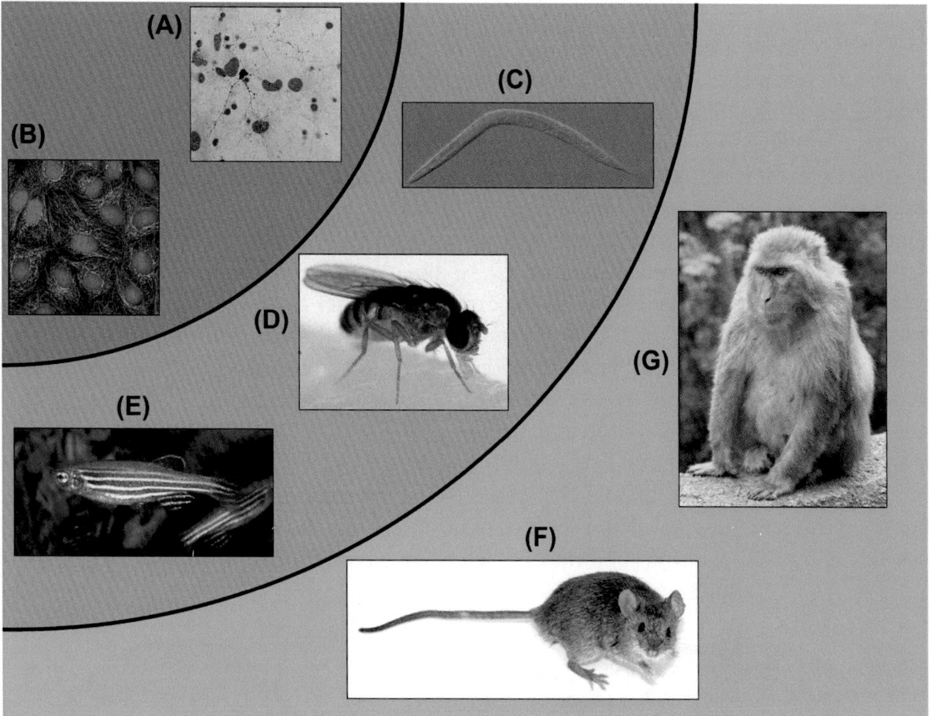

FIGURE 1.7

Cell and animal model systems for neurodegeneration. A summary of some of the model systems used to investigate neurodegenerative diseases. (A) Human neurons derived from induced pluripotent stem cells. (B) Immortalized human cancer cell lines, including the HeLa cell line shown here. (C) The simple nematode worm, *Caenorhabditis elegans*. (D) The fruit fly, *Drosophila melanogaster*. (E) The zebra fish, *Danio rerio*. (F) Rodent models, such as the mouse (*Mus musculus*). (G) Nonhuman primates such as the macaque, *Macaca mulatta*.

Images modified from Wikimedia collection using a Creative Commons Licence.

molecule development is when the majority of targets and drugs fail to progress, often well before they get anywhere near human trials. The nature of this preclinical development varies depending on the precise nature of the disease and favored models but can include analysis of a specific target in cells and tissues relevant to the disorder followed by development of small molecules or biological drugs (such as artificially generated antibodies) focusing on the chemistry, pharmacokinetic profile, and pharmacological activity. These drugs would then be tested in cellular and animal models for disease to evaluate both safety and efficacy in systems as close to humans as possible. If a drug yields positive results in all these stages, then it will enter into the clinical trial system.

The clinical trial process itself is a highly regulated and formalized series of tests in human volunteers to assess whether the putative drug therapy is safe for humans, and to examine whether it is successful in targeting the disease process it aims to mitigate. Clinical trials are, for the most part, major undertakings in terms of time, logistics, and funding and are regulated by a number of governmental or supragovernmental agencies. The two largest, and most important, of these are the European Medicines

Agency, covering the European Union, and the Food and Drug Administration in the United States of America. These provide a framework and legal oversight of drug testing as well as recording the progress of trials (documented at http://www.ema.europa.eu/ema/ and www.clinicaltrials.gov, respectively). The basic stages of the clinical trial evaluation process are displayed in Fig. 1.8 [44,45]. The initial stage, a phase 1 clinical trial, is focused on testing safety in healthy human volunteers, providing them with the drug or therapy and monitoring for adverse drug reactions in this population. The second and third stages of the clinical trial evaluation concentrate on efficacy, the ability of the treatment to modify the symptoms or disease process in humans with a beneficial clinical outcome. As can be seen from Fig. 1.8, the three stages of clinical trial evaluation take a long time (normally a minimum period of 6 years, and frequently much longer), are extremely expensive, and suffer from a high rate of attrition. All this conspires to limit the number of drugs that make it through from preclinical development to the approved drug entering into clinical practice, and there are numerous examples of drugs that have been approved for human use only for serious side effects to become apparent once they are being used by hundreds of thousands or even millions of patients.

For any disease, the clinical trial evaluation is an arduous process. For neurodegenerative diseases, however, there are particular, additional challenges [46]. First and foremost is the nature of these disorders. For the most part, these are slow progressive disorders that develop over many years. The symptoms can be highly heterogeneous and may be acting on physiological processes or capacities where

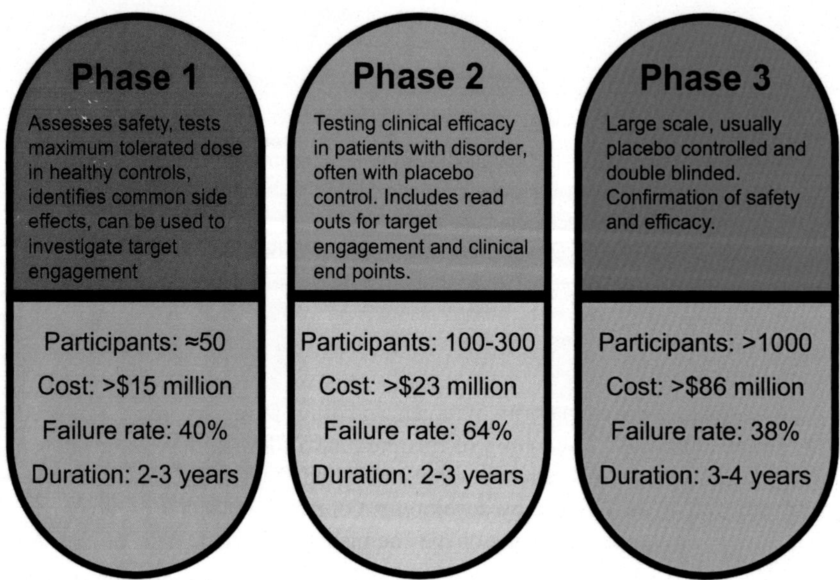

FIGURE 1.8

The anatomy of a clinical trial. A diagrammatic representation of the stages of a clinical trial, moving from smaller scale phase 1 safety trials through to phase 3 efficacy trials that may involve many thousands of participants and cost hundreds of millions of dollars.

Data from DiMasi JA, Grabowski HG, Hansen RW. Innovation in the pharmaceutical industry: new estimates of R&D costs. J Health Econ 2016;47:20–33.

there is a very variable baseline, a good example being cognitive ability. As such, trials for efficacy require large sample sizes measured over a long period with high sensitivity. It is also increasingly apparent that in most neurodegenerative diseases, significant damage has already been inflicted on the central nervous system long before the onset of clinical symptoms. This being the case, it is likely to be very challenging to intervene in these diseases at a late stage where the majority of neuronal cell loss has already occurred. In addition, it is well recognized that targeting the central nervous system with drugs is intrinsically difficult, partly owing to the privileged nature of the brain and the likely requirement that drugs will have to cross the blood–brain barrier to have an impact.

1.8 SUMMARY

Neurodegenerative diseases present a unique and complex challenge to biomedicine. They are multifactorial, involving a mix of genetic and environmental causes, and afflict the most complex organ in the human body. The huge challenges in understanding these disorders, not to mention the hurdles to be overcome in developing drugs that can be used to treat them, are matched only by their alarming and increasing cost. As stated at the start of this chapter, this textbook seeks to provide an overview of the current state of our knowledge of neurodegeneration across some of the most common and intriguing forms of these diseases.

FURTHER READING

Crossman AR, Neary D. Neuroanatomy: an illustrated colour text. Elsevier; 2005. ISBN: 0443100365. An overview of the anatomy of the human brain.

Donaghy M, editor. Brain's diseases of the nervous system. Oxford University Press; 2009. ISBN: 0198569386. A comprehensive guide to neurological disease.

Kandel ER, Schwartz JH, Jessell TM, Siegelbaum SA, Hudspeth AJ, editors. The principles of neural science. Elsevier; 2012. ISBN: 0071390111. A leading textbook on the function of the nervous system and of neural cells.

Ng R. Drugs: from discovery to approval. Wiley-Blackwell; 2015. ISBN: 1118907272. This textbook covers the drug discovery process in detail, from the early stages of target validation through to clinical trials.

Wood NW. Neurogenetics: a guide for clinicians. Cambridge University Press; 2012. ISBN: 0521154372X. An introduction to neurogenetics, providing a summary of how genetics can and has been used to inform our understanding of neurological disease.

REFERENCES

[1] Gustavsson A, Svensson M, Jacobi F, Allgulander C, Alonso J, Beghi E, Dodel R, Ekman M, Faravelli C, Fratiglioni L, Gannon B, Jones DH, Jennum P, Jordanova A, Jonsson L, Karampampa K, Knapp M, Kobelt G, Kurth T, Lieb R, Linde M, Ljungcrantz C, Maercker A, Melin B, Moscarelli M, Musayev A, Norwood F, Preisig M, Pugliatti M, Rehm J, Salvador-Carulla L, Schlehofer B, Simon R, Steinhausen HC, Stovner LJ, Vallat JM, Van den Bergh P, van Os J, Vos P, Xu W, Wittchen HU, Jonsson B, Olesen J, Group CD. Cost of disorders of the brain in Europe 2010. Eur Neuropsychopharmacol 2011;21:718–79.

[2] DiLuca M, Olesen J. The cost of brain diseases: a burden or a challenge? Neuron 2014;82:1205–8. Epub 2014/06/20.

[3] Brookmeyer R, Johnson E, Ziegler-Graham K, Arrighi HM. Forecasting the global burden of Alzheimer's disease. Alzheimers Dement 2007;3:186–91.

[4] In: Michael Donaghy, editor. Brain's diseases of the nervous system. 12th ed. Oxford: Oxford University Press; 2009.

[5] Prayson RA. Neuropathology. 2nd ed. Philadelphia, PA: Elsevier/Saunders; 2012.

[6] Schapira AHV, Wszolek ZK, Dawson TM, Wood NW. Neurodegeneration. John Wiley & Sons:New York; 2017.

[7] Crossman AR, Neary D. Neuroanatomy: an illustrated colour text. 3rd ed. Edinburgh: Churchill Livingstone; 2005.

[8] Brodmann K. Vergleichende Lokalisationslehre der Grosshirnrinde in ihren Prinzipien dargestellt auf Grund des Zellenbaues: Barth. 1909.

[9] Zilles K, Amunts K. Centenary of Brodmann's map–conception and fate. Nat Rev Neurosci 2010;11:139–45.

[10] Jones AR, Overly CC, Sunkin SM. The Allen Brain Atlas: 5 years and beyond. Nat Rev Neurosci 2009;10:821–8. Epub 2009/10/15.

[11] Mikula S, Denk W. High-resolution whole-brain staining for electron microscopic circuit reconstruction. Nat Methods 2015;12:541–6.

[12] Filippi M. Oxford textbook of neuroimaging. 1st ed. 2015.

[13] Helmstaedter M. Cellular-resolution connectomics: challenges of dense neural circuit reconstruction. Nat Methods 2013;10:501–7.

[14] Amunts K, Ebell C, Muller J, Telefont M, Knoll A, Lippert T. The human brain project: creating a European research infrastructure to decode the human brain. Neuron 2016;92:574–81.

[15] Alivisatos AP, Chun M, Church GM, Greenspan RJ, Roukes ML, Yuste R. The brain activity map project and the challenge of functional connectomics. Neuron 2012;74:970–4.

[16] Ballabh P, Braun A, Nedergaard M. The blood–brain barrier: an overview: structure, regulation, and clinical implications. Neurobiol Dis 2004;16:1–13.

[17] Allen NJ, Barres BA. Neuroscience: glia – more than just brain glue. Nature 2009;457:675–7.

[18] Lange W. Cell number and cell density in the cerebellar cortex of man and some other mammals. Cell Tissue Res 1975;157:115–24.

[19] Azevedo FA, Carvalho LR, Grinberg LT, Farfel JM, Ferretti RE, Leite RE, Jacob Filho W, Lent R, Herculano-Houzel S. Equal numbers of neuronal and nonneuronal cells make the human brain an isometrically scaled-up primate brain. J Comp Neurol 2009;513:532–41.

[20] De Carlos JA, Borrell J. A historical reflection of the contributions of Cajal and Golgi to the foundations of neuroscience. Brain Res Rev 2007;55:8–16.

[21] Ascoli GA, Donohue DE, Halavi M. NeuroMorpho.Org: a central resource for neuronal morphologies. J Neurosci Off J Soc Neurosci 2007;27:9247–51.

[22] Kandel ER, Schwartz JH, Jessell TM, Siegelbaum S, Hudspeth AJ. Principles of neural science. 5th ed. 2017

[23] Eccles JC, Llinas R, Sasaki K. The excitatory synaptic action of climbing fibres on the Purkinje cells of the cerebellum. J Physiology 1966;182:268–96.

[24] Charcot J. Lectures on the diseases of the nervous system. 2nd ed. Philadelphia: Henry Lea; 1879.

[25] Gowers WR. A manual of diseases of the nervous system. London: J & A Churchill; 1886.

[26] Frisoni GB, Fox NC, Jack Jr CR, Scheltens P, Thompson PM. The clinical use of structural MRI in Alzheimer disease. Nat Rev Neurol 2010;6:67–77.

[27] Wood NW. Neurogenetics: a guide for clinicians. Cambridge: Cambridge University Press; 2012.

[28] Strimbu K, Tavel JA. What are biomarkers? Curr Opin HIV AIDS 2010;5:463–6.

[29] Hardy J, Gwinn-Hardy K. The relationship between nosology, etiology and pathogenesis in neurodegenerative diseases. Handb Clin Neurol 2008;89:189–92.

[30] Guerreiro RJ, Gustafson DR, Hardy J. The genetic architecture of Alzheimer's disease: beyond APP, PSENs and APOE. Neurobiol Aging 2012;33:437–56.
[31] Gandhi S, Wood NW. Genome-wide association studies: the key to unlocking neurodegeneration? Nat Neurosci 2010;13:789–94. Epub 2010/06/29.
[32] Bras J, Guerreiro R, Hardy J. Use of next-generation sequencing and other whole-genome strategies to dissect neurological disease. Nat Rev Neurosci 2012;13:453–64.
[33] Devine MJ, Gwinn K, Singleton A, Hardy J. Parkinson's disease and alpha-synuclein expression. Mov Disord Off J Mov Disord Soc 2011;26(12):2160–8.
[34] Berg JMTJL, Gatto GJ, Stryer L. Biochemistry. 8th ed. Palgrave Macmillan; 2015.
[35] Banaszak LJ. Foundations of structural biology. Elsevier; 2000.
[36] Langston JW. The MPTP story. J Park Dis 2017;7:S11–22.
[37] Selkoe DJ. Cell biology of protein misfolding: the examples of Alzheimer's and Parkinson's diseases. Nat cell Biol 2004;6:1054.
[38] Gitler AD, Dhillon P, Shorter J. Neurodegenerative disease: models, mechanisms, and a new hope. The Company of Biologists Ltd.; 2017.
[39] Bateson P. Ethics and behavioral biology. Adv Study Behav 2005;35:211–33.
[40] Enna SJ, Coyle JT. Pharmacological management of neurological and psychiatric disorders. New York; London: McGraw-Hill; 1998.
[41] Ng RA. Drugs: from discovery to approval. 3rd ed. 2015.
[42] Dahlin JL, Inglese J, Walters MA. Mitigating risk in academic preclinical drug discovery. Nat Rev Drug Discov 2015;14:279–94.
[43] Cogo S, Greggio E, Lewis PA. Leucine rich repeat Kinase 2: beyond Parkinson's and beyond kinase inhibitors. Expert Opin Ther Targets 2017;21:751–3.
[44] Van Norman GA. Drugs, devices, and the FDA: Part 1: an overview of approval processes for drugs. JACC Basic Transl Sci 2016;1:170–9.
[45] DiMasi JA, Grabowski HG, Hansen RW. Innovation in the pharmaceutical industry: new estimates of R&D costs. J Health Econ 2016;47:20–33.
[46] Gribkoff VK, Kaczmarek LK. The need for new approaches in CNS drug discovery: why drugs have failed, and what can be done to improve outcomes. Neuropharmacology 2017;120:11–9.

CHAPTER

ALZHEIMER'S DISEASE AND DEMENTIA

2

CHAPTER OUTLINE

The Molecular and Clinical Pathology of Neurodegenerative Disease. https://doi.org/10.1016/B978-0-12-811069-0.00002-1

2.1 INTRODUCTION

The term dementia derives from the Latin *demens*, meaning literally to be out of one's mind. As discussed in more detail below (Section 2.2), dementia is now used in a medical context to define a broad collection of chronic neurological diseases where the main presenting symptoms are associated with a loss of mental capacity. In particular, where there is loss of cognitive ability, impacting on memory, learning, language, and other higher functions of the brain. The most common form of dementia is Alzheimer's disease, a clinical entity defined by progressive cognitive decline and specific neuropathological changes in the brain—notably the loss of neurons and the accumulation of protein aggregates made up of the amyloid beta peptide and the microtubule associated protein tau. As the leading cause of dementia and a growing burden on health-care systems across the globe, Alzheimer's disease has been the focus of huge research efforts over the last 50 years and will be the main focus of this chapter. There are, however, many other causes of dementia with relevance for understanding the molecular etiology of both Alzheimer's and neurodegeneration more broadly, some of which will be covered in this chapter and some of which—for example, the prion diseases and Huntington's disease—are dealt with in stand-alone chapters. It is also important to note that there is an increasing appreciation of the overlap between different neurological diseases involving neurodegeneration, most notably between Alzheimer's disease and Parkinson's disease, and between frontal temporal dementia and some forms of motor neuron disease. This overlap can be pathological, as with Alzheimer's disease and Parkinson's disease sharing Lewy body accumulation in the brain, or genetic as is the case with frontal temporal dementias and motor neuron disease (see Chapter 4). As such, it is critical to consider the content of this chapter in the context of the wider field of neurodegeneration.

Dementia, encompassing the erosion of many of the aspects of brain function that define our personality and character, has a history that stretches back many thousands of years [1]. Early on in the development of medical thought, Greek and Roman writers started making the distinction between temporary alterations in mental state and the slow decline in capacity frequently found in older age [2]. Retrospective analysis of a number of texts, both fictional and nonfictional, throughout the medieval and early modern period highlight examples of the portrayal of symptoms that would display some of the characteristics that would now be associated with dementia—an example of which is the portrayal of mental decline in William Shakespeare's play. The Tragedy of King Lear [3]. The concept of dementia began to change during the 19th century as the fields of psychiatry and neurology took shape, with a significant turning point occurring in the first few years of the 20th century. A physician called Alois

BOX 2.1 ALOIS A. AND AUGUSTE D.

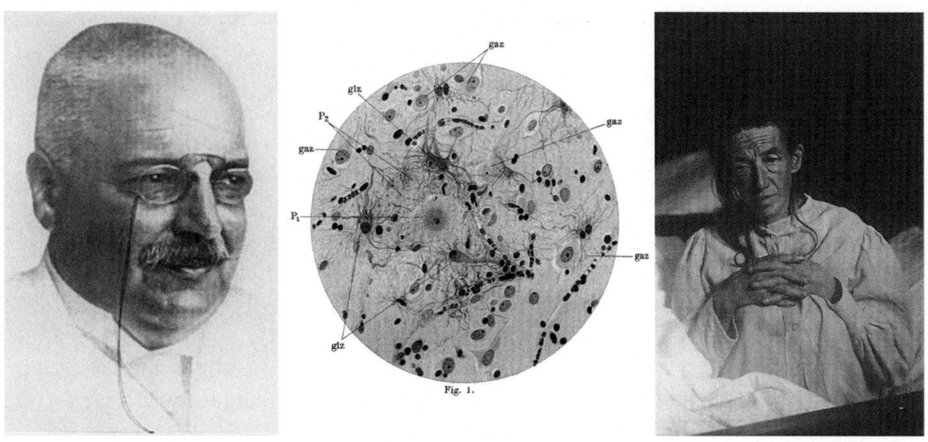

Alois Alzheimer (left-hand panel) met and studied a patient at a Frankfurt asylum, Auguste Dieter (right-hand panel) who had been committed to the care of the hospital due to severe changes in her mental state. These descriptions, along with a pathological analysis of her brain postmortem (an example of which is shown in the middle panel, revealing an amyloid plaque in the middle and activated astrocytes surrounding this), were the first to bring together the symptoms and changes in the brain that we now recognize as Alzheimer's disease.

Alzheimer, working at a hospital in Frankfurt, started working with a patient who had suffered a rapid decline in cognitive ability [4]. The patient, Auguste Dieter, had presented with personality change, loss of memory, and confusion and had eventually been admitted to hospital as an inpatient. Alzheimer carefully documented Auguste Dieter's case and after her death in 1906 carried out a detailed pathological analysis of her brain (Box 2.1) [5]. His pathological examination, partly using staining techniques developed by his colleague Franz Nissl, revealed extensive neuronal loss in the cerebral cortex and the presence of peculiar accumulations which we now recognize as amyloid plaques and neurofibrillary tangles (see Section 2.3). He reported his observations at a meeting in Tübingen in 1906, later publishing a description of the patient and the pathology [6,7], with further publications from other researchers following soon after. In these publications, and subsequent reports on similar patients by both Alzheimer and others [8], the foundations were laid for the establishment of the clinical entity that we now call Alzheimer's disease (an eponym that was first applied by Alzheimer's eminent colleague Emil Kraepelin in 1909) [9]. It is important to note that a number of other physicians and scientists working around the same time were observing and characterizing similar changes in the brain and symptoms, for example, Arnold Pick working in Prague describing a presenile form of dementia in 1891 [10]. A key change driven by the research of Alzheimer, Pick, and others was the gradual differentiation between the chronic, degenerative disorders of cognition that are now recognized as dementia, and other psychiatric disorders such as schizophrenia and psychosis [11]. Another important realization was that neurodegeneration leading to dementia was distinct from the normal process of aging. Taken together, and combined with advances in assessing cognitive function, neuroimaging and genetics, this leads to our current understanding of dementia.

2.2 CLINICAL PRESENTATION

As noted above, dementia is the umbrella term given to a group of conditions that cause a progressive decline in cognition that is of sufficient severity to disrupt activities of daily living. It is a major global health problem, with a prevalence of about 10%–15% at the age of 80 [12,13]. It is currently estimated that more than 40 million people have dementia worldwide, and these numbers are expected to continue to rise, doubling every 20 years, as the global population ages [14]. To take the example of one country, there are approximately 850,000 people in the United Kingdom living with dementia and the condition costs the UK £26 billion per year [15], highlighting the huge challenge that the dementias present to health systems. This is reflected in the urgency with which governments, charities, and pharmaceuticals are acting to develop novel treatments across the spectrum of dementia—in 2013 the G8 stated that dementia should be a major global priority and set an ambition to identify a cure or disease-modifying therapy for dementia by 2025.

Although dementia is most commonly thought of in the context of memory impairment, it is important to recognize the other cognitive domains that can be affected including language, praxis (the learning of novel skills), visual perception, and executive function [16].

There are many different causes of dementia, and (as befits a textbook on neurodegeneration) in this chapter the focus will be on the neurodegenerative forms of these diseases. Cognitive impairment may, however, also occur as a result of various systemic illnesses including renal and liver failure, nutritional deficiency, and infections such as HIV and syphilis. Although these causes are rare, accounting for <1% of new cases of dementia, it is important that they are considered, as they are potentially treatable and reversible—a key distinction when compared with neurodegenerative causes of dementia [17,18].

2.2.1 ALZHEIMER'S DISEASE

Alzheimer's disease is the most common neurodegenerative cause of dementia accounting for about 50% of all cases [19]. The prevalence increases with age, with rates doubling every 5 years after 65 [20]. Young-onset Alzheimer's disease, defined as onset before the age of 65, is rare, but it is still the most common cause of dementia in this age group. When Alzheimer's is presented in a younger patient, a genetic cause should be considered [21]. The condition was first described by Alois Alzheimer in 1906 when he described a 50-year-old patient who developed rapid memory loss (see Box 2.1). He examined her brain postmortem and described the neurofibrillary tangles and amyloid plaques that are characteristic of the disease [22]. Difficulties with memory are the most common initial symptoms of Alzheimer's disease. Declarative memory, the memory of events, is generally profoundly affected, whereas procedural memory, the memory of how to do things, is relatively well preserved, at least in the early stages of the disease. Relatives may have noticed the patient asking repetitive questions, forgetting tasks and getting lost in familiar environments. Memory problems alone are not sufficient for a diagnosis of Alzheimer's as other cognitive domains must be affected. If only memory is affected, the patient is diagnosed with amnestic mild cognitive impairment. Typically, patients with amnestic mild cognitive impairment convert to frank Alzheimer's at a rate of 10%–15% per year [23]. With time as pathological process extends beyond the medial temporal lobes there is more obvious involvement of other cognitive domains including praxis, language, and visuospatial function. Unlike frontotemporal dementia (see below), Alzheimer's disease is not usually associated with significant changes in personality, although behavioral problems and symptoms such as apathy, depression, and anxiety become more prominent later in the disease. Myoclonus and seizures may also occur later in the disease process. A number of atypical presentations of Alzheimer's disease are recognized. In these syndromes, amnestic symptoms may not be the first presentation but rather language, visual dysfunction, or difficulties with praxis are prominent early features of the disease.

2.2.2 VASCULAR DEMENTIA

Cerebrovascular disease is an increasingly recognized cause of dementia and accounts for around 15% of cases [24]. There are a number of different mechanisms by which cerebrovascular disease can lead to dementia. Multiple cortical infarcts are a well-recognized cause, but even a single infarct can cause cognitive impairment if it is in a strategic location such as the thalamus. Small vessel disease results from infarcts of the deep white matter and accounts for up to 50% of vascular dementia cases [25].

Cerebral hemorrhage is also associated with dementia—the risk of dementia after spontaneous intracerebral hemorrhage is 14% at 1 year. The greatest is risk for lobar intracerebral hemorrhage, which together with superficial siderosis, and microbleeds may be associated with amyloid angiopathy [26,27]. Poststroke dementia is a recognized syndrome affecting up to 10% of patients after a first stroke and 30% after multiple strokes [28]. The pathophysiology of post stroke dementia is unclear, and it has sometimes been considered as a separate subtype of dementia. Some hereditary disorders are associated with vascular dementia such as cerebral autosomal dominant arteriopathy with subcortical infarcts and leukoencephalopathy (CADASIL) [29].

The clinical features of vascular dementia can be variable and depend on the structures and the pathomechanisms involved. In the case of a single strategic infarct, the clinical syndrome will depend on the location of that infarct. Patients with subcortical vascular disease typically have difficulty with attention, executive function, and processing speed. Noncognitive symptoms such as depression, psychosis, and apathy are frequently seen, but memory impairment may not be a significant feature [30].

The management of vascular dementia centers around supportive care and management of comorbid conditions. Acetylcholinesterase inhibitors have been used, but their benefit seems to be less than that seen in Alzheimer's disease [24].

2.2.3 POSTERIOR CORTICAL ATROPHY

Posterior cortical atrophy is a rare dementia syndrome characterized predominantly by a progressive decline in visual processing skills. The early symptoms may be nonspecific complaints of visual blurring, but symptoms progress to difficulty reading, difficulties perceiving textures and patterns, and an inability to piece together visual scenes to create a global picture that makes sense (simultagnosia) [31]. Most cases of posterior cortical atrophy are related to Alzheimer's pathology, but other pathologies have described too, including Lewy body dementia, corticobasal degeneration, and prion disease [32,33].

2.2.4 LOGOPENIC APHASIA

Logopenic aphasia (from the Greek for lack of words) describes a syndrome with paucity of verbal output with sparing of grammar and motor speech. Clinically, patients have difficulty with naming and repetition of sentences [34]. In contrast to other forms of primary progressive aphasia (see below), logopenic aphasia has been shown to be due to Alzheimer's pathology.

2.2.5 DYSEXECUTIVE OR FRONTAL VARIANT

A subset of patients with Alzheimer's disease may present with profound executive impairment that may resemble behavioral variant frontotemporal dementia (see below). These patients tend to have a more rapid progression, and the difference in symptoms is believed to reflect a disproportionate burden of pathology affecting the frontal lobes [35].

2.2.6 DIAGNOSIS OF ALZHEIMER'S DISEASE

A detailed clinical assessment is of paramount importance, with both the patient and their relatives questioned carefully about the nature of their symptoms. A number of simple, questionnaire-style, systematic, cognitive assessment tests have been developed to provide an initial insight into cognitive ability, for example the mini-mental state exam [36]. These tests, however, provide only an initial insight into cognitive performance and require further detailed evaluation and combination with other diagnostic tools [37].

Neuroimaging using structural magnetic resonance imaging will show generalized and focal atrophy, in particular, affecting the hippocampi and medial temporal lobe [38]. Functional imaging with fluoro-2-deoxy-D-glucose positron emission tomography scans or single-photon emission computed tomography scans will show corresponding areas of low metabolism or hypoperfusion, respectively [39]. Direct imaging of the pathological hallmarks of Alzheimer's disease, in particular, is possible using Pittsburgh compound B, a modified amyloid stain that specifically labels amyloid plaques (see Section 2.3 for further discussion of this) [40]. Importantly, a combination of these approaches can be used to provide longitudinal analysis of brain structure and function, thereby facilitating identification of rate of change and degeneration [38].

Analysis of cerebrospinal fluid is increasingly used in the assessment of Alzheimer's disease. Measurement of cerebrospinal fluid amyloid beta and tau, and in particular, the amyloid beta:tau ratio reflect the neuropathological hallmarks of Alzheimer's disease [41,42] (see Section 2.3). These biomarkers can be used to support the diagnosis of Alzheimer's, particularly in early stages or in atypical cases, and are of increasing importance in efforts to develop novel treatments for dementia [43].

A definitive diagnosis of Alzheimer's is, however, pathological and hence achieved only at autopsy, with demonstration of the accumulation of extracellular amyloid plaques and intracellular neurofibrillary tangles.

2.2.7 MANAGEMENT

The management of Alzheimer's disease must be holistic, addressing not only the needs of the patient but also taking into consideration their carers, which means that social support should form a large part of the care package [44]. In terms of medical treatment, there is currently no disease-modifying or neuroprotective treatment. There are, however, a variety of symptomatic treatments available. Acetylcholinesterase inhibitors are a group of drugs that were specifically developed for Alzheimer's and were first used in the late 1990s. Their use is based on the neurochemical finding that the brains of patients with Alzheimer's display a reduction in the neurotransmitter acetylcholine. The benefit of these drugs is relatively modest; double-blind, placebo-controlled trials have demonstrated that donepezil, rivastigmine, and galantamine have the potential to mildly improve cognition, ability to carry out activities in daily living and behavior in patients with mild-to-moderate Alzheimer's for periods of between 6 and 18 months [45]. The other disease-specific treatment is memantine, which antagonizes glutamate at the NMDA receptor. It has been shown to improve cognitive function in moderate to severe disease [46]. A more detailed description of currently available treatments for Alzheimer's disease is included in Section 2.5.

2.2.8 FRONTOTEMPORAL DEMENTIA

Frontotemporal dementia is an umbrella term that describes a clinically, genetically, and pathologically heterogeneous group of disorders characterized by deficits in behavior, executive function, personality, or language. It is the second most common cause of early onset dementia [47], with a prevalance in the UK of 3.5 per 100,000 person years in the age range 50–64 [48]. It was first described in 1892 when Arnold Pick wrote about a patient with progressive aphasia and lobar atrophy [49]. In frontotemporal dementia, there is preferential involvement of the frontal and/or temporal cortices, which accounts for the presentation. Under the umbrella term of frontotemporal dementia, there are distinct clinical syndromes, the behavioral variant and the speech variants—also known as semantic dementia and progressive nonfluent aphasia [50].

2.2.9 BEHAVIORAL VARIANT FRONTOTEMPORAL DEMENTIA

The behavioral variant is the most common subtype, accounting for half of the cases of frontotemporal dementia. It is characterized by personality changes with apathy, disinhibition, and socially inappropriate behavior. Behavioral disinhibition can manifest in different ways including urinating in public, inappropriate sexual behavior, and the use of offensive or inappropriate language. Patients may also exhibit risky or impulsive behavior such as dangerous driving, gambling, substance abuse, or even criminal activity [51]. In addition, patients may become hyperoral with binge eating and a preference for sweet foods. Executive function is impaired; ritualistic or stereotyped behaviors can be seen; but memory and visuospatial skills are typically preserved initially. Patients typically lack insight and in the initial stages, the changes may be subtle and mistaken for psychiatric illness [52]. Neuroimaging typically shows focal frontal or temporal atrophy that is often asymmetrical with a right-sided predominance [53].

2.2.10 LANGUAGE VARIANT

Primary progressive aphasia is a subtype of frontotemporal dementia that causes an insidious decline in language function [52]. There are two main types; semantic dementia and progressive nonfluent aphasia. Semantic dementia is characterized by word finding difficulty and impaired word comprehension. Fluency and grammar are not affected, at least in the initial stages, but the speech content is often empty and circumlocutory. The main pathology starts asymmetrically in the left anterior temporal lobe and amygdala, but as the disease progresses to involve the frontal lobes behavioral changes may become apparent. A remarkable feature of some cases of semantic dementia is the emergence of hitherto unrecognized artistic talent. It has been postulated that this possibly releases a visually orientated right hemisphere with a greater capacity for art [54].

Patients with progressive nonfluent aphasia in contrast, develop slow effortful speech with agrammatism and difficulty with articulation. There may also be apraxia of orofacial movements, with a failure to yawn or cough on command. Word meaning is retained, and behavioral features are not seen until later in the disease process. Neuroimaging demonstrates atrophy of the left perisylvian fissure.

There is an overlap between motor neuron disease and frontotemporal dementia, particularly the behavioral variant. About 15% of patients with amyotrophic lateral sclerosis, a form of motor neuron disease, will have features of frontotemporal dementia, and more will have subtle features revealed upon cognitive testing [55]. Similarly, about 12.5% of patients with behavioral variant frontotemporal dementia will develop motor neuron disease [56]. Other neurological symptoms may also be seen in patients with frontotemporal dementia. Up to 20% of patients will have evidence of parkinsonism or an overlap with progressive supranuclear palsy due to tau pathology [57].

Unlike Alzheimer's disease, trials have not demonstrated clinical benefit for the use of cholinesterase inhibitors in frontotemporal dementia and have been shown occasionally to worsen symptoms [58]. Treatment is focused on management of the symptoms, particularly management of behavioral symptoms.

2.2.11 DOWN SYNDROME AND TRISOMY 21

It has been recognized for many decades that the clinical spectrum of Down syndrome, caused by the presence of an additional copy of chromosome 21 (trisomy 21), includes the onset of presenile dementia [59,60]. This is driven by the accumulation of amyloid plaque and neurofibrillary tangle formation in the brain, caused by an additional copy of the *APP* gene on chromosome 21 (see Section 2.4) [61]. The clinical presentation of dementia in individuals with trisomy 21 is complicated by the multifactorial nature of Down syndrome; however, it is notable that the age at onset for cognitive decline in such cases is in the fifth and sixth decade of life, significantly younger onset as compared with classical Alzheimer's disease [60].

2.2.12 DEMENTIA WITH LEWY BODIES

Dementia with Lewy bodies is a distinct dementia disease entity that shares many clinical aspects with the more common Alzheimer's disease [62]. It can be differentiated from Alzheimer's by neuropathological examination (see Section 2.3), where there is an absence of amyloid plaque or neurofibrillary tangle pathology. Importantly, dementia with Lewy bodies forms part of a clinical spectrum of disease with Parkinson's disease and Parkinson's disease dementia, classified based on the initial presenting symptom (that is, whether these are centered on declining cognition or on disorder of movement).

2.2.13 CHRONIC TRAUMATIC ENCEPHALOPATHY

Cognitive alterations were first documented in individuals with repeated head trauma in the 1920s, with the description of *dementia pugilistica* (originally described as being "punch drunk") in boxers [63,64]. Detailed studies of a number of contact sports [65,66], as well as other sources of traumatic brain injury, have highlighted head trauma as a major risk factor for dementia and other forms of neurodegenerative disease (see Box 2.2). These are now grouped together under the designation chronic traumatic encephalopathy [67]. In terms of clinical presentation, chronic traumatic encephalopathy is heterogeneous, with a range of symptoms that span cognitive decline through movement disorders similar to those seen in Parkinson's disease and amyotrophic lateral sclerosis (see Chapters 3 and 5) [68].

BOX 2.2 DEMENTIA PUGILISTICA AND CHRONIC TRAUMATIC ENCEPHALOPATHY

The impact of repeated head trauma on cognition and other brain functions has been recognized for many years, most notably in the identification of *dementia pugilistica* in boxers (as shown in *Club Night,* by George Bellows). Over the past two decades, it has become clear that these changes in brain function are not limited to boxers but can occur in a wide range of individuals who suffer repeated head trauma. This was highlighted by the case of Mike Webster, a National Football League professional football player who developed cognitive changes and was diagnosed with chronic traumatic encephalopathy and whose case was made famous by the neuropathologist Bennet Omalu.

2.2.14 OTHER DEMENTIA SYNDROMES

Corticobasal syndrome is classically described as a progressive asymmetrical movement disorder. Typically patients present with a "useless" arm due to a combination of symptoms including rigidity, dystonia, tremor, myoclonus, and apraxia [69]. The alien limb phenomenon is also seen when patients experience their limb moving without conscious control. About one-fifth of patients present with a gait disorder. Although corticobasal syndrome was originally thought of as a movement disorder, cognitive features are now recognized as a prominent part of the disease with difficulties with executive function, language, visuospatial function, and memory reported [70]. Corticobasal syndrome is the term used to describe the clinical phenotype and can be associated with a number of different pathologies, including tau, TDP 43, Alzheimer's pathology, and prion disease. Corticobasal degeneration is a distinct pathological entity characterized by tau inclusions in neurons and glia with gliosis and ballooned achromatic astrocytes. As well as the corticobasal syndrome, the pathology of corticobasal degeneration can be seen in patients with different clinical phenotypes including progressive supranuclear palsy, behavioral variant frontotemporal dementia, and primary progressive aphasia. A number of other neurodegenerative disorders include dementia as part of their spectrum of symptoms, notable examples being the prion disorders and Huntington's disease [71,72]. These are considered at greater length in Chapters 4 and 6, respectively.

2.3 PATHOLOGY

The pathological profile of dementia is multifaceted, involving a number of different key neuropathological hallmarks in addition to the cell loss indicative of neurodegeneration. Part of the complexity of the neuropathology of dementia stems from the range of different disorders that can be classified as dementia, and indeed the pathological profile of the different subtypes of dementia plays a critical role in the final diagnosis of individual patients' postmortem [73].

2.3.1 NEURONAL LOSS

As is to be expected for a neurodegenerative disorder, a defining feature of dementia is the loss of neurons in the brain. This loss of neurons is extensive and clearly visible at a macroscopic level in the end stages of the disease. The precise distribution of this cell loss varies depending on the associated clinical presentation—although a common theme is degeneration of the cerebral cortex. In Alzheimer's disease, there is widespread degeneration across the cortex and hippocampus, regions that are involved in higher cognitive functions including memory and spatial orientation (Fig. 2.1A) [74,75]. Cholinergic neurons are the main, but not exclusive, victim of this degeneration [76,77]. Specific dementia variants can have more localized loss of cells. Posterior cortical atrophy, as the name suggests, is characterized by marked neuronal loss in the occipital lobe of the cerebral cortex (located to the posterior of the brain) [32]. Frontal temporal dementia, in contrast, is characterized predominantly by neuronal loss in the frontal lobe [78]. In both cases, the spatial distribution of this cell loss correlates with the symptoms observed in these patients—visual alterations in posterior cortical atrophy, behavioral symptoms in

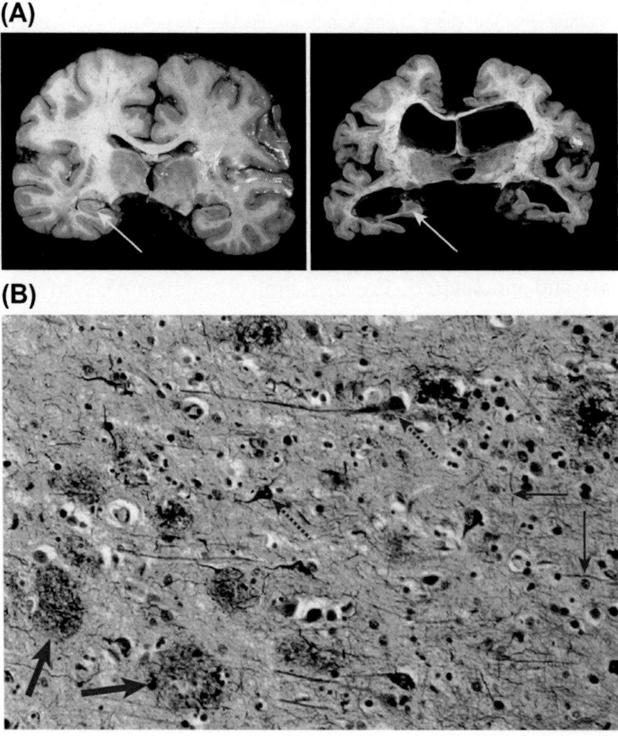

FIGURE 2.1

The neuropathology of Alzheimer's disease. (A) Coronal section of human brain showing macroscopic loss of brain mass due to neurodegeneration (right panel) contrasted with healthy brain (left panel). The hippocampus is highlighted in both cases with a *yellow arrow*. (B) Microscopic pathology of Alzheimer's disease in the human brain, showing amyloid plaques (*solid arrows*) and neurofibrillary tangles (*dotted arrows*).

Image adapted from Memory Loss, Alzheimer's Disease, and Dementia (Second Edition) A Practical Guide for Clinicians.

frontotemporal dementia. Importantly, and as noted above, these changes in regional volume within the brain can now be measured longitudinally using a variety of imaging techniques [38]. When these techniques have been applied to genetically determined forms of dementia, this has revealed that pathological changes occur many years before the onset of symptoms—casting new light on the disease process in dementia and highlighting a long, prodromal stage of disease [79].

2.3.2 AMYLOID PLAQUES AND NEUROFIBRILLARY TANGLES

Another defining neuropathological feature of dementia is the presence of abnormal aggregates of a range of proteins in the brain. The key role of the aggregates in the pathology of the dementias became apparent very early in the development of these disorders as clinical entities with the application of staining techniques in the late 19th and early 20th centuries, revealing the presence of extracellular plaques that became heavily stained when exposed to dyes that could bind to starch—resulting in them being named as amyloid plaques (*amylum* being the Latin for starch), and intracellular aggregates, given the name neurofibrillary tangles by Alzheimer [80,81]. Both amyloid plaques and neurofibrillary tangles are pathognomonic for Alzheimer's disease, with the presence of both being required for a categorical diagnosis of the disease [16,82]. Amyloid plaques, also known as neuritic plaques, are dense accumulations of the amyloid beta peptide. This small peptide, around 40 amino acids in length, is a product of the proteolytic processing of the integral membrane amyloid precursor protein (see Section 2.4 for a more detailed description of the production and role of amyloid beta in the pathogenesis of Alzheimer's disease) [83]. In individuals with Alzheimer's, this peptide adopts a beta sheet–rich structure and forms aggregated fibrils in the extracellular space. The plaques made up of this protein are roughly spherical, with a dense inner core surrounded by a halo of aggregated material (Fig. 2.1B) [84], although there is a degree of morphological diversity in amyloid beta deposition with some patients presenting with less dense accumulations, dubbed cotton wool plaques [85]. Within the brain, amyloid plaques are closely associated with activated glial cells—first noted by Alzheimer himself (see Box 2.1 and Section 2.3.4). Neurofibrillary tangles are aggregations of the microtubule-associated protein tau, assembled into paired helical filaments (Fig. 2.1B) [81]. These are found in the cytoplasm of neurons in the brains of individuals with Alzheimer's disease but are also associated with a range of other neurodegenerative disorders including chronic traumatic encephalopathy, corticobasal degeneration, progressive supranuclear palsy, and some forms of frontotemporal dementia [86].

For both plaque and tangle pathology in Alzheimer's disease, it has been proposed that there is a stereotypical progression of their occurrence through the brain—for example, through the Braak staging model for Alzheimer's [87]. This has been the subject of much debate within the neuropathology community, in particular, as to how representative the staging of pathology in different regions of the brain is and how generalizable this is across disease [88]. An even more controversial area of debate is the nature of the mechanisms underpinning this progression, with some suggestion that there is a spreading of protein aggregates analogous to an infectious process (see Section 2.4.5 and Chapter 4) [89]. That there is progression of pathology within the brain is, however, clear from analysis of early stage mild cognitive impairment compared with end-stage Alzheimer's disease. Indeed, it is also clear that these changes within the brain—both neuronal loss and protein aggregation—precede the onset of clinical symptoms by many years (Fig. 2.2) [38], an observation that has important implications for the development of treatments for dementia (see Section 2.5). With regard to the correlation of plaque and tangle formation, and the process of neuronal death and

(A)

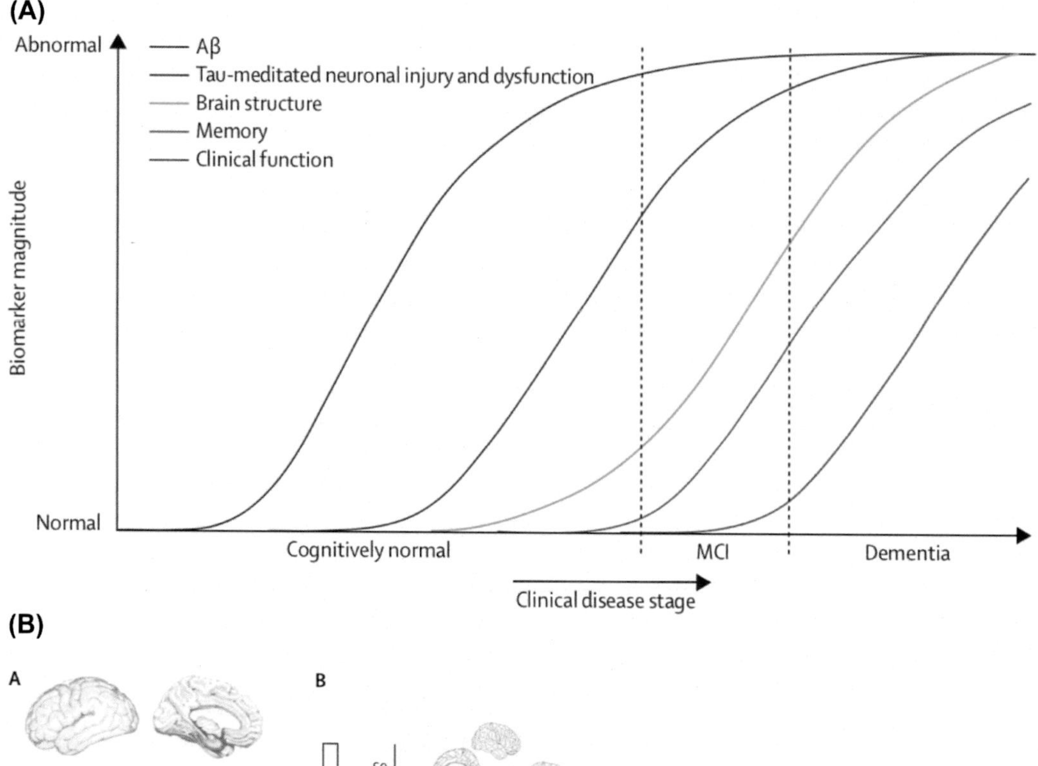

(B)

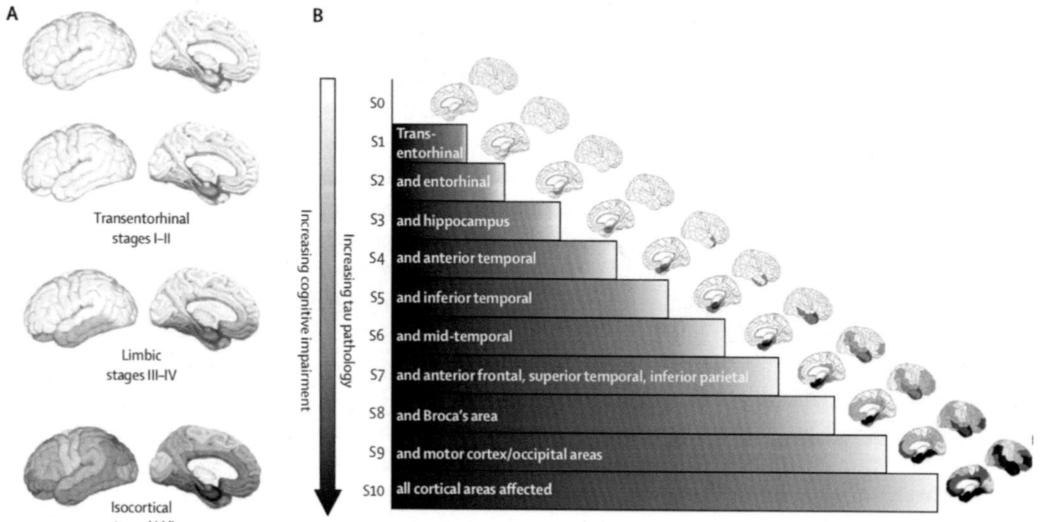

FIGURE 2.2

Neuronal loss and pathological markers through time (A) a model for the time dependent changes in the pathological markers for Alzheimer's disease, with symptomatic stage noted on the x axis. (B) The Braak staging model for the spread of tau pathology in the Alzheimer's brain.

Image adapted from Jack Jr, Clifford R, et al. Hypothetical model of dynamic biomarkers of the Alzheimer's pathological cascade. The Lancet Neurology 2010;9.1:119–8. (upper panel) and Villemagne, Victor L, et al. Tau imaging: early progress and future directions. The Lancet Neurology 2015;14.1:114–4 (lower panel).

cell loss, an observation with great relevance for our understanding of the molecular mechanisms of disease is that the burden of neurofibrillary tangles is much more closely aligned with neuronal loss than amyloid plaque number and location [90].

2.3.3 LEWY BODIES

Lewy bodies, first described in the brains of individuals with Parkinson's disease by Frederick Lewy in the 1920s [91], are another common protein aggregate inclusion frequently observed in dementia. Lewy bodies are composed predominantly of aggregates of the protein alpha synuclein—mutations which are associated with Parkinson's disease and dementia with Lewy bodies [92]. There are two main links between Lewy bodies and dementia. First, as indicated above, Lewy bodies are pathognomonic for dementia with Lewy bodies [62]. In this disorder, Lewy bodies and neuronal loss within the cerebral cortex (and in the absence of amyloid plaques and neurofibrillary tangles) are tightly correlated with the development of progressive cognitive dysfunction. A second key link relates to Alzheimer's disease. A substantial proportion of individuals with a pathological diagnosis of Alzheimer's disease, that is displaying plaque and tangle pathology, also harbor Lewy bodies within their brains [93]. The precise role that Lewy bodies play in the pathological events that result in neuronal loss in Alzheimer's disease is unclear, as is the specific nature of the interaction between alpha synuclein, tau, and amyloid beta, but it is clear that the Lewy body is part of the neuropathological spectrum found in the Alzheimer's disease brain, and acts to link Alzheimer's to Parkinson's disease.

2.3.4 GLIAL PATHOLOGY

A common feature of the brains of individuals with dementia is the presence of activated glial cell populations [94]. Both activated astrocytes and microglia are found in close proximity to the pathology found in the brains of dementia patients, excellently illustrated by the exquisite drawings presented by Alois Alzheimer in his original description of the disorder that bears his name (see Box 2.1). This reactive astrogliosis and microglial response is thought to be part of the brain's response to the destruction being wrought by the neurodegenerative process; however, there is increasing evidence that this reaction may exacerbate the underlying process of cell death (discussed in further detail in Section 2.4.6).

2.3.5 TAUOPATHIES

In addition to being a central pathological marker for the development of Alzheimer-type dementia, neurofibrillary tangles made up of the microtubule-associated protein tau are found in a number of other diseases that include dementia as part of their phenotype. Corticobasal degeneration, chronic traumatic encephalopathy, progressive supranuclear palsy, and some forms of frontotemporal dementia present with tau inclusions in the absence of amyloid pathology [86].

2.3.6 OTHER PATHOLOGIES

As will be clear from Section 2.2, the dementias are a heterogeneous group of disorders, and this is reflected in the pathologies that are observed in the brains of patients suffering with these forms of neurodegeneration. In addition to the pathologies noted above, there are a number of other specific changes and protein aggregates found in subtypes of dementia. In the prion diseases, dealt with at greater length in Chapter 4, dementia is accompanied by spongiform degeneration of the brain and by the occurrence of extracellular plaques composed of the prion protein [95]. In Huntington's disease, where dementia is part of the symptomatic spectrum of the disorder and described in detail in Chapter 6, intranuclear aggregates of huntingtin (the protein which is mutated in Huntington's disease) are found in the areas affected by neurodegeneration [96]. One very rare, but mechanistically very revealing, further example is provided by the protein deposited in the brains of a handful of families carrying mutations in the *ITM2B* gene [97]. Mutations in this gene result in the production of an abnormal peptide, distinct in sequence but similar in properties to the amyloid beta peptide, and leads to the development of an inherited dementia (originally described by Charles Worster-Drought in the 1930s) [98]. Importantly, this peptide is deposited in extracellular plaques that share many characteristics of those observed in the brains of individual with amyloid beta pathology—suggesting that the process connecting protein aggregation, amyloid (or amyloid-like) plaque formation, and neuronal death does not depend on the precise sequence of the peptide or protein involved.

2.4 MOLECULAR MECHANISMS OF DEGENERATION

Since Alzheimer first described the symptoms and pathology he observed in Auguste D in 1907, a huge amount has been learned about the molecular causes of Alzheimer's disease and the broader spectrum of the dementias. These advances, similar to those seen in most neurodegenerative disorders over the past century, have been driven by a combination of clinical characterization, pathological investigation, and genetic analysis. The last of these, in particular, has revolutionized our understanding of the causes of dementia across the whole spectrum of disease, providing direct evidence of the role of specific cellular processes in the underlying etiology of disease. Notably, advances in our genetic understanding of dementia have proceeded in step with insights from pathology and model systems to feed into drug development for novel treatments for these disorders (dealt with in further detail in Section 2.5).

2.4.1 NEUROGENETICS

The majority of patients developing dementia display no clear evidence of their disease being caused directly by inherited mutations. For Alzheimer's, approximately 90% of cases fall into this category, variously described as idiopathic or sporadic disease (that is, disease with no clear identified cause). It has been equally clear for a number of decades that a subset of patients with dementia develop cognitive dysfunction with a familial pattern of inheritance, displaying segregation of disease according to an autosomal dominant or recessive pattern. For some rare diseases that include dementia as part of their clinical presentation, a much higher proportion of cases are caused by inherited mutations—a key example of which is Huntington's disease, which is a purely genetic disorder (see Chapter 6 for further details). For Alzheimer's disease, between 1% and 5% of cases exhibit a Mendelian pattern of inheritance [99]. In addition to these cases, research over the past two decades has uncovered both strong risk factors that substantially alter your lifetime risk of developing disease (while not resulting in certain disease), and common variation in the human genome, which can result in a small, but statistically significant, increase in risk [100,101]. This is also the case for many rarer forms of dementia, and overlapping genetic architectures for both Alzheimer's disease and dementia more generally are now become clear.

2.4.1.1 Mendelian Forms of Alzheimer's—APP, PSEN1, PSEN2

Mutations in three genes have been identified to date that are directly causative for Alzheimer's disease. All three are inherited in an autosomal dominant fashion (with a few specific exceptions) and fall at a molecular level into a common pathway. The first of these to be described was mutations in the amyloid precursor protein gene located on chromosome 21 [102]. The identification of mutations in *APP* provides an excellent example of the intimate and reciprocal relationship between research into the pathology and genetics of the dementias and the contribution of molecular cell biology to joining these up. The *APP* gene codes for a 770 amino acid type I transmembrane protein (that is, a transmembrane protein with its N terminus in the extracellular space and its C terminus in the cytoplasm), a protein that is subject to a series of proteolytic cleavage events producing a range of peptides (Fig. 2.3). One of these protein products, the amyloid beta peptide, had first been linked with Alzheimer's disease in the early 1980s after its identification as a major component of the amyloid plaques found in the brains of individuals with Alzheimer's disease [103]. Intriguingly, and importantly from a genetics perspective, at the same time the investigators responsible for this report highlighted that the same peptide was found in the pathology observed in the brains of individuals with trisomy 21 [104]. Subsequent immunohistochemical analysis using antibodies specific for this peptide revealed that it could be identified in both the amyloid plaque pathology found in the Alzheimer's brain and in amyloid deposits surrounding the blood supply to the brain [105]. The gene from which this peptide was derived was later identified, localized to chromosome 21 and given the designation the amyloid precursor protein gene [106]. A clear implication of these studies was that genetic variation in this gene could be linked to the development of amyloid plaques (and other amyloid pathology) in the brain and therefore the underlying etiology of dementia and Alzheimer's disease [107]. This was supported by the increase in gene dosage identified in trisomy 21; however, the additional copy of the *APP* gene in Down's syndrome is accompanied by additional copies of most, if not all, genes found in chromosome 21 (Box 2.3) [61]. A major breakthrough came in 1990 with the identification of point mutations in the *APP* gene associated with cerebral amyloid angiopathy in a Dutch family [108]. This was soon followed by the identification of point mutations in the *APP* gene in familial Alzheimer's disease, initially in a family from London [109] and subsequently in a number of families with inherited, early-onset Alzheimer's—providing clear genetic evidence that the *APP* gene and the amyloid beta peptide were causatively linked to the onset of dementia in Alzheimer's disease. There is now an extensive list of point mutations in the *APP* gene, identified in multiple families from across the globe. These mutations cluster in the section of the gene coding for the amyloid beta peptide fragment, and in particular, around the cleavage sites for the alpha, beta, and gamma secretase enzymatic proteolytic activities (Fig. 2.4) and favor the production of either great amounts of the amyloid beta peptide or more aggregation-prone peptides (see Section 2.4.2 for a more detailed discussion of this). The data from these rare familial cases of Alzheimer's disease and cerebral amyloid angiopathy, combined with the Alzheimer's type pathology and symptoms observed in trisomy 21, raised the possibility that further copy number variations in the human genome could also be associated with Alzheimer's disease, and indeed these were identified in 2006, with gene duplications of the *APP* locus linked to autosomal dominant disease [110]. An intriguing further twist to the *APP* story is that a coding variant (A673T) in the *APP* gene was identified in 2012 by DeCode, the Icelandic genomics biotechnology company, as protecting against dementia and age-related cognitive decline [111].

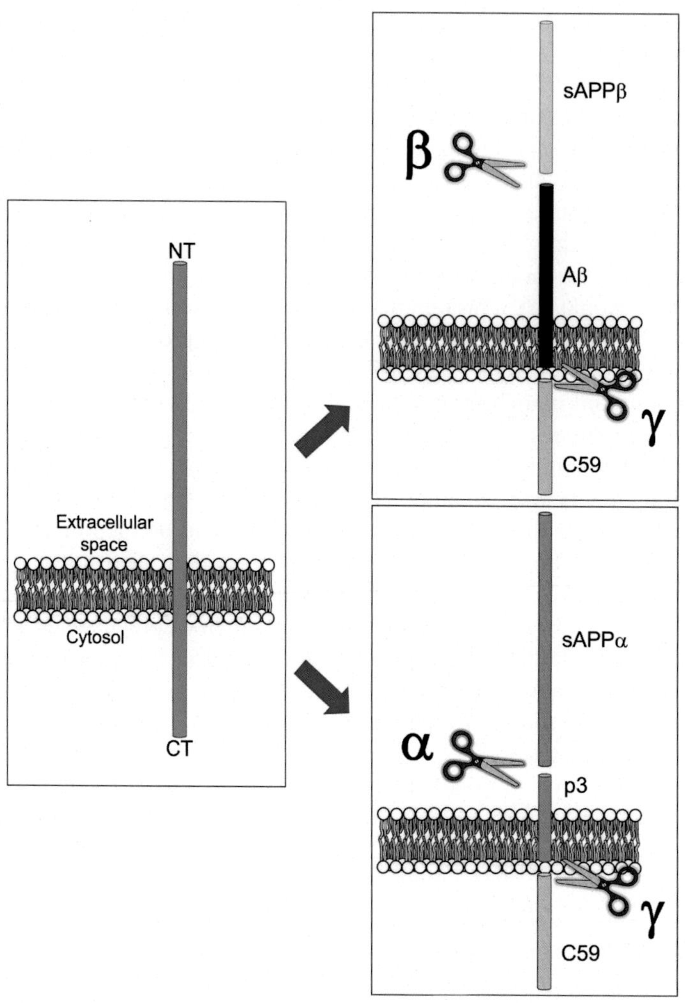

FIGURE 2.3

Amyloid precursor protein-processing summary. The processing of the amyloid precursor protein through the amyloidogenic pathway (upper panel cleavage by beta and gamma secretases) or the nonamyloidogenic pathway (lower panel, cleavage by alpha and gamma secretases).

The *APP* gene is, therefore, causatively linked to Alzheimer's through a number of different mechanisms. First, point mutations in the gene that alter the amino acid sequence of the protein result in early onset Alzheimer's disease/cerebral amyloid angiopathy. Second, increasing the production of the amyloid precursor protein due to gene or chromosomal copy number increase results in either Alzheimer's disease or Alzheimer's pathology/symptoms in trisomy 21. Less frequently, variants in the *APP* gene can protect against risk of dementia. Combined with the pathological importance of amyloid plaques in Alzheimer's, these data place the amyloid beta peptide at the heart of the pathological process leading to neurodegeneration.

BOX 2.3 TRISOMY 21 AND DEMENTIA

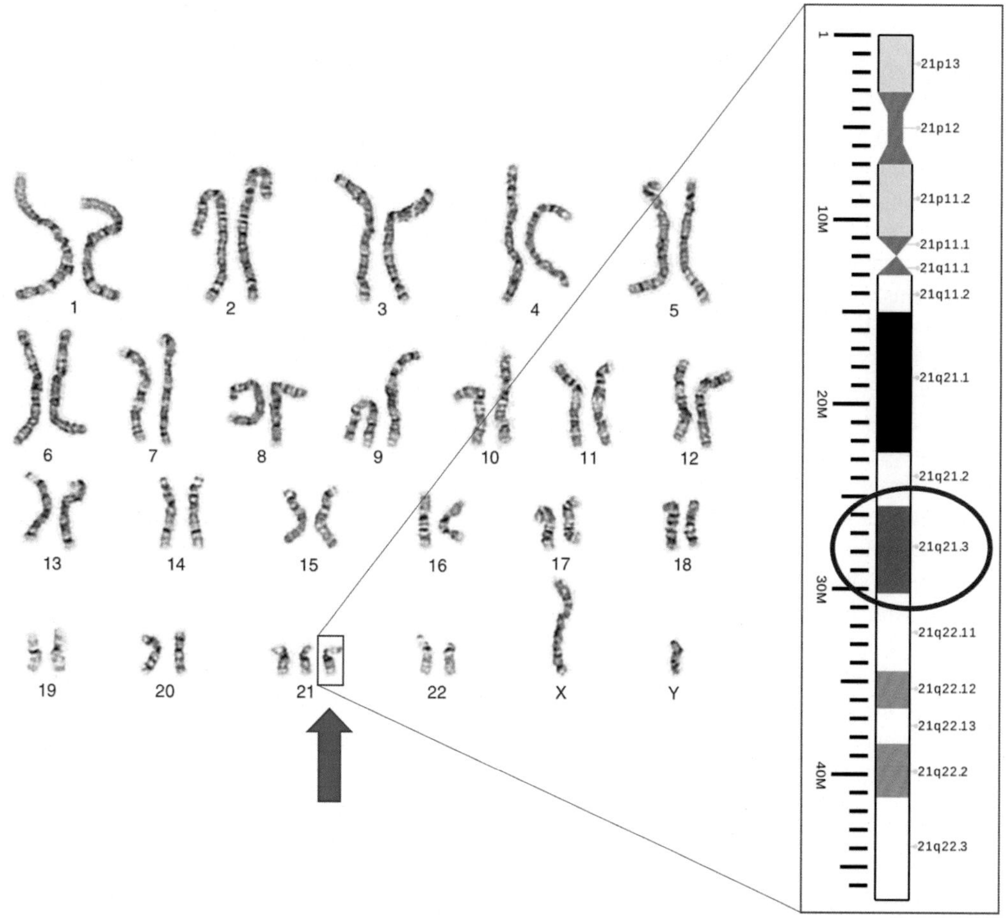

Trisomy 21 is a genetic condition where an additional copy of chromosome 21 is present, causing Down's syndrome. The additional copy of chromosome 21 includes the *APP* gene, coding for the amyloid precursor protein, and so individuals with trisomy 21 have 50% more amyloid beta peptide in their brains than those with two copies of this chromosome. As a consequence, people with Down's syndrome develop severe cognitive decline in the fifth and sixth decade of life and eventually will be diagnosed with Alzheimer's disease. *Images adapted from Wikimedia commons under a creative commons licence.*

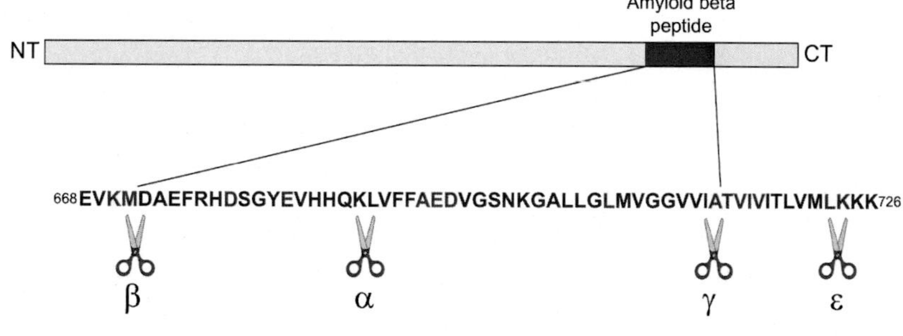

FIGURE 2.4

Mutations in *APP*. A summary of point mutations in the *APP* gene, highlighting secretase cleavage points.

Importantly, it was noted on the identification of mutation in the *APP* gene that these accounted for only a small percentage of cases of familial Alzheimer's disease—suggesting that other genes could also be playing a role in the genetic etiology of this disease. A second breakthrough in this area came in 1995, with the identification of point mutations segregating with autosomal dominant Alzheimer's disease in the *PSEN1* gene on chromosome 14 [112]. This was soon followed by the identification of mutations in a close ortholog of Presenilin 1, Presenilin 2 (encoded by the *PSEN2* gene on chromosome 1), also segregating with Alzheimer's disease [113]. The Presenilin genes code for large integral membrane proteins with multiple transmembrane domains and, at the time of their identification, uncertain function. What became clear after ensuing mechanistic studies was that point mutations in Presenilin 1 or Presenilin 2 caused an alteration in the production of amyloid beta peptide from the amyloid precursor protein, with an increase in longer fragments that were more prone to aggregation. As an intriguing postscript, it has been suggested that Auguste Dieter, the original dementia case described by Alzheimer in 1906 (Box 2.1) may have carried a mutation in *PSEN1* based on genetic analysis of preserved tissue samples from her brain [114]. These data, discussed in more detail in Section 2.4.2, linked together *APP*, *PSEN1*, and *PSEN2* into a pathway for the processing and production of the amyloid beta peptide. A more recent discovery has been the identification of coding variants in the *ADAM10* gene, coding for the enzyme responsible for the nonamyloidogenic secretary pathway for the amyloid precursor protein, associated with increased risk of Alzheimer's disease. These variants result in reduced proteolytic activity, favoring the production of amyloid beta and consistent with the impact of mutations in *APP*, *PSEN1*, and *PSEN2* [115,116]. The insights from human genetics in this area have provided a key rationale for the huge volume of subsequent cellular and animal model investigations into the biology of amyloid beta in Alzheimer's disease and dementia, and through this for drug discovery efforts and clinical trials [117].

2.4.1.2 Other Mendelian Genes

In parallel to genetic analysis of inherited Alzheimer's disease, similar approaches have been used to identify genes linked to a number of other dementias where there is a clear genetic component. These studies have provided crucial insights into the molecular causes of dementia and indeed have made major contributions to our understanding of Alzheimer's dementia. One of the most important discoveries in this area was the identification of mutations in the *MAPT* gene in 1998 associated with frontotemporal dementia [118]. Tau had been previously identified as a key player in the pathogenesis of Alzheimer's and a group of related neurodegenerative disorders based on the presence of neurofibrillary tangles in the brains of individuals with Alzheimer's, progressive supranuclear palsy, corticobasal degeneration, and *dementia pugilistica*/chronic traumatic encephalopathy; however, the causative link between tau and disease was unclear until this point. Mutations in tau include coding variants (for example, the P301L variant) as well as splice site acceptor mutations that alter the expression of tau, favoring the production of a larger form with increased inclusion of exon 10 [119]. As noted below (Section 2.4.2), these mutations alter the propensity of tau to associate with microtubules and to form aggregates within the brain and establish tau dysfunction as a primary cause of neurodegeneration. Beyond tau, mutations in the *SNCA* gene, coding for alpha synuclein, were found to cause Parkinsonism and dementia with Lewy bodies in a family of Greek and Italian ancestry—the Contursi kindred—in 1997 [120]. This discovery led to the identification of alpha synuclein as the main component of Lewy bodies later that year [121], a key discovery with implications across Alzheimer's disease, dementia with Lewy bodies, and Parkinson's disease (see Chapter 3 for further details relating to alpha synuclein).

Mutations in the *HTT* gene, located on chromosome 4, are inextricably linked to the development of Huntington's chorea (dealt with in greater detail in Chapter 6) [96]. Dementia is a core component of the disease phenotype in Huntington's, and so the identification of mutations in the *HTT* gene causing a polyglutamine repeat expansion that results in increased aggregation [122] reinforces the role of protein aggregates in the disease process leading to dementia.

As noted in Section 2.3.6, the identification of protein aggregates in the brains of individuals with British and Danish familial dementia with a similar etiology to Alzheimer's amyloid plaques, but divergent protein constituents provided a fascinating insight into the processes driving disease in the brains of individuals with dementia. This research was derived from the identification of a mutation in the *ITM2B* gene, and the subsequent molecular characterization of the impact of these mutations on the proteolytic product of the *ITM2B* protein (causing an extension of the peptide released by the protease furin in both cases) [97,123].

A final example of a disease with a prominent dementia component and genetic antecedents, and again one that has provided significant mechanistic insight into the process driving dementia, is Creutzfeldt–Jakob disease and related transmissible spongiform encephalopathies (also known as the prion diseases, the subject of Chapter 5). These disorders have a complex, tripartite etiology where disease can be sporadic in origin, driven by environmental exposure to infectious protein aggregates or by the presence of autosomal dominant mutations in the *PRNP* gene located on chromosome 20. These mutations provide a genetic link between protein aggregation, propagation of these aggregates, and a quasi-infectious process that has had implications across neurodegenerative disease and beyond (see Section 2.4.4 and Chapter 4) [124].

2.4.1.3 Risk Factors

In addition to very rare mutations in genes that are directly causative for dementia, the rapid development of human genetic analysis over the last 30 years has facilitated the identification of a large number of variants in the human genome that alter lifetime risk of developing disease.

The earliest of these to be described, and still one of the most important from both a clinical and mechanistic perspective, is variation in the *APOE* gene. This gene, located on chromosome 19, was first associated with risk of Alzheimer's in a case–control study from 1993 and codes for apolipoprotein ε [125,126]. Apolipoprotein ε is part of a family of proteins that act to coordinate metabolism of lipids in the body and has a particular role in the trafficking and metabolism of cholesterol. The *APOE* gene has four common allelic variants: apolipoprotein ε2, ε3, and ε4. These are distinguished by coding changes in the amino acid sequence at positions 112 and 158 (coding for either cysteine or arginine). The 1993 study described an overrepresentation of the ε4 allelic variant—in which both residues are arginine—in late-onset Alzheimer's disease, larger studies since then have both replicated and refined this finding [127]. An individual who is heterozygous for the ε4 allele (that is, one copy of ε4 and one copy of another allelic variant) has an odds ratio of 3.2 compared with those with the ε3 allele—this equates to having over three times the risk of disease. For individuals who are homozygous for ε4 this risk increases to an odds ratio of 14.9—with a 90% probability of developing disease over their lifetime. *APOE* status also has an impact on mean age at onset of symptoms for those who develop disease, with a younger onset for homozygous individuals versus heterozygotes versus noncarriers. Interestingly, possessing the ε2 allele results in a lower odds ratio compared with ε3 homozygotes, suggesting that this form of apolipoprotein can protect against the onset of dementia. The significance of *APOE* variation becomes clear when the frequency of the ε4 allele in the general population is considered—at more than 13% this is a variant that is a key genetic contributor to the incidence of Alzheimer's disease. This is emphasized when the extreme rarity of the Mendelian forms of disease is contrasted with that of *APOE* ε4.

Despite the large number of people potentially impacted by this genetic variant, the precise mechanism driving the association of the ε4 allele with the onset of Alzheimer's disease is not completely clear. There is evidence of a close link between the function of apolipoprotein and the clearance of amyloid beta within the brain, with the ε4 allele resulting in reduced amyloid clearance [128]. The role of the *APOE* gene in the transport of cholesterol has also been highlighted as a potential link to risk for Alzheimer's, as a number of studies have demonstrated that cholesterol levels in the cell membrane can act to modulate the processing of the amyloid precursor protein [129]. Regardless of precise mechanism, the sheer number of individuals carrying the dementia-associated allelic variant of *APOE* make this one of the most important—and clinically relevant—genetic associations with dementia.

A second major risk gene was identified in 2012 by two groups—one studying the genetics of the Icelandic population and one examining a rare form of dementia coupled with bone cysts called Nasu-Hakola disease. Both studies identified coding variants in the *TREM2* gene, including an R47H amino acid change, as significantly increasing risk of developing dementia [130,131]. Although not as frequently found in the population as the *APOE* risk alleles, the identification of *TREM2* as a risk gene for Alzheimer's has had a very important functional consequence in that it raised the profile of glial cell biology in genetic forms of dementia. *TREM2* is predominantly expressed in the brain in microglial cells and plays an important role in innate immune response. Combined with other, more recent gene variants linked to innate immunity (see below) [132], this has opened up a new front in mechanistic studies of the cell biology of Alzheimer's.

An area where major strides have been made over the past decade is that of genome-wide association of variants linked to lifetime risk of dementia. As noted in Chapter 1, the rapid development of cheap and robust methods to assess the human genome has made possible large-scale genetic analysis of patients with dementia and control populations. This, in turn, has led to the identification of a large number of loci scattered throughout the human genome associated with increased risk of developing dementia. A key strength of these studies is that they do not focus on powerful, but rare, familial mutations. Genome-wide association studies, instead, examine common variation, variants that are found in large numbers of the individuals, but that may have only a modest impact on the lifetime risk of developing disease [133]. Importantly, this means that genetic insights garnered from these studies can have implications for the more common idiopathic forms of dementia. The scale of genome-wide association studies can be breathtaking. For Alzheimer's disease, which (as the most common form of dementia) represents the neurodegenerative disease with the largest pool of patients to draw from, the most recent metaanalysis of genome-wide association included data from more than 25,000 individuals with Alzheimer's and more than 40,000 controls [134]. This resulted in the identification of 20 loci that reached genome-wide significance for increased lifetime risk of developing Alzheimer's disease. Notably, this included the *APOE* locus and the coding variants found in the ε4 allele. Beyond *APOE*, a number of genes linked to the metabolism and trafficking of the amyloid precursor protein have been identified as being associated with increased lifetime risk of Alzheimer's. An example of this is *SORL1*, a gene that is involved in the trafficking and endocytosis of the amyloid precursor protein—a gene that had previously been implicated in Alzheimer's disease [135]. In addition, a number of genes linked to immune response were identified, again highlighting the role of immunity and inflammation in the etiology of dementia. This study represents only one of the many that have been conducted for

Alzheimer's disease [136], studies that have been matched by equivalent analysis for dementia with Lewy bodies [137] and frontotemporal dementia [138]. These investigations represent an unprecedented attempt to decode the genetic architecture of these disorders and are particularly revealing when used to compare and contrast the pathways highlighted in clinical-related disorders—such as Alzheimer's and dementia with Lewy bodies—that have been revealed to be genetically divergent.

It is important to note, however, that these major undertakings have significant drawbacks. Most notably, genome-wide association studies do not, in and of themselves, necessarily identify genes that are linked to risk of disease. They pinpoint, instead, regions of the genome, which may (or may not) directly correspond to a particular variant in a particular coding region that alters the function of a specific gene. A major challenge, therefore, is to identify precisely what biological function is linked to the genetic risk highlighted by this type of study, an undertaking that is not trivial [139]. It is also critical to remember that the majority of variants identified by genome-wide association in the field of dementia account for small (>1%) increases in lifetime risk—an increase that has very little relevance in terms of predictive power in a clinical setting.

Taken together with studies of familial disease, however, genome-wide association studies provide an important part of the genetic information that has allowed scientists to construct a genetic architecture for dementia—a process that is most advanced for Alzheimer's disease (Fig. 2.5) [140]. An architecture such as this spans rare, but highly penetrant or directly causative, inherited forms of disease through to much more common, but phenotypically weaker, variants linked to disease. Combined, the genetic framework provided by such studies has provided a key part of the rationale for functional studies in dementia, studies that have led to significant insights into disease pathology and are a major driver of drug development for these disorders.

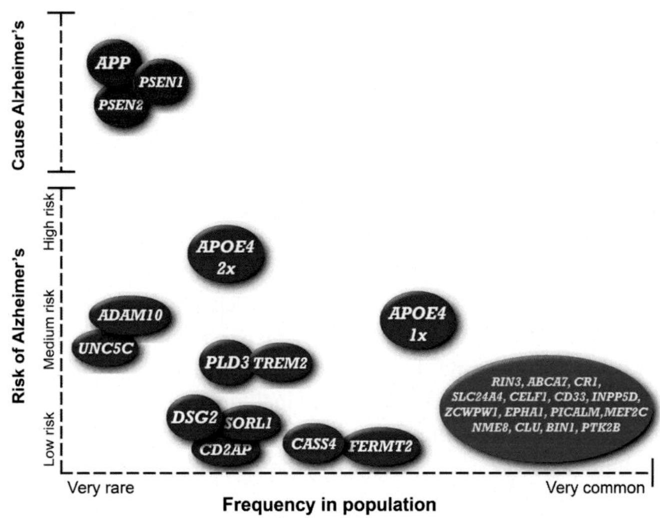

FIGURE 2.5

Genetics of Alzheimer's disease. A genetic architecture for Alzheimer's disease, showing high-risk, low-frequency variants (in red, top left corner), medium- to low-risk variants (blue) and genome-wide associated hits with high frequency but low lifetime risk of disease increase (green).

2.4.2 ANIMAL AND CELLULAR MODELS FOR DISEASE

Genetic insights into the causes of dementia have been translated into animal and cellular models of disease, using genetic engineering (and more recently genome editing) to express or modulate the genes that have been identified. A huge amount of effort has been invested in generating rodent models for dementia, with the majority of these being mouse models. Soon after the identification of mutations in the *APP* gene, attempts were made to generate transgenic mice overexpressing *APP* or mutant variants of this gene [141]. Initially, these were relatively simple—with the earliest examples expressing either wild-type *APP* or one of the familial mutations PDAPP [142]. Over time, however, these became progressively more sophisticated—for example, with transgenic mice generated to overexpress human *APP* carrying five different mutations in the gene [143]. Overexpressing *APP*, regardless of how many mutations it contains, results in an overall boost to amyloid beta production in the central nervous system of the mice, resulting (in most cases) in the deposition of the amyloid beta peptide in pathological inclusions similar to those observed in the brains of individuals with Alzheimer's disease. Notably, however, this occurs without widespread neuronal loss and in the absence of tau pathology. Although some alterations in behavior were observed in a number of these mouse models, it is unclear how closely this mimics the human condition. It is also important to note that overexpression of *APP* results in increased levels of all of the products of amyloid precursor protein cleavage. This has made it technically challenging to examine, for example, the pathological impact of just amyloid beta 1-42—although mice expressing a hybrid of the *ITM2B* gene spliced with amyloid beta have allowed this to be partially addressed [144]. After the identification of mutations in the *MAPT* gene, transgenic mice were generated expressing a range of different forms of mutated tau, including the P301L point mutation [145]. These developed tangle-like inclusions, providing potential insight into the mechanisms of the tauopathies, and led to the generation of multigene transgenic mice expressing mutations in *APP*, *MAPT* and in *PSEN1* [146]. Taken in combination, these triple transgenics replicate many of the pathological features of Alzheimer's disease, including amyloid plaque-like and neurofibrillary tangle-like pathology. As such, they have provided a valuable (if controversial) tool for understanding the biology of dementia and for drug development [147]. The final protein in the triad of pathologies observed in Alzheimer's is alpha synuclein, and again discoveries in human disease genetics have resulted in the generation of mice overexpressing human genes carrying mutations linked to disease (in this case dementia with Lewy bodies) [148]. One interesting twist in relation to alpha synuclein transgenic mice, highlighting the potential pitfalls of using mice to model complex human neurodegenerative disease, is that the murine *SNCA* gene (which codes for alpha synuclein) already harbors the original mutation identified in alpha synuclein from the Contursi kindred [149]. This complicates the interpretation of this model, and indeed many of the models for alpha synuclein dysfunction.

A large number of nonmammalian model systems have also been used in an attempt to dissect the events that result in neurodegeneration and dementia, including the nematode worm *Caenorhabditis elegans* [150], simple yeast cells [151], and the fruitfly *Drosophila melanogaster* [152]. Despite evolutionary distance from humans, these have provided important insights into the underlying biology of many of the genes linked to dementia.

None of the models described thus far, however, investigate dementia in the context of human neurons or other brain cells. Technical and ethical challenges prevented such studies, until the recent development of induced pluripotent stem cell technology allowed the generation of stem cells capable of differentiating into a range of human cell types [153]. These include both neurons and glial cells such as astrocytes and have been generated for both wild-type controls and from individuals carrying mutations in some of the genes closely linked to dementia, such as those in *APP* [154]. These have revealed cellular dysfunction linked to these mutations, as well as providing a potentially invaluable tool for drug discovery [155]. Cellular models such as these, however, must by their very nature reduce the complexity of the system being observed compared with that found in the human brain. In an attempt to address this, recent experiments have attempted to weld together the organismal complexity presented by mouse models for dementia and the human-specific benefits of neuronal cells derived from induced pluripotent stem cells by transplanting human cells into mice [156]. Intriguingly, this resulted in the human neuronal cells generating pathology similar to that observed in the brains of humans carrying the equivalent mutations.

2.4.3 PROTEIN AGGREGATION AND DISEASE

A recurring theme in the etiology of neurodegeneration is that of protein misfolding and aggregation. This is especially true of dementia, where the aggregation of amyloid beta, tau, alpha synuclein, huntingtin, and the prion protein are just some examples of proteins where the accumulation of misfolded aggregates is directly linked to the initiation and progression of disease.

In Alzheimer's disease, the centrality of the amyloid plaque to the pathology and definition of the disorder has long made protein aggregation one of the most important aspects of the underlying causes of degeneration. Indeed, the interplay between neuropathology, molecular etiology, and genetics was critical in establishing amyloid beta as a key part of the pathways that lead to Alzheimer's. Amyloid beta first came to light with two seminal studies carried out during the early 1980s, where researchers at the University of California San Diego identified a short peptide from plaques isolated from the brains of individuals with Alzheimer's disease and trisomy 21 [103,104]. The subsequent identification of the *APP* gene on chromosome 21 [106], and then the discovery of mutations in this gene linked to familial Alzheimer's disease [109], provided conclusive proof that the amyloid beta peptide was central to the disease mechanisms driving the development of Alzheimer's disease. Detailed mechanistic cellular analysis of the processing of the amyloid precursor protein, initially focused on the impact of mutations in *APP*, but soon coupled to those found in *PSEN1* and *PSEN2*, resulted in the realization that a shift in the production of amyloid beta was the underlying insult in the Mendelian forms of Alzheimer's disease. To understand this, it is critical to understand the proteolytic events that result in the production of amyloid beta from its precursor protein (as shown in Fig. 2.3). This is a complex process, with cleavage by alpha or beta secretase activities at specific amino acids in the extracellular portion of the amyloid precursor protein in tandem with a sequence-independent intramembranous cleavage event mediated by the gamma secretase enzymatic activity—resulting in a variable-length peptide with predominant forms of 40 or 42 amino acids [83]. As noted in Section 2.3, the majority of mutations in *APP* linked to Alzheimer's disease increase the processing events that lead to the production of amyloid beta ahead of the nonamyloidogenic pathways. Gene duplication, either of just the *APP* locus or due to an extra copy of chromosome 21, drives a 50% increase in production of all forms of the products of amyloid precursor protein processing—thereby increasing both amyloid beta 40 and 42. A critical insight came with the identification of mutations in Presenilin 1 and 2 [112,113], and the demonstration that the majority of these mutations shifted the balance of amyloid beta peptide production from shorter forms of the peptide to longer forms, and in particular amyloid beta 1-42 [164,165]. These reinforced efforts to understand the biochemical behavior of the amyloid beta peptide, and in particular, its propensity to form aggregates within the brain. The dense accumulation of amyloid beta in plaques, as well as the characterization of the fibrils that make up these plaques [84], had previously highlighted that misfolding and aggregation of amyloid beta were occurring in the brains of individuals who developed Alzheimer's disease [166]. The aggregation of amyloid beta into ordered multimeric states follows a clear kinetic pathway that mirrors that observed across amyloid-forming proteins—a large number of which have now been described [167]. Importantly, in vitro analysis of the aggregation of amyloid beta peptides clearly indicated that the longer forms of the protein have a greater propensity to aggregate than the shorter forms, implying that the events that favor the formation of these aggregates are central to the early etiology of Alzheimer's disease [168]. Further evidence that the kinetics of aggregation of amyloid beta are a critical factor in Alzheimer's disease was the identification of a coding mutation in the amyloid precursor protein that did not alter the processing of the protein per se but did impact on the rate of aggregation of the peptide produced [169].

In terms of the characterization of the molecular nature of the processing of the amyloid precursor protein, significant advances were made in the late 1990s and early 2000s—in particular, the identification of the proteases responsible for the alpha, beta, and gamma secretase cleavage events [83]. The enzyme responsible for the beta secretase cleavage of the amyloid precursor protein was identified by several groups working in parallel in 1999, with the protease in question dubbed BACE (for beta site amyloid precursor protein cleaving enzyme) [170–172], the structure of which was revealed in 2000 (Fig. 2.6A) [173]. The nature of gamma secretase, very clearly critical for the processing steps associated with the pathogenesis of Alzheimer's disease, proved enigmatic for many years. This was partly due to the atypical nature of the proteolytic event that it carried out—most proteases have clear sequence specificity, and do not operate in a hydrophobic intramembrane environment. Gamma secretase, in contrast, has poorly defined sequence specificity and operates intramembranously. Given the impact of mutations in the Presenilins on gamma secretase activity, these proteins were strongly implicated as being closely linked to this proteolytic event but did not share homology with any known protease. Over a number of years, however, and through the work of a large number of research groups, data began to emerge that suggested that the Presenilin proteins were integral components, supplying key residues for

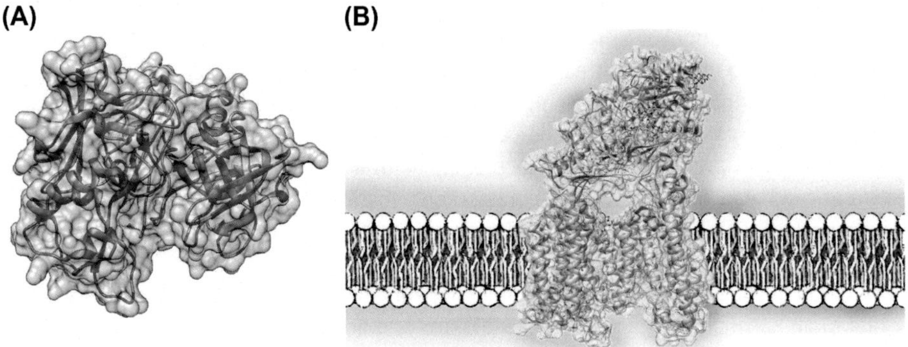

FIGURE 2.6

Atomic resolution structures for the secretases. (A) Crystal structure for BACE-1, responsible for the beta secretase cleavage of the amyloid precursor protein. (B) Cryo-electron microscopy structure for gamma secretase, made up of Presenilin 1, APH1-A, PEN-2, and nicastrin. Structural images created using Chimera software, based on coordinates from PDB 1M4H and 5A63 [173,175].

the active site, of a multiprotein enzymatic complex capable of carrying out gamma secretase cleavage of the amyloid precursor protein [174]. This complex, consisting of a Presenilin protein and a number of other transmembrane proteins including nicastrin, aph1, and Pen-2, has now been characterized to an atomic level using cryoelectron microscopy [175] (Fig. 2.6B). The non-amyloidogenic alpha secretase cleavage event can be carried out by a number of structurally related members of the ADAM family of proteases [176], with the key player being ADAM10 [177]. The role of ADAM10 has been further emphasized by the identification of genetic variants in the gene coding for this protease linked to risk of developing Alzheimer's, variants that reduce ADAM10 activity [115,116].

Finally, more recent research has identified a fourth cleavage event—nominated as the epsilon secretase cleavage. This occurs to the c terminal of the amyloid precursor protein, beyond the gamma secretase cleavage and is mediated by the same multiprotein complex that carries out the gamma cleavage [178]. It is thought that the epsilon cleavage of the amyloid precursor protein acts as a priming event for some gamma secretase cleavage events.

Combined, we now have a comprehensive overview of the enzymes involved in the processing of the amyloid precursor protein and an increasingly high-resolution picture of the precise molecular events that drive the production of the amyloid beta peptide [179]. This information has been critical for efforts to develop a disease-modifying treatment for Alzheimer's (see Section 2.5.2).

Thus far, only the production of the amyloid beta peptide and how this might influence aggregation has been discussed. Of equal importance to how amyloid beta is produced, however, is how this peptide is degraded and removed. Here, once again, human genetics has provided pointers as to important pathways relating to this and relevant to the development of dementia. Most notably, this has been through the identification of genes associated with risk of dementia and that have been closely implicated in cellular events linked to the clearance of amyloid beta—key examples being apolipoprotein ε (identified through case–control analyses) and complement receptor 1 (identified through genome-wide association studies for Alzheimer's disease) [180,181]. There are also extensive data linking insulin-degrading enzyme and the clearance of the amyloid beta peptide [182]. When considering the clearance of amyloid beta peptides, and this hold true for other aggregates found in the degenerating brain, it is important to examine both the cellular clearance pathways that act to remove waste in the cell (including the cellular processes of the ubiquitin proteasome system and autophagy) [183,184] and the clearance mechanism that collect up aggregates from the extracellular space—notably phagocytosis [185], a process closely linked to immune response in the brain (see Section 2.4.6). With regard to the former, there is increasing interest in the general governance of proteostasis in the brain and brain cells, and the possibility of system-wide disruption of this contributing to the build of protein in dementia [186]. In all of these cases, disruption of the mechanisms to degrade and remove amyloid beta peptide will have significant consequences on the absolute levels of this peptide in the brain. This, in turn, will increase the likelihood of further aggregation events and subsequent degeneration. Many of the pathways regulating protein turnover, for example autophagy, have become the focus of drug discovery efforts for neurodegenerative disease [187]. To date, however, these have not progressed to clinical trials.

The second major protein player in aggregation linked to Alzheimer's disease is the microtubule-associated protein tau. This was first implicated in the disorder after the isolation and characterization of the constituent proteins making up neurofibrillary tangles [188], with subsequent investigations noting abnormally high phosphorylation of this protein as being associated with the disease state in Alzheimer's disease [189,190]. A critical discovery was that mutations in the *MAPT* gene (coding for tau) cause frontotemporal dementia [118]. This shifted the balance of research toward understanding

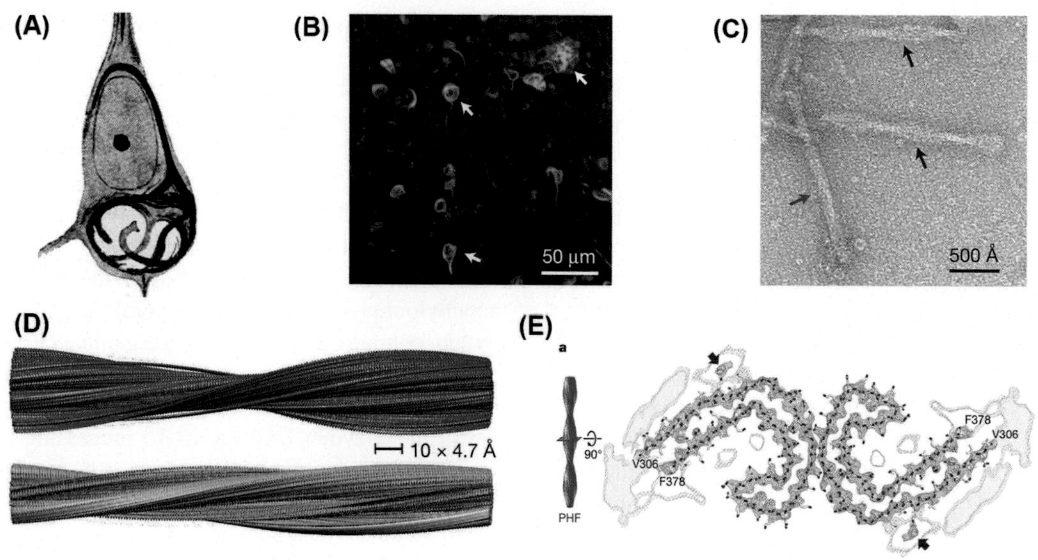

FIGURE 2.7

The molecular structure of tau from the human brain. (A) Drawing of neurofibrillary tangle by Alois Alzheimer. (B) Neurofibrillary tangles (*yellow arrows*) and amyloid plaque plaque (*white arrow*) in the human brain. (C) Electron micrograph of tau-paired helical filaments (*blue arrows*) and straight filaments (*green arrow*) can be seen. (D) The atomic structure of paired helical filaments (blue) straight filaments (green) revealed by cryo-electron microscopy reconstruction. (E) Cross-section through the core of a paired helical filament showing molecular interactions.

Images (B–E) taken from Fitzpatrick AWP, Falcon B, He S, Murzin AG, Murshudov G, Garringer HJ, Crowther RA, Ghetti B, Goedert M, Scheres SHW. Cryo-EM structures of tau filaments from Alzheimer's disease. Nature 2017;547:185–90.

tau dysfunction and accumulation as a cause of neurodegeneration in dementia. The *MAPT* gene, located on chromosome 17, is a complicated locus and produces a number of different mRNA transcripts produced through alternative splicing of the several exons in the gene. Splicing of exon 10 in particular has important relevance for disease, as the first mutations identified in *MAPT* alter splicing of this exon and increase the proportion of expressed protein containing the amino acids coded for by exon 10 (after this a range of point mutations have been described). The normal role of tau is to bind to and stimulate the assembly of microtubules made up of tubulin [119]. Inclusion of exon 10, and hyperphosphorylation of tau protein, favors the dissociation of tau from microtubules and increases the concentration of free tau in the cytoplasm. Here, under the right conditions, tau can begin to form small oligomers and eventually develop into mature fibrils—forming the paired helical filaments that sit at the heart of neurofibrillary tangles. Recent research using cryoelectron microscopy has revealed the structure of these fibrils in exquisite atomic detail, beginning to shed light on the precise molecular characteristics of such aggregates (Fig. 2.7) [191].

A final protein to consider in detail with regard to protein aggregation in dementia is alpha synuclein. Although most closely associated with Parkinson's disease (see Chapter 3), alpha synuclein and the Lewy bodies that it forms in the brains of individuals with Parkinson's, Alzheimer's, and dementia with Lewy bodies, are also important contributors to the degenerative process in dementia. Indeed, the 140 amino acid alpha synuclein protein was first identified as being linked to neurodegeneration by its isolation from preparations of amyloid material from Alzheimer's brain—being described as the non-amyloid component of plaques (or NACP) at the time [192,193]. The subsequent identification of mutations in the *SNCA* gene [120], coding for alpha synuclein, and the realization that this protein was the major constituent of Lewy bodies [121], placed this protein center stage in the events leading to cell death in the brain. Importantly, and echoing results examining the pathobiology of amyloid beta and tau, it is clear that the aggregation of alpha synuclein is a key event in the pathway leading to neurodegeneration in individuals with Lewy bodies and disease. Coding mutations in alpha synuclein, such as the A53T mutation identified in the Contursi kindred, all act to modify the aggregation of this protein, favoring the formation of amyloid aggregates or of oligomeric intermediates [194]. *SNCA* gene duplications or triplications, both of which are very clearly causative for disease and lead to widespread Lewy body pathology, also act to increase the critical mass of alpha synuclein within brain cells—and increase the propensity of the protein to aggregate. This latter contribution to disease has direct parallels with the increased production of amyloid beta in cases with *APP* gene duplication or trisomy 21 and has led to suggestions that this is a unifying theme in neurodegeneration [195].

2.4.4 THE AMYLOID CASCADE

The insights from human genetics and models, both cellular and organismal, relating to dementia highlight a number of events that are central to our understanding of the etiology of these disorders. Several attempts have been made to synthesize these into pathways or overarching hypotheses for disease—the most influential of which has been the amyloid cascade hypothesis. This hypothesis was proposed in 1992 by John Hardy and Gerald Higgins [157] and stated that the accumulation and aggregation of the amyloid beta peptide was the initiating event in a toxic cascade that resulted in cell death and, eventually, dementia in individuals with Alzheimer's disease and trisomy 21. The hypothesis has been revisited a number of times and has evolved over time to incorporate new discoveries and insights [158,159]. A simplified summary of the amyloid cascade hypothesis is that alterations in the production or clearance of amyloid beta results in the accumulation of either amyloid beta 1-40 and 1-42, or a shift in production to the longer form. These peptides begin to aggregate within the brain, forming small oligomers that may have a direct toxic impact (see Section 2.4.5), but also drive dysfunction of tau, resulting in it dissociating from microtubules and becoming hyperphosphorylated. This, in turn, results in the aggregation of tau eventually leading to the formation of neurofibrillary tangles, the pathway to which involves the generation of directly toxic species of tau aggregates which result in cytotoxicity (Fig. 2.8). Feeding into this cascade, and an area where there is increasing interest in terms of the underlying mechanisms driving degeneration, is an immune response within the brain and inflammatory processes. These secondary events may act to magnify and exacerbate the cell death initiated by exposure to protein aggregates. It should be emphasized that the amyloid cascade hypothesis is exactly what it is says: a hypothesis. Although there is significant evidence suggesting that it describes at least some of the events that form the pathway leading to degeneration in Alzheimer's dementia, the veracity of the hypothesis has been subject to criticism for a number of years [160–162].

A key test of the hypothesis is the removal of one or more steps in the cascade resulting in the slowing or halting of the disease process—a test that has been the underpinning driving significant efforts in drug discovery over the past several decades [163]. Such efforts have, to date, failed to yield any

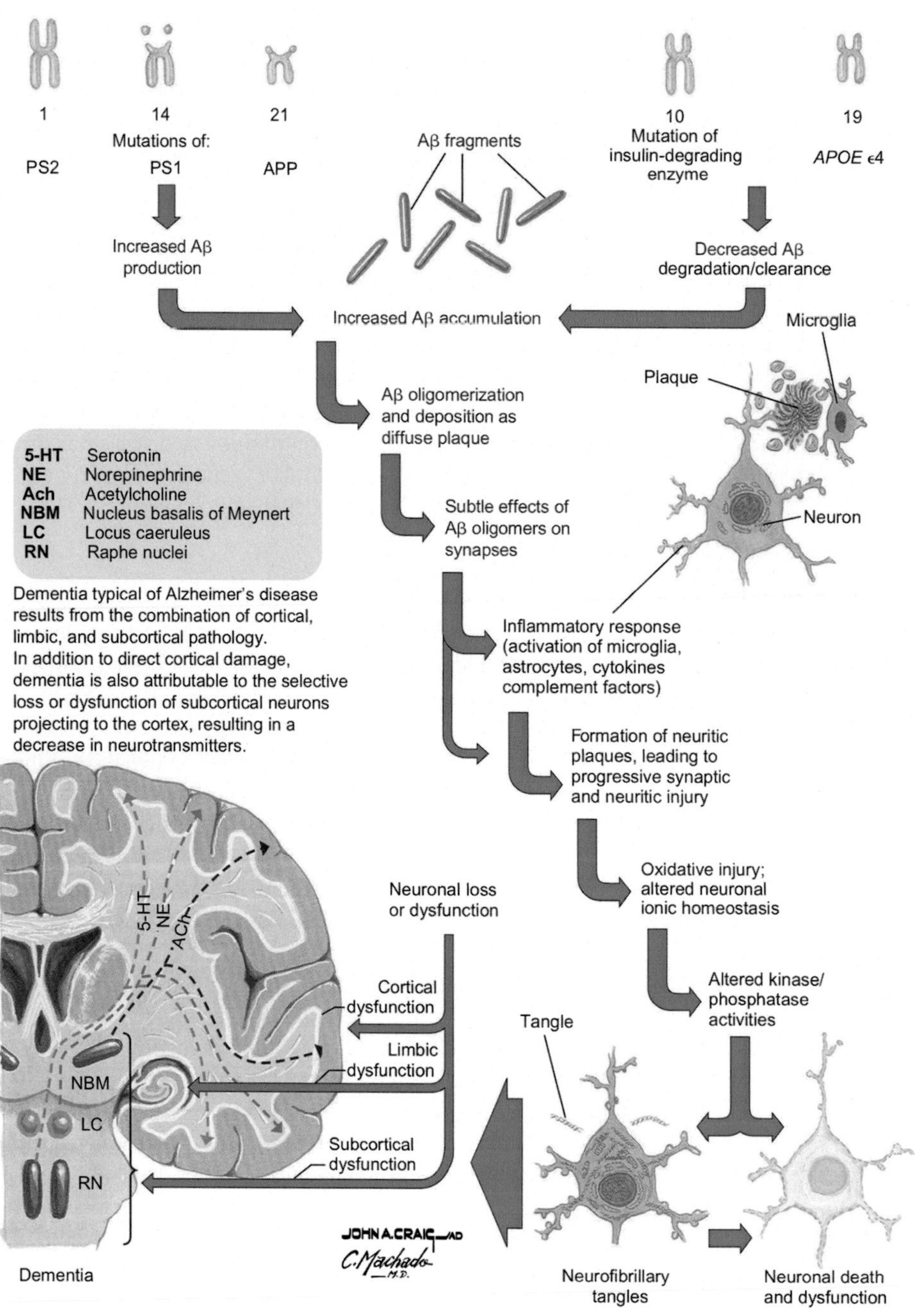

FIGURE 2.8

The amyloid cascade hypothesis. The hypothesis states that the accumulation of amyloid beta in the brain results in the buildup of oligomeric aggregates of this protein. These stimulate the aggregation of tau, which in turn leads to cytotoxicity and neurodegeneration.

Image reproduced with permission from Netter Images.

significant clinical benefits in human trials (see Section 2.5) —an outcome that has led to discussion of how physiologically relevant the amyloid cascade hypothesis is. It is also important to note that there are significant gaps in our understanding of how the events that form the amyloid cascade hypothesis link together. For example, although it is clear that tau dysfunction and accumulation is a consequence of amyloid beta accumulation in the central nervous system, it is not clear precisely how this occurs. Equally, it is unclear as to the exact mechanism that connects tau aggregation to cytotoxicity.

2.4.5 MOLECULAR MECHANISMS DRIVING NEURODEGENERATION

Regardless of the sequence of events that results in cell death within the brains of people with dementia, there is now a great deal of evidence linking the formation of protein aggregates within the human brain and the loss of neurons that characterize dementia. A number of possible mechanisms have been proposed as to why this happens, with much of the interest now centered on small oligomeric assemblies of proteins such as amyloid beta and tau. Oligomers (at least in the context of neurodegeneration and disease) are small, multicomponent regular aggregates of proteins that accumulate in the disease state. These share generic properties, most notably a propensity to adopt a beta-pleated sheet conformation and shared kinetics of formation and disaggregation. Indeed, it has been suggested that the formation of oligomers, and through these higher-order aggregates, may be a generic property of all proteins [196]. What is clear is that the accumulation of small aggregates, both within and without cells, has serious consequences in terms of cellular damage and eventual cytotoxicity. This has been examined for a number of proteins involved in dementia, including the amyloid beta peptide [197], tau [198], and alpha synuclein [199]. Research across these disorders, also including studies of proteins that form aggregates but are not involved in neurodegeneration, has highlighted that common mechanisms of toxicity may be involved [200]. The precise nature of this toxicity remains a matter of much debate, although it is notable that the formation of small oligomers of beta-sheet-folded proteins or peptides results in the exposure of hydrophobic elements of protein structure that can interact and damage cellular structures [201]. There is also some evidence that peptide oligomers can form porelike structures, which can punch holes in cellular membranes resulting in significant dysfunction [202]. A major caveat to many of these studies is that the protein aggregates examined are frequently formed from protein produced under nonphysiological conditions, for example, extracted from bacterial cells that have been modified to express the protein in question. How the behavior of these aggregates relates to the situation in the human central nervous system is unclear, although there are a number of studies that have isolated oligomeric aggregates from human brain samples consisting of amyloid beta [203] and tau [204]. The past several years have seen major advances in our understanding of the structural organization of both amyloid beta and tau aggregates, with cryo-electron microscopy yielding atomic resolution structures for both proteins [191,205]. These studies, however, are currently limited to larger aggregates and mature fibrils. It is unclear how these larger aggregates relate to the toxic intermediaries that are thought to be more closely associated with cell death, and by their very nature (as transitory oligomeric states that lead to the formation of fibrils) it is extremely challenging to isolate and carry out a structural characterization of these forms [206].

The posttranslational modification of tau and alpha synuclein, in particular by phosphorylation, has been strongly linked to the pathological behavior of these proteins in dementia and beyond. It was recognized in the 1980s that tau was endogenously phosphorylated [207], with further research indicating that this phosphorylation could be increased in a disease situation [208]. It is now clear that hyperphosphorylation of tau is intimately linked to the aggregation properties of this protein and its accumulation in neurofibrillary tangles [86]. Importantly, given the link to the pathological events in dementia, a number of candidate kinases have been identified that can phosphorylate tau under disease conditions—most notably glycogen synthase kinase 3 [209]. After the identification of alpha synuclein as the main constituent of Lewy bodies, researchers in Tokyo identified a phosphorylation event on serine 129 of the protein that was tightly linked to its presence in Lewy bodies [210].

There is also evidence for synergistic interactions between protein aggregates linked to dementia—notably amyloid beta, tau, and alpha synuclein [211]. This encompasses evidence from in vivo models [212] as well as the in vitro aggregation behavior of these proteins [213]. Again, whether this is relevant to their behavior in the human brain has not yet been established, and is notable that the different protein aggregates found in the demented brain can occur together or independently (see Section 2.3).

An important aspect of the study of protein aggregates in dementia is the correlation between aggregation state and neurodegeneration. For example, and as noted above, there is a weak correlation between amyloid plaque burden and neuronal cell loss in the human brain—although there may be a correlation with smaller amyloid aggregates [214]. This is mirrored in rodent models, where large-scale amyloid beta accumulation can be tolerated by brain cells [147].

An important insight into the balance between small aggregates, larger inclusions, and neuronal toxicity was provided by a seminal study examining how cultured neurons tolerate expression of the protein involved in Huntington's disease, huntingtin [215]. In this experiment, the long-term survival of neurons expressing expanded repeat huntingtin was assessed using longitudinal, repeated imaging of cells. Strikingly, the cells that survived longest were those that formed visible inclusions—suggesting that the cells that die are those where the aggregates remain as smaller protein assemblies (possibly oligomers). This has raised the possibility that the large-scale aggregates observed in the human brain across a wide range of neurodegenerative disorders, including amyloid plaques and neurofibrillary tangles, may actually represent a protective mechanism. In this model, these large macromolecular ordered assemblies act to sequester reactive, damaging smaller oligomers and reduce the toxic impact of these [216].

2.4.6 PROTEIN AGGREGATE SPREAD

One of the most exciting, and controversial, areas of research into the mechanisms of dementia relates to the possibility that the protein aggregates observed in the brains of individuals with dementia can spread from cell to cell. This is a theme that cuts across many areas of neurodegeneration, from Alzheimer's disease through Parkinson's to the prion disease—with the latter being defined by the propagation of infectious aggregates from cell to cell and from person to person [217]. The prion diseases, including Creutzfeldt–Jakob disease and Kuru (dealt with in greater detail in Chapter 4), were the first neurodegenerative disorders where clear evidence for protein aggregate spread and transmission was uncovered [218]. The overlap in pathology and clinical presentation between some of the prion disorders and more common forms of dementia resulted in early efforts to assess whether Alzheimer's disease could be transmitted from humans to nonhuman primates, but with no success [219]. Efforts to investigate this continued through the 1980s and began to show some evidence that protein aggregates could be propagated in mammalian models for disease [220,221]. A resurgence of interest came, however, with data from pathological studies of the dementias and the proposal that amyloid beta, tau, and alpha synuclein pathology spread throughout the brain in a stereotypical fashion [87]. One possible explanation for this was that aggregates were literally spreading from cell-to-cell or between brain regions, inducing the further aggregation of proteins in an autocatalytic fashion [222]. This interest was magnified by the demonstration that alpha synuclein aggregates could propagate from diseased tissue into fetal implants, introduced in an effort to treat Parkinson's disease, and possible evidence that propagation of protein aggregates relevant to dementia could occur in the human brain [223,224]. Subsequent to this, there have been extensive efforts to test whether amyloid beta, tau, and alpha synuclein aggregates and pathology can spread in cellular and animal models.

For amyloid beta, experiments using artificially produced recombinant aggregates and material from patients with Alzheimer's disease have demonstrated injection of these protein seeds into the brains of mouse models can induce the formation of amyloid plaque pathology, suggesting that some form of transmission can occur [225–227]. These data are consistent with earlier studies in nonhuman primate models [220,221]. This has also been investigated extensively for tau, with both animal and cellular models for the tauopathies revealing evidence that tau pathology can be seeded and propagated both in vivo and in vitro [228–232]. Alpha synuclein, the central pathological player in Parkinson's disease but also found in dementia with Lewy bodies and in a large number of Alzheimer's disease patients, has been demonstrated to possess some aspects of transmissibility in animal models—both with recombinant protein and with aggregates derived from human brain [233,234]. Data from cellular models also suggest that alpha synuclein aggregates possess prionlike properties [235]. These results have added to evidence from fetal transplants to raise the suggestion that Parkinson's disease could be a prion disorder [236].

As noted above, it has been documented for many years that the pathology observed in the brains of patients with dementia is linked very closely to aggregated, misfolded protein—and that these aggregates can misfold and propagate in a cell-free environment [168]. It has been proposed that this templating could underpin the propagation of misfolded protein between cells and in vivo, in a manner directly analogous to that observed in the prion diseases [237].

With all of these experimental data, there is increasing evidence that the protein aggregates found in the brains of people with dementia can exhibit some form of transmissible behavior. Intriguingly, however, there has not been unequivocal epidemiological evidence that Alzheimer's dementia, the tauopathies, or any of the Lewy body disorders, can be transmitted from person to person—evidence that is definitive for the true prion diseases. What evidence there is comes from cases where a prion disease, Creutzfeldt–Jakob disease, has been transmitted through iatrogenic mechanisms. These include cases where disease has been transmitted by corneal transplants and human growth hormone extracted postmortem [238,239]. In a small number of these cases, in addition to the transmission of prion pathology and disease there is evidence for amyloid beta pathology [240–242]. Whether such transmission can occur with just amyloid pathology, and whether this extends to tau and/or alpha synuclein pathology, remains to be determined.

It is important to note that the proposition that neurodegenerative disease pathology (other than for the prion disorders) can be transmitted or propagated remains a controversial topic. Indeed, this was previously also the case for the prion diseases—especially as evidence accrued that this was occurring as a protein-only phenomenon. There are a number of potential explanations as to why pathology can be staged in Alzheimer's disease and other dementias, explanations that may not necessarily depend on a prion-like mechanism [89].

2.4.7 NEUROINFLAMMATION AND IMMUNE RESPONSE

The role of the immune system and inflammatory responses to cellular damage in the brains of people with dementia represents an area of research of increasing interest and importance [243]. A number of observations have accelerated this, deriving from pathological and genetic studies. With regard to genetics, the identification of risk variants in the *TREM2* gene have made a major impact on the field, highlighting a role for microglial cells and the immune response that they partially coordinate within the brain [244]. Intriguingly, there is now evidence that TREM2 can act as a cellular receptor for amyloid beta and that this can stimulate microglial activation [245]. There are also data from animal models of tau [246] and amyloid beta [247] suggesting that TREM2 can modulate the pathology of these cells. Taken together with pathological data revealing the presence of activated microglia and astrocytes in the Alzheimer's brain, this strongly supports an important role for immune response and neuroinflammation in the disease processes that lead to neuronal cell death.

2.4.8 INTERPLAY BETWEEN GENETICS, PATHOLOGY, AGING, AND SPORADIC DISEASE

A key question in the dementia field is how closely the studies of proteins and pathways implicated in the etiology of dementia by genetic studies map onto the more common idiopathic forms of dementia. Alzheimer's disease is a standout example of this: can conclusions drawn from the study of mutations in *APP*, *PSEN1*, and *PSEN2* be applied to idiopathic Alzheimer's? This is more than an academic question, as many of the disease-modifying drug therapies currently under development are at least partly based on hypotheses or models derived from familial forms of disease. There are a number of areas of biology that are thought to contribute to the age-dependent risk of idiopathic Alzheimer's disease, some of which are thought to link to the mechanisms revealed through the study of the familial form of the disease. These include an age-dependent increase in oxidative stress in the brain, a consequence of which may include aberrant folding of proteins, including the amyloid beta peptide [248]. There is also significant evidence for a decrease in global mechanisms for maintaining proteostasis in the aging brain, resulting in decreased degradation of amyloid beta [249]. Finally, there is evidence for close links between alterations in metabolism as the brain ages and the changes in the level of aggregation state of amyloid beta [250].

2.5 THERAPIES

The dementias present exceptionally challenging targets for modern medicine. As noted in the introductory chapter, the combination of multifactorial disease processes, a heterogeneous patient population and the substantial challenges of engaging targets within the central nervous system make neurodegenerative disorders difficult to develop drugs for, and the dementias are no exception. Because of its status as the most common neurodegenerative disorder, Alzheimer's has been the focus of huge efforts attempting to develop both drugs that mitigate some of the symptoms of the disorder and disease-modifying therapies. For the former, there has been some limited success—with several drugs approved for human use. For the latter, the past two decades have witnessed large-scale drug development projects based on our understanding of the molecular pathogenesis of Alzheimer's fail repeatedly in clinical trials.

2.5.1 EXISTING THERAPIES

Existing therapies for dementia are exclusively symptomatic, acting to partially mitigate some of the symptoms of disease on a temporary basis. Those that have successfully gone through clinical trials are focused on neurochemical dysfunction that occurs as a consequence of neuronal cell loss in dementia. As noted above (Section 2.3), in Alzheimer's disease a disproportionate number of the neurons lost during the disease are cholinergic, using the neurotransmitter acetylcholine to communicate with neighboring cells. The first class of drugs licensed for use in Alzheimer's target this acetylcholine deficit, by acting to inhibit one of the

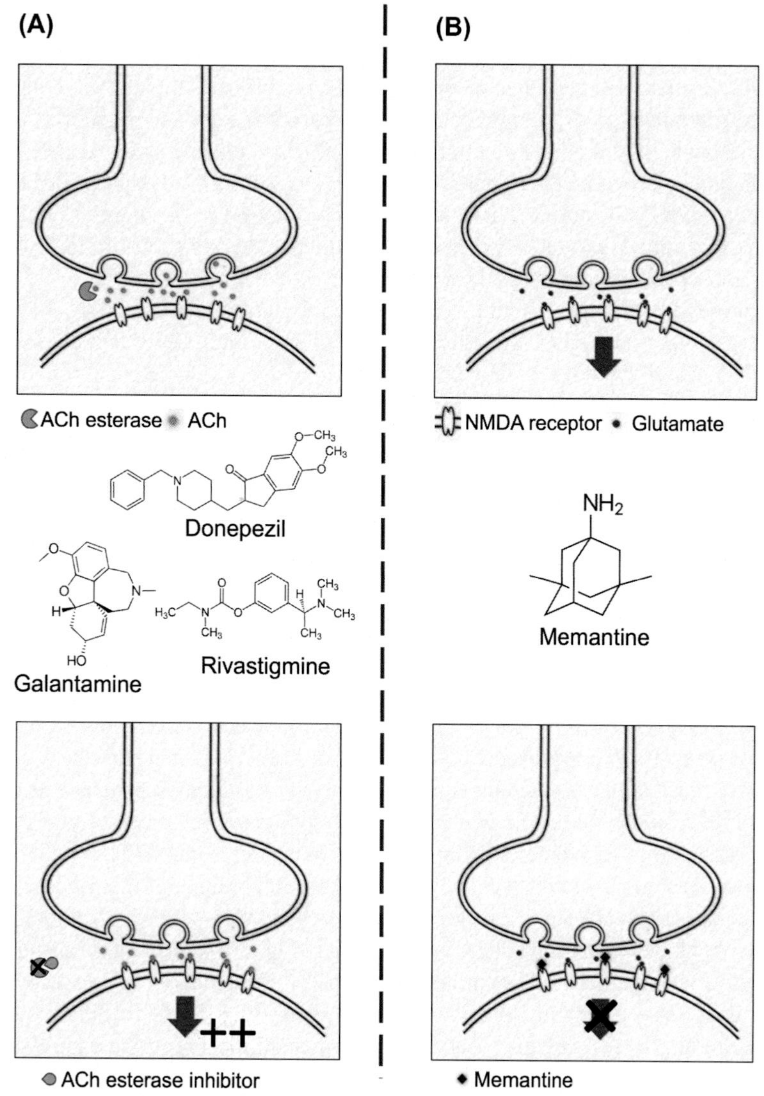

FIGURE 2.9

Current drug therapies for Alzheimer's disease. (A) Acetylcholinesterase inhibitors such as galantamine, donepezil, and rivastigmine act to potentiate acetylcholine based signaling in the brain. (B) The NMDA receptor antagonist memantine acts to block glutamate excitotoxicity.

enzymes responsible for the catabolism of this neurotransmitter—acetylcholine esterase. Reversible inhibitors of acetylcholine esterase include donepezil, galantamine, and rivastigmine (Fig. 2.9A), which act to partially block the breakdown of acetylcholine at the synaptic cleft thereby potentiating signals within the brain using this neurotransmitter. This provides a small but significant improvement in cognitive function for some patients in the early stages of Alzheimer's disease [251,252].

The second class of drug licensed for use as a symptomatic treatment in Alzheimer's targets alterations in glutamatergic signaling within the central nervous system. Memantine (Fig. 2.9B), which acts as an antagonist to n-methyl-D-aspartate receptors within the brain, is thought to reduce excitotoxicity caused by excess glutamatergic signaling that occurs as part of the disease process in Alzheimer's. The precise mechanism whereby this results in clinical benefit is unclear; however, there is evidence of a small but significant improvement in cognition, mood, and ability to carry out tasks in moderate to severe Alzheimer's [253]. Despite the implication of memantine in a potential neuroprotective role (that is, sparing cells from excitotoxicity), there is no evidence that it is able to modify the trajectory of degeneration in individuals living with Alzheimer's.

Both memantine and the acetylcholine esterase inhibitors used in Alzheimer's have been assessed for potential use in other dementias, for example, in dementia with Lewy bodies and for individuals with Parkinson's disease who have developed a significant dementia component to their clinical symptoms. These studies suggest that there are some clinical benefits observed in these patients, although these are of a similar magnitude to those observed in Alzheimer's disease [254,255].

Crucially, none of the drugs currently licensed for use in dementia have been demonstrated to have a disease-modifying impact and therefore make no difference to the trajectory of decline or the duration of disease.

2.5.2 DRUG DISCOVERY

The absence of a disease-modifying therapy for any of the dementias is a key unmet need in the field of neurology. As the burden of dementia on individuals, health services and society has grown over the past several decades, ever greater efforts have been made across academic, pharmaceutical, and charitable research bodies to develop novel approaches and compounds that act to halt or slow the degenerative process in dementia. As for much of the research into the dementias, this has focused on Alzheimer's disease due to the prevalence of this disorder, although there have been important developments in rarer forms of disease that have direct relevance to Alzheimer's. Drug discovery in the dementia field has built on the substantial advances that have been made in understanding the fundamental biology of the degenerative process that occurs in dementia described in Section 2.4 [117]. This has resulted in a large number of clinical trials across a wide spectrum of pathological mechanisms, summarized in Fig. 2.10 [256]. Pharmaceutical companies in particular have invested significant effort in targeting aspects of the amyloid cascade hypothesis and the putative downstream consequences of amyloid beta peptide generation and aggregation, and tau dysfunction (Fig. 2.11). The premise is a simple one: that by removing a component of the pathological cascade that results in cell death, neurodegeneration can be averted. A number of candidate drugs looking to reduce the production of amyloid beta, remove amyloid peptide from the central nervous system or to directly target tau have reached phase III clinical trials, although to date none of these have resulted in clinically beneficial outcomes. It is not yet clear whether this is due to fundamental flaws in the amyloid cascade hypothesis or due to the precise approaches and methodologies adopted for these trials.

2.5.2.1 Targeting Amyloid Beta Production

Starting at the top of the cascade, the first broad category of drug development targeting the amyloid cascade hypothesis has focused on attempting to reduce the production of the amyloid beta peptide at source. This area of research derives from several decades of work characterizing the enzymatic activities that result in the processing of the amyloid precursor protein into amyloid beta and other fragments,

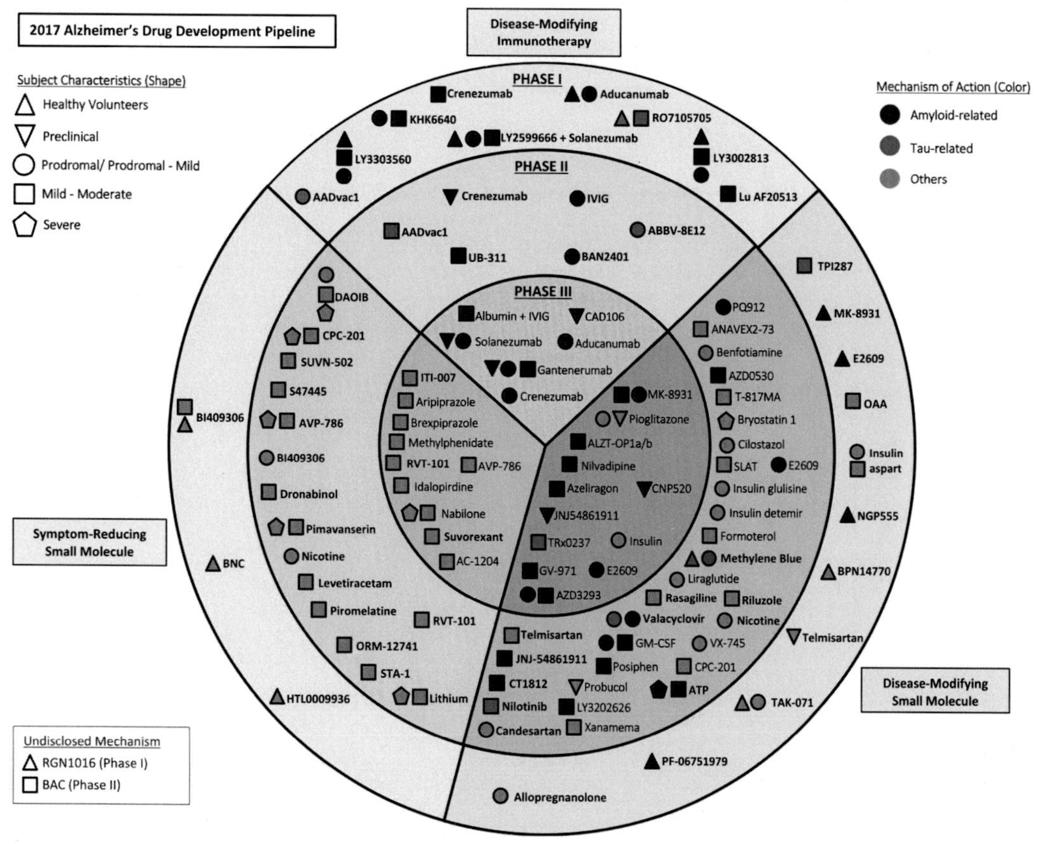

FIGURE 2.10

Clinical trials summary. A snapshot of ongoing clinical trials as of 2017, displayed by mechanism of action, target, stage of trial, and subject population.

Taken from Cummings J, Lee G, Mortsdorf T, Ritter A, Zhong K. Alzheimer's disease drug development pipeline. Alzheimers Dement (NY) 2017;2017(3):367–84.

and in particular, the beta and gamma secretase activities that result in amyloid beta. As the amyloid precursor protein is processed by multiple activities, by applying inhibitors that specifically reduce BACE1 or gamma secretase (Fig. 2.12) activity, it is possible to shift the balance of processing from the amyloidogenic amyloid beta pathway towards the nonamyloidogenic alpha secretase (ADAM10) pathway [257]. Several pharmaceutical companies have developed compounds that achieve just this and have advanced them to the point that in-human trials have occurred. For BACE1, these have proceeded to the point of phase II trials [258,259]; however, a recent phase II/III trial was halted [260]. For gamma secretase, inhibitors such as Semagacestat have progressed all the way through to large-scale phase III clinical trials for efficacy. Despite showing considerable promise in animal models for amyloid deposition, the results of clinical trials have, thus far, been disappointing—showing no clinical benefits resulting from the inhibition of gamma secretase [261,262].

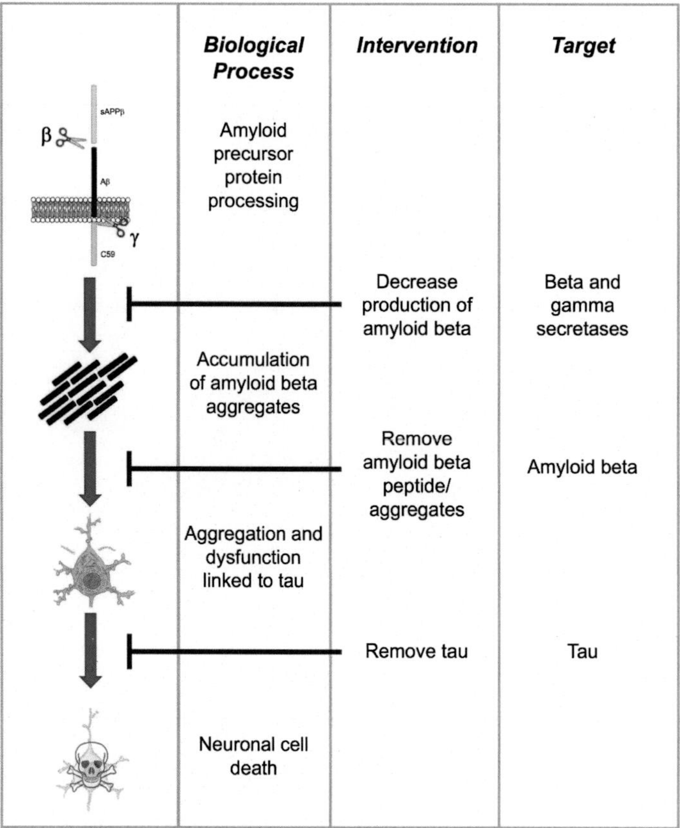

FIGURE 2.11

Strategies for targeting the amyloid cascade. These are divided into biological processes targeted, the mechanism of intervention and the target.

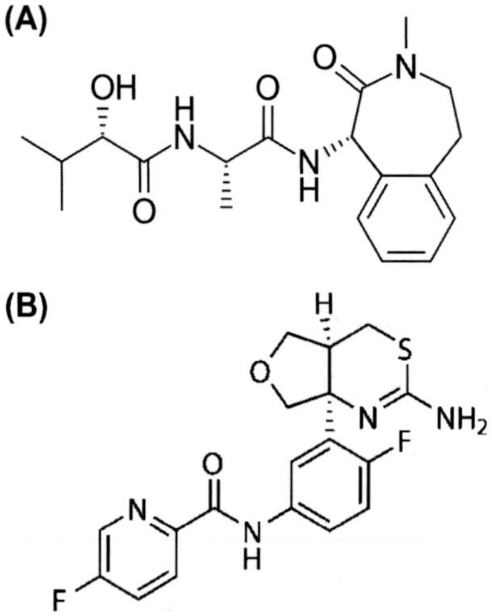

(A)

(B)

FIGURE 2.12

Beta and gamma secretase inhibitors. Chemical structures for (A) Semagacestat (gamma secretase inhibitor) and (B) LY2886721 (BACE-1 inhibitor).

A final direct strategy to alter amyloid precursor protein processing as a means to reduce amyloid beta production is to boost the nonamyloidogenic processing pathway, an approach reinforced by the identification of genetic variation at the *ADAM10* locus, coding for the protein responsible for the alpha secretase cleavage of the amyloid precursor protein, associated with increased lifetime risk of developing Alzheimer's. This has been achieved in cellular and animal models by increasing the activity of alpha secretase [263,264]. To date, however, this approach has not been developed to the point of clinical trials in humans [265].

Data from cellular and animal models of amyloid precursor protein processing suggested that an indirect approach to modulating amyloid beta peptide production in the Alzheimer's brain could be to alter cholesterol levels in the brain. A number of trials looking to test this using statins, drugs that are widely used in humans to reduce cholesterol levels, have been carried out, however none resulted in clinical benefits [266].

2.5.2.2 Removing Amyloid Beta

Moving beyond the production of amyloid beta, the next point at which intervention is possible is by removing the peptide once it has been produced and/or starts to form aggregates. This has proved to be one of the most active areas of drug discovery over the past several decades, driven by exciting preclinical data from mouse models demonstrating that both active and passive immunization targeting the amyloid beta peptide could significantly reduce amyloid burden within the brain [267,268]. Phase II trials during the early 2000s with active immunization, that is using fragments of the amyloid beta peptide to instigate an immune response from a patient's own immune system, were terminated due to a subset of patients (6% of more than 300 person trials) developing acute meningoencephalitis [269]. This presents as an inflammation of the meninges, the membrane covering the surface of the brain, with very serious and potentially life-threatening consequences. Intriguingly, follow-up postmortem analysis of patients enrolled in the trial who had died from other causes revealed that there were changes in the amyloid deposition in the brains of some of these individuals—suggesting that active immunization was able to drive a change in amyloid burden in humans [270]. An important caveat to these results was that there was no evidence of an improvement in cognition concomitant with this reduction in amyloid pathology, suggesting a decoupling of amyloid plaque pathology and cognitive deficits.

In parallel to the development of active immunization approaches, a number of pharmaceutical companies have developed passive immunization as a strategy to target amyloid beta in the central nervous system. Passive immunization relies on the in vitro generation of antibodies to a specific target, followed by the use of these antibodies as a biologic therapy—an approach that has demonstrated important clinical benefit in the oncology field [271]. Again, promising preclinical data from mouse models for disease prompted pharmaceutical companies to launch clinical trials [272]. These successfully cleared preclinical and human safety trials and demonstrated some promise in phase II trials. To date, however, large-scale phase III trials of passive immunization for amyloid beta—using biologic therapies such as Solanezumab (developed by Eli Lilly) or Bapineuzumab (developed jointly by Pfizer and Johnson & Johnson) have not demonstrated any clinical benefit in mild to moderate Alzheimer's disease [273,274]. There are, however, ongoing trials across this area of therapeutic development, for example, with other antiamyloid beta antibodies such as Crenezumab, manufactured by Roche [275].

2.5.3 TARGETING TAU

The final component of the amyloid cascade hypothesis to be targeted is that of the dysfunction and aggregation of the microtubule-associated protein tau. There is robust pathological evidence that suggests that tau accumulation is more tightly correlated with neuronal cell death than is the case for amyloid plaque pathology, and so the proposition of intervening at the stage of tau dysfunction is an attractive one as this would appear to be an event that is proximal to cytotoxicity and neurodegeneration. The underlying process of tau aggregation [276] shares many properties with those exhibited by amyloid beta (albeit occurring in the cytoplasm of cells rather than in the extracellular space), and so some of the drug strategies that have been adopted by biotech companies and the pharmaceutical industry are similar to those targeting amyloid. This is most notable in the development of immunization strategies to try to remove tau and reduce aggregation [277]. An active immunization strategy, using a fragment of tau has passed phase I safety evaluation and is currently undergoing phase II trials for efficacy [278], and passive immunization efforts targeting tau are currently the subject of phase I safety trials and phase II efficacy trials [279,280]. An illuminating illustration of the multifaceted approaches that pharmaceutical companies are taking to tackling tau is provided by the latter. A passive immunization therapy developed jointly between Bristol-Myers Squibb and Biogen is currently in parallel phase II trials, one in Alzheimer's disease and one in progressive supranuclear palsy (a movement disorder where tau is the predominant pathology, and in the absence of amyloid plaques—Chapter 3). The rationale for testing in progressive supranuclear palsy is that this disorder is driven by a less complex disease mechanism, solely—or at least overwhelmingly—associated with tau [281]. Testing efficacy in this disorder, which is also more rapid in progression than Alzheimer's, has the potential to provide a proof of concept that tau can be engaged as a target with clinical benefits before moving into the more complex disease etiology of Alzheimer's disease (while also providing a desperately needed therapy for a devastating disorder).

A small molecule approach to targeting tau aggregation has also been adopted, with phase III trials running to assess whether preventing the accumulation of monomeric tau into larger structural assemblies can reduce the cytotoxicity and neuronal loss associated with the development of tau pathology [282].

A distinctive approach to targeting tau has been to modulate the phosphorylation of this protein. As noted in Section 5.4, the phosphorylation of tau is a component of the biochemical changes in this protein that are thought to lead to increased aggregation and the eventual formation of neurofibrillary tangles. A number of kinases have been identified as being able to phosphorylate tau, a prominent example of which is GSK3β, and several clinical trials have taken place to test whether decreasing phosphorylation of tau can provide clinical benefits. Despite progressing to phase III in some cases, all of these have failed [283,284].

2.5.4 INTERVENING IN NEUROINFLAMMATION AND THE IMMUNE RESPONSE

There is a substantial body of evidence implicating neuroinflammation and immune response as part of the etiology of Alzheimer's disease, and this has led to a number of clinical trials looking to dampen inflammation in the brain [243]. The most prominent of these has been to test whether nonsteroidal antiinflammatory drugs, the use of which has been demonstrated to be associated with a lower incidence of Alzheimer's disease, can be used as an intervention to prevent progression of neurodegeneration. Assessment of cyclooxygenase I and II inhibitors, compounds that are already in widespread clinical use for suppression of inflammation, in Alzheimer's disease resulted in no benefit. As discussed more extensively below (Section 2.5.6), it is unclear as to whether this represents a failure of the hypothesis that neuroinflammation is a tractable component of the disease etiology in Alzheimer's, or whether this represents a failure to intervene early enough in the disease process to yield clinical benefits.

2.5.5 OTHER STRATEGIES

As illustrated in Fig. 2.10, there is a range of other therapeutic approaches currently undergoing clinical trials in the dementia field. A number of these are being carried out in cohorts of patients with rare forms of dementia or neurodegeneration, for example targeting expression of huntingtin, the protein associated with Huntington's disease, using RNA-mediated gene silencing [285]. The important role of alpha synuclein aggregate formation in dementia, and beyond dementia in Parkinson's disease, has resulted in immunotherapy approaches analogous to those applied to amyloid beta and tau being tested in human trials [286]. Other strategies to target disease progression in Alzheimer's include modifying metabolism in the brain, based on changes in metabolic pathways that have been identified in patients [287], as well as general neuroprotective approaches that provide neurons in the Alzheimer's brain with greater resistance to the pathological changes that are driving cell death [288]. To date, none of these has yielded clinical benefits, although it should be noted that several of these are only now entering phase II/III trials.

2.5.6 WHY ARE DRUGS FAILING?

Given the extensive list of failed phase III trials over the past decade, there has been a considerable amount of reflection across the pharmaceutical sector and in academia as to why so many trials are failing to demonstrate any clinical benefits in dementia [289,290]. The reasons for this are manifold and reflect the complex etiology of dementia, our evolving understanding of this, and the long-standing challenges of targeting the central nervous system.

A central issue is the level of degeneration that occurs in the central nervous system before the onset of symptoms (as indicated in Fig. 2.2). Based on a model where the symptomatic stage of Alzheimer's disease represents the end stage of the disorder, by definition, interventions that occur after the onset of symptoms will find it more challenging to modify the disease course than intervening earlier on in the disease process. This raises a paradox that is common to most neurodegenerative diseases: clinical trials for dementia are at present limited to recruiting patients who actively exhibit symptoms of disease, as we do not have robust procedures to identify individuals who are certain to go on to develop symptoms. As such, the clinical endpoints assessed in trials rely on alterations in symptomatic disease trajectory, or using proxies for this such as alterations in brain volume measured by magnetic resonance imaging or amyloid burden as measured by Pittsburgh compound B signal. If it were possible to confidently, and accurately, predict the onset of disease then delay of this could be used as a clinical endpoint in and of itself, providing a more definitive read out for disease modification. It is of note that this is increasingly a route that clinical trials are moving toward in dementia.

This is consistent with an increasingly important role for precision medicine—carefully characterizing patients before treatment or clinical trials—in drug development [291,292]. As our understanding of the genetics and risk factors that lead to dementia increase, there is hope that this approach can be applied to Alzheimer's disease [293] and yield benefits similar to those that have been realised in the cancer field over the past decade [294]. Related to this is the concept of patient stratification, ensuring that the patients (or presymptomatic individuals) entering into clinical trials are those that stand a realistic chance of benefiting from the mechanism of action under scrutiny [295]. A key issue in identifying patients ahead of the onset of the symptoms that are normally associated with dementia is the development of biomarkers that can successfully predict disease, as well as react to changes in disease state as a read out for efficacy [296]. Although much progress has been made in this area, for example in the use of cerebrospinal fluid amyloid beta and tau measurements, there is still significant research required.

A particular challenge for drugs targeting the processing of the amyloid precursor protein is that the enzymes/enzymatic activities that are involved in this, most notably ADAM10 and gamma secretase, are promiscuous and carry out a large number of proteolytic processing events in human cells [297–299]. This has the potential for significant issues relating to on target toxicity, as inhibiting (for gamma secretase) or stimulating (in the case of ADAM10) will have an impact on a much wider range of processes than just the cleavage of the amyloid precursor protein. A standout example of this is the role of gamma secretase in the proteolytic processing of Notch, a protein that has a critical function during development [300].

A more fundamental issue is the possibility that the majority of drug trials in Alzheimer's disease in particular share a flawed hypothetical basis—and that the amyloid cascade hypothesis is incorrect. Although there is a substantial body of evidence supporting a cascade of events driven by amyloid beta accumulation, the status of the amyloid cascade hypothesis has long been a subject of debate [162,301], and it remains as a hypothesis for the pathological events that drive neurodegeneration in Alzheimer's. Indeed, it has been proposed that the critical test of the amyloid cascade hypothesis is what happens when you remove components of that cascade [302]. To date, results in clinical trials have not borne out the hypothesis—although there are many caveats to these trials.

There is no one clear path to addressing the many challenges facing drug trials in dementia. One route that is being followed is to identify and treat patients with candidate drugs before the onset of symptoms, focusing on presymptomatic familial mutation-carrying patients [303]. Such trials have been embarked on but are by their nature of long duration [256].

2.6 CONCLUSIONS

The dementias hold particular fascination and horror as they strike at many of the brain functions that define us as individuals. Significant strides have been made over the past three decades in understanding the changes that occur in the brains of individuals who develop dementia; however, there is as yet no disease-modifying therapy available. As the numbers of individuals afflicted by Alzheimer's disease and other dementias increase, this will become a bigger and bigger challenge to health-care systems across the world—highlighting the continued importance of developing a greater understanding of these diseases.

REFERENCES

[1] Boller F, Forbes MM. History of dementia and dementia in history: an overview. J Neurol Sci 1998;158:125–33.

[2] Berchtold NC, Cotman CW. Evolution in the conceptualization of dementia and Alzheimer's disease: Greco-Roman period to the 1960s. Neurobiol Aging 1998;19:173–89.

[3] Matthews BR. Portrayal of neurological illness and physicians in the works of shakespeare. Front Neurol Neurosci 2010;27:216–26.

[4] Hippius H, Neundorfer G. The discovery of Alzheimer's disease. Dialogues Clin Neurosci 2003;5:101–8.

[5] Maurer K, Volk S, Gerbaldo H. Auguste D and Alzheimer's disease. Lancet 1997;349:1546–9.

[6] Alzheimer A. Uber einen eigenartigen schweren Erkrankungsprozess der Hirninde. Neurol Cent 1906;25:1134.

[7] Alzheimer A. Uber eine eigenartige Erkrankung der Hirnrinde. Allg Z Psychiatr 1907;64:146–8.

[8] Perusini G. Über klinisch und histologisch eigenartige psychische Erkrankungen des späteren Lebensalters. Histologische und Histopathologische Arbeiten. Jena: Verlag G Fischer; 1909. p. 297–351.

[9] Kraepelin E. Psychiatrie, Bd. 1–4. Leipzig: Barth; 1909. p. 1915.

[10] Pick A. Ueber primare chronische Demenz (so. Dementia praecox) im jugendlichen Alter. Prag Med Wochenschr 1891;16:312–5.

[11] Berrios GE. Alzheimer's disease: a conceptual history. Int J Geriatr Psychiatry 1990;5:355–65.

[12] Alexander M, Perera G, Ford L, Arrighi HM, Foskett N, Debove C, Novak G, Gordon MF. Age-stratified prevalence of mild cognitive impairment and dementia in European populations: a systematic review. J Alzheimer's Dis 2015;48:355–9.

[13] Ferri CP, Prince M, Brayne C, Brodaty H, Fratiglioni L, Ganguli M, Hall K, Hasegawa K, Hendrie H, Huang Y, Jorm A, Mathers C, Menezes PR, Rimmer E, Scazufca M. Alzheimer's Disease I. Global prevalence of dementia: a Delphi consensus study. Lancet 2005;366:2112–7.

[14] Prince M, Bryce R, Albanese E, Wimo A, Ribeiro W, Ferri CP. The global prevalence of dementia: a systematic review and metaanalysis. Alzheimer's Dement J Alzheimer's Assoc 2013;9:63–75. e2.

[15] Lewis F, Karlsberg Schaffer S, Sussex J, O'Neill P, Cockcroft L. The trajectory of dementia in the UK-making a difference. Office Health Econ Consult Rep 2014.

[16] McKhann GM, Knopman DS, Chertkow H, Hyman BT, Jack Jr CR, Kawas CH, Klunk WE, Koroshetz WJ, Manly JJ, Mayeux R, Mohs RC, Morris JC, Rossor MN, Scheltens P, Carrillo MC, Thies B, Weintraub S, Phelps CH. The diagnosis of dementia due to Alzheimer's disease: recommendations from the National Institute on Aging-Alzheimer's Association workgroups on diagnostic guidelines for Alzheimer's disease. Alzheimers Dement 2011;7:263–9.

[17] McGinnis SM. Infectious causes of rapidly progressive dementia. Semin Neurol 2011;31:266–85.

[18] Paterson RW, Takada LT, Geschwind MD. Diagnosis and treatment of rapidly progressive dementias. Neurol Clin Pract 2012;2:187–200.

[19] Lobo A, Launer L, Fratiglioni L, Andersen K, Di Carlo A, Breteler M, Copeland J, Dartigues J, Jagger C, Martinez-Lage J. Prevalence of dementia and major subtypes in Europe: a collaborative study of population-based cohorts. Neurology 2000;54:S4.

[20] Plassman BL, Langa KM, Fisher GG, Heeringa SG, Weir DR, Ofstedal MB, Burke JR, Hurd MD, Potter GG, Rodgers WL. Prevalence of dementia in the United States: the aging, demographics, and memory study. Neuroepidemiology 2007;29:125–32.

[21] Rossor MN, Fox NC, Mummery CJ, Schott JM, Warren JD. The diagnosis of young-onset dementia. Lancet Neurol 2010;9:793–806.

[22] Alzhelmer A. Uber eine eigenartige Erkrankung der Hirnrinde. Allg Z Psychiatr 1907;64:146–8.

[23] Petersen RC. Early diagnosis of Alzheimer's disease: is MCI too late? Curr Alzheimer Res 2009;6:324–30.

[24] O'Brien JT, Thomas A. Vascular dementia. Lancet 2015;386:1698–706.

[25] Kalaria RN, Erkinjuntti T. Small vessel disease and subcortical vascular dementia. J Clin Neurol 2006;2:1–11.

[26] Moulin S, Labreuche J, Bombois S, Rossi C, Boulouis G, Hénon H, Duhamel A, Leys D, Cordonnier C. Dementia risk after spontaneous intracerebral haemorrhage: a prospective cohort study. Lancet Neurol 2016;15:820–9.

[27] Gregoire S, Smith K, Jäger H, Benjamin M, Kallis C, Brown M, Cipolotti L, Werring D. Cerebral microbleeds and long-term cognitive outcome: longitudinal cohort study of stroke clinic patients. Cerebrovasc Dis 2012;33:430–5.

[28] Pendlebury ST, Rothwell PM. Prevalence, incidence, and factors associated with pre-stroke and post-stroke dementia: a systematic review and meta-analysis. Lancet Neurol 2009;8:1006–18.

[29] Chabriat H, Joutel A, Dichgans M, Tournier-Lasserve E, Bousser M-G. Cadasil Lancet Neurol 2009;8:643–53.

[30] O'Brien JT, Thomas A. Vascular dementia. Lancet 2015;386:1698–706.

[31] Beh SC, Muthusamy B, Calabresi P, Hart J, Zee D, Patel V, Frohman E. Hiding in plain sight: a closer look at posterior cortical atrophy. Pract Neurol 2014. practneurol-2014–20000883.

[32] Tang-Wai DF, Graff-Radford N, Boeve BF, Dickson DW, Parisi JE, Crook R, Caselli RJ, Knopman DS, Petersen RC. Clinical, genetic, and neuropathologic characteristics of posterior cortical atrophy. Neurology 2004;63:1168–74.

[33] Renner J, Burns J, Hou C, McKeel D, Storandt M, Morris J. Progressive posterior cortical dysfunction a clinicopathologic series. Neurology 2004;63:1175–80.

[34] Leyton CE, Piguet O, Savage S, Burrell J, Hodges JR. The neural basis of logopenic progressive aphasia. J Alzheimer's Dis 2012;32:1051–9.

[35] Dickerson BC, Wolk DA. Initiative AsDN. Dysexecutive versus amnesic phenotypes of very mild Alzheimer's disease are associated with distinct clinical, genetic and cortical thinning characteristics. J Neurol Neurosurg Psychiatry 2011;82:45–51.

[36] Folstein MF, Folstein SE, McHugh PR. "Mini-mental state". A practical method for grading the cognitive state of patients for the clinician. J Psychiatr Res 1975;12:189–98.

[37] Arevalo-Rodriguez I, Smailagic N, Roque IFM, Ciapponi A, Sanchez-Perez E, Giannakou A, Pedraza OL, Bonfill Cosp X, Cullum S. Mini-Mental State Examination (MMSE) for the detection of Alzheimer's disease and other dementias in people with mild cognitive impairment (MCI). Cochrane Database Syst Rev 2015:CD010783.

[38] Frisoni GB, Fox NC, Jack Jr CR, Scheltens P, Thompson PM. The clinical use of structural MRI in Alzheimer disease. Nat Rev Neurol 2010;6:67–77.

[39] Tartaglia MC, Rosen HJ, Miller BL. Neuroimaging in dementia. Neurotherapeutics 2011;8:82–92.

[40] Cohen AD, Klunk WE. Early detection of Alzheimer's disease using PiB and FDG PET. Neurobiol Dis 2014;72(Pt A):117–22.

[41] Tapiola T, Alafuzoff I, Herukka SK, Parkkinen L, Hartikainen P, Soininen H, Pirttila T. Cerebrospinal fluid {beta}-amyloid 42 and tau proteins as biomarkers of Alzheimer-type pathologic changes in the brain. Arch Neurol 2009;66:382–9.

[42] Ritchie C, Smailagic N, Noel-Storr AH, Ukoumunne O, Ladds EC, Martin S. CSF tau and the CSF tau/ABeta ratio for the diagnosis of Alzheimer's disease dementia and other dementias in people with mild cognitive impairment (MCI). Cochrane Database Syst Rev 2017;3:CD010803.

[43] Blennow K, Dubois B, Fagan AM, Lewczuk P, de Leon MJ, Hampel H. Clinical utility of cerebrospinal fluid biomarkers in the diagnosis of early Alzheimer's disease. Alzheimers Dement 2015;11:58–69.

[44] Livingston G, Sommerlad A, Orgeta V, Costafreda SG, Huntley J, Ames D, Ballard C, Banerjee S, Burns A, Cohen-Mansfield J, Cooper C, Fox N, Gitlin LN, Howard R, Kales HC, Larson EB, Ritchie K, Rockwood K, Sampson EL, Samus Q, Schneider LS, Selbaek G, Teri L, Mukadam N. Dementia prevention, intervention, and care. Lancet 2017;390:2673–734.

[45] Birks JS. Cholinesterase inhibitors for Alzheimer's disease. Cochrane Libr 2006;25(1):CD005593.
[46] van Marum RJ. Update on the use of memantine in Alzheimer's disease. Neuropsychiatr Dis Treat 2009;5:237.
[47] Vieira RT, Caixeta L, Machado S, Silva AC, Nardi AE, Arias-Carrión O, Carta MG. Epidemiology of early-onset dementia: a review of the literature. Clin Pract Epidemiol Ment Health CP EMH 2013;9:88.
[48] Mercy L, Hodges J, Dawson K, Barker R, Brayne C. Incidence of early-onset dementias in Cambridgeshire, United Kingdom. Neurology 2008;71:1496–9.
[49] Pick A. Ueber die Beziehungen der senile Hirnatrophie zur Aphasie. Prag Med Wochenschr 1892;17:165–7.
[50] Woollacott IO, Rohrer JD. The clinical spectrum of sporadic and familial forms of frontotemporal dementia. J Neurochem 2016;138(Suppl. 1):6–31.
[51] Diehl-Schmid J, Perneczky R, Koch J, Nedopil N, Kurz A. Guilty by suspicion? Criminal behavior in frontotemporal lobar degeneration. Cogn Behav Neurol 2013;26:73–7.
[52] Bang J, Spina S, Miller BL. Frontotemporal dementia. Lancet 2015;386:1672–82.
[53] Rohrer JD. Structural brain imaging in frontotemporal dementia. Biochim Biophys Acta Mol Basis Dis 2012;1822:325–32.
[54] Miller BL, Hou CE. Portraits of artists: emergence of visual creativity in dementia. Arch Neurol 2004;61:842–4.
[55] Murphy JM, Henry RG, Langmore S, Kramer JH, Miller BL, Lomen-Hoerth C. Continuum of frontal lobe impairment in amyotrophic lateral sclerosis. Arch Neurol 2007;64:530–4.
[56] Burrell JR, Kiernan MC, Vucic S, Hodges JR. Motor neuron dysfunction in frontotemporal dementia. Brain J Neurol 2011;134:2582–94.
[57] Espay AJ, Litvan I. Parkinsonism and frontotemporal dementia: the clinical overlap. J Mol Neurosci 2011;45:343–9.
[58] Mendez MF, Shapira JS, McMurtray A, Licht E. Preliminary findings: behavioral worsening on donepezil in patients with frontotemporal dementia. Am J Geriatr Psychiatry 2007;15:84–7.
[59] Burger PC, Vogel FS. The development of the pathologic changes of Alzheimer's disease and senile dementia in patients with Down's syndrome. Am J Pathol 1973;73:457–76.
[60] Ballard C, Mobley W, Hardy J, Williams G, Corbett A. Dementia in Down's syndrome. Lancet Neurol 2016;15:622–36.
[61] Wiseman FK, Al-Janabi T, Hardy J, Karmiloff-Smith A, Nizetic D, Tybulewicz VL, Fisher EM, Strydom A. A genetic cause of Alzheimer disease: mechanistic insights from Down syndrome. Nat Rev Neurosci 2015;16:564–74.
[62] Walker Z, Possin KL, Boeve BF, Aarsland D. Lewy body dementias. Lancet 2015;386:1683–97.
[63] Winterstein C. Head injuries attributable to boxing. Lancet 1937;230:719–22.
[64] Martland HS. Punch drunk. J Am Med Assoc 1928;91:1103–7.
[65] Omalu BI, DeKosky ST, Minster RL, Kamboh MI, Hamilton RL, Wecht CH. Chronic traumatic encephalopathy in a national football League player. Neurosurgery 2005;57:128–34. discussion -34.
[66] Blennow K, Brody DL, Kochanek PM, Levin H, McKee A, Ribbers GM, Yaffe K, Zetterberg H. Traumatic brain injuries. Nat Rev Dis Prim 2016;2:16084.
[67] Hay J, Johnson VE, Smith DH, Stewart W. Chronic traumatic encephalopathy: the neuropathological legacy of traumatic brain injury. Annu Rev Pathol 2016;11:21–45.
[68] Jordan BD. The clinical spectrum of sport-related traumatic brain injury. Nat Rev Neurol 2013;9:222–30.
[69] Parmera JB, Rodriguez RD, Studart Neto A, Nitrini R, Brucki SMD. Corticobasal syndrome: a diagnostic conundrum. Dement Neuropsychol 2016;10:267–75.
[70] Graham NL, Bak TH, Hodges JR. Corticobasal degeneration as a cognitive disorder. Mov Disord 2003;18:1224–32.
[71] Kubler E, Oesch B, Raeber AJ. Diagnosis of prion diseases. Br Med Bull 2003;66:267–79.
[72] Eddy CM, Parkinson EG, Rickards HE. Changes in mental state and behaviour in Huntington's disease. Lancet Psychiatry 2016;3:1079–86.

[73] Elahi FM, Miller BL. A clinicopathological approach to the diagnosis of dementia. Nat Rev Neurol 2017;13:457–76.

[74] Schuff N, Woerner N, Boreta L, Kornfield T, Shaw LM, Trojanowski JQ, Thompson PM, Jack Jr CR, Weiner MW. Alzheimer's Disease Neuroimaging I. MRI of hippocampal volume loss in early Alzheimer's disease in relation to ApoE genotype and biomarkers. Brain J Neurol 2009;132:1067–77.

[75] Sabuncu MR, Desikan RS, Sepulcre J, Yeo BT, Liu H, Schmansky NJ, Reuter M, Weiner MW, Buckner RL, Sperling RA, Fischl B. Alzheimer's Disease Neuroimaging I. The dynamics of cortical and hippocampal atrophy in Alzheimer disease. Arch Neurol 2011;68:1040–8.

[76] Whitehouse PJ, Price DL, Clark AW, Coyle JT, DeLong MR. Alzheimer disease: evidence for selective loss of cholinergic neurons in the nucleus basalis. Ann Neurol 1981;10:122–6.

[77] Mufson EJ, Counts SE, Perez SE, Ginsberg SD. Cholinergic system during the progression of Alzheimer's disease: therapeutic implications. Expert Rev Neurother 2008;8:1703–18.

[78] Seelaar H, Rohrer JD, Pijnenburg YA, Fox NC, van Swieten JC. Clinical, genetic and pathological heterogeneity of frontotemporal dementia: a review. J Neurol Neurosurg Psychiatry 2011;82:476–86.

[79] Welsh-Bohmer KA. Defining "prodromal" Alzheimer's disease, frontotemporal dementia, and Lewy body dementia: are we there yet? Neuropsychol Rev 2008;18:70–2.

[80] Sipe JD, Cohen AS. Review: history of the amyloid fibril. J Struct Biol 2000;130:88–98.

[81] Iqbal K, Novak M. From tangles to tau protein. Bratisl Lek Listy 2006;107:341–2.

[82] Hyman BT. The neuropathological diagnosis of Alzheimer's disease: clinical-pathological studies. Neurobiol Aging 1997;18:S27–32.

[83] Haass C, Kaether C, Thinakaran G, Sisodia S. Trafficking and proteolytic processing of APP. Cold Spring Harb Perspect Med 2012;2:a006270.

[84] Terry RD, Gonatas NK, Weiss M. Ultrastructural studies in Alzheimer's presenile dementia. Am J Pathol 1964;44:269–97.

[85] Houlden H, Baker M, McGowan E, Lewis P, Hutton M, Crook R, Wood NW, Kumar-Singh S, Geddes J, Swash M, Scaravilli F, Holton JL, Lashley T, Tomita T, Hashimoto T, Verkkoniemi A, Kalimo H, Somer M, Paetau A, Martin JJ, Van Broeckhoven C, Golde T, Hardy J, Haltia M, Revesz T. Variant Alzheimer's disease with spastic paraparesis and cotton wool plaques is caused by PS-1 mutations that lead to exceptionally high amyloid-beta concentrations. Ann Neurol 2000;48:806–8.

[86] Lee VM, Goedert M, Trojanowski JQ. Neurodegenerative tauopathies. Annu Rev Neurosci 2001;24:1121–59.

[87] Bancher C, Braak H, Fischer P, Jellinger KA. Neuropathological staging of Alzheimer lesions and intellectual status in Alzheimer's and Parkinson's disease patients. Neurosci Lett 1993;162:179–82.

[88] Burke RE, Dauer WT, Vonsattel JP. A critical evaluation of the Braak staging scheme for Parkinson's disease. Ann Neurol 2008;64:485–91. Epub 2008/12/11.

[89] Walsh DM, Selkoe DJ. A critical appraisal of the pathogenic protein spread hypothesis of neurodegeneration. Nat Rev Neurosci 2016;17:251–60.

[90] Nelson PT, Alafuzoff I, Bigio EH, Bouras C, Braak H, Cairns NJ, Castellani RJ, Crain BJ, Davies P, Del Tredici K, Duyckaerts C, Frosch MP, Haroutunian V, Hof PR, Hulette CM, Hyman BT, Iwatsubo T, Jellinger KA, Jicha GA, Kovari E, Kukull WA, Leverenz JB, Love S, Mackenzie IR, Mann DM, Masliah E, McKee AC, Montine TJ, Morris JC, Schneider JA, Sonnen JA, Thal DR, Trojanowski JQ, Troncoso JC, Wisniewski T, Woltjer RL, Beach TG. Correlation of Alzheimer disease neuropathologic changes with cognitive status: a review of the literature. J Neuropathol Exp Neurol 2012;71:362–81.

[91] Lewy F. Zur pathologischen Anatomie der Paralysis agitans. Dtsch Z Nervenheilk 1913;50:50–5.

[92] Goedert M, Spillantini MG, Del Tredici K, Braak H. 100 years of Lewy pathology. Nat Rev Neurol 2013;9:13–24.

[93] Toledo JB, Gopal P, Raible K, Irwin DJ, Brettschneider J, Sedor S, Waits K, Boluda S, Grossman M, Van Deerlin VM, Lee EB, Arnold SE, Duda JE, Hurtig H, Lee VM, Adler CH, Beach TG, Trojanowski JQ. Pathological alpha-synuclein distribution in subjects with coincident Alzheimer's and Lewy body pathology. Acta Neuropathol 2016;131:393–409.

[94] Sarlus H, Heneka MT. Microglia in Alzheimer's disease. J Clin Invest 2017;127:3240–9.
[95] Iwasaki Y. Creutzfeldt-Jakob disease. Neuropathology 2017;37:174–88.
[96] A novel gene containing a trinucleotide repeat that is expanded and unstable on Huntington's disease chromosomes. The Huntington's Disease Collaborative Research Group. Cell 1993;72:971–83.
[97] Vidal R, Frangione B, Rostagno A, Mead S, Revesz T, Plant G, Ghiso J. A stop-codon mutation in the BRI gene associated with familial British dementia. Nature 1999;399:776–81. Epub 1999/07/03.
[98] Worster-Drought C, Hill TR, McMenemey WH. Familial presenile dementia with spastic paralysis. J Neurol Psychopathol 1933;14:27–34.
[99] Karch CM, Goate AM. Alzheimer's disease risk genes and mechanisms of disease pathogenesis. Biol Psychiatry 2015;77:43–51.
[100] Nicolas G, Charbonnier C, Campion D. From common to rare variants: the genetic component of Alzheimer disease. Hum Hered 2016;81:129–41.
[101] Rosenberg RN, Lambracht-Washington D, Yu G, Xia W. Genomics of Alzheimer disease: a review. JAMA Neurol 2016;73:867–74.
[102] Tcw J, Goate AM. Genetics of beta-amyloid precursor protein in Alzheimer's disease. Cold Spring Harb Perspect Med 2017:7.
[103] Glenner GG, Wong CW. Alzheimer's disease: initial report of the purification and characterization of a novel cerebrovascular amyloid protein. Biochem Biophys Res Commun 1984;120:885–90.
[104] Glenner GG, Wong CW. Alzheimer's disease and Down's syndrome: sharing of a unique cerebrovascular amyloid fibril protein. Biochem Biophys Res Commun 1984;122:1131–5.
[105] Wong CW, Quaranta V, Glenner GG. Neuritic plaques and cerebrovascular amyloid in Alzheimer disease are antigenically related. Proc Natl Acad Sci USA 1985;82:8729–32.
[106] Kang J, Lemaire HG, Unterbeck A, Salbaum JM, Masters CL, Grzeschik KH, Multhaup G, Beyreuther K, Muller-Hill B. The precursor of Alzheimer's disease amyloid A4 protein resembles a cell-surface receptor. Nature 1987;325:733–6.
[107] Rumble B, Retallack R, Hilbich C, Simms G, Multhaup G, Martins R, Hockey A, Montgomery P, Beyreuther K, Masters CL. Amyloid A4 protein and its precursor in Down's syndrome and Alzheimer's disease. N Engl J Med 1989;320:1446–52.
[108] Levy E, Carman MD, Fernandez-Madrid IJ, Power MD, Lieberburg I, van Duinen SG, Bots GT, Luyendijk W, Frangione B. Mutation of the Alzheimer's disease amyloid gene in hereditary cerebral hemorrhage, Dutch type. Science 1990;248:1124–6.
[109] Goate A, Chartier-Harlin MC, Mullan M, Brown J, Crawford F, Fidani L, Giuffra L, Haynes A, Irving N, James L, et al. Segregation of a missense mutation in the amyloid precursor protein gene with familial Alzheimer's disease. Nature 1991;349:704–6.
[110] Sleegers K, Brouwers N, Gijselinck I, Theuns J, Goossens D, Wauters J, Del-Favero J, Cruts M, van Duijn CM, Van Broeckhoven C. APP duplication is sufficient to cause early onset Alzheimer's dementia with cerebral amyloid angiopathy. Brain J Neurol 2006;129:2977–83.
[111] Jonsson T, Atwal JK, Steinberg S, Snaedal J, Jonsson PV, Bjornsson S, Stefansson H, Sulem P, Gudbjartsson D, Maloney J, Hoyte K, Gustafson A, Liu Y, Lu Y, Bhangale T, Graham RR, Huttenlocher J, Bjornsdottir G, Andreassen OA, Jonsson EG, Palotie A, Behrens TW, Magnusson OT, Kong A, Thorsteinsdottir U, Watts RJ, Stefansson K. A mutation in APP protects against Alzheimer's disease and age-related cognitive decline. Nature 2012;488:96–9.
[112] Sherrington R, Rogaev EI, Liang Y, Rogaeva EA, Levesque G, Ikeda M, Chi H, Lin C, Li G, Holman K, Tsuda T, Mar L, Foncin JF, Bruni AC, Montesi MP, Sorbi S, Rainero I, Pinessi L, Nee L, Chumakov I, Pollen D, Brookes A, Sanseau P, Polinsky RJ, Wasco W, Da Silva HA, Haines JL, Perkicak-Vance MA, Tanzi RE, Roses AD, Fraser PE, Rommens JM, St George-Hyslop PH. Cloning of a gene bearing missense mutations in early-onset familial Alzheimer's disease. Nature 1995;375:754–60.

[113] Rogaev EI, Sherrington R, Rogaeva EA, Levesque G, Ikeda M, Liang Y, Chi H, Lin C, Holman K, Tsuda T, et al. Familial Alzheimer's disease in kindreds with missense mutations in a gene on chromosome 1 related to the Alzheimer's disease type 3 gene. Nature 1995;376:775–8.
[114] Muller U, Winter P, Graeber MB. A presenilin 1 mutation in the first case of Alzheimer's disease. Lancet Neurol 2013;12:129–30.
[115] Suh J, Choi SH, Romano DM, Gannon MA, Lesinski AN, Kim DY, Tanzi RE. ADAM10 missense mutations potentiate beta-amyloid accumulation by impairing prodomain chaperone function. Neuron 2013;80:385–401.
[116] Kim M, Suh J, Romano D, Truong MH, Mullin K, Hooli B, Norton D, Tesco G, Elliott K, Wagner SL, Moir RD, Becker KD, Tanzi RE. Potential late-onset Alzheimer's disease-associated mutations in the ADAM10 gene attenuate {alpha}-secretase activity. Hum Mol Genet 2009;18:3987–96.
[117] Karch CM, Cruchaga C, Goate AM. Alzheimer's disease genetics: from the bench to the clinic. Neuron 2014;83:11–26.
[118] Hutton M, Lendon CL, Rizzu P, Baker M, Froelich S, Houlden H, Pickering-Brown S, Chakraverty S, Isaacs A, Grover A, Hackett J, Adamson J, Lincoln S, Dickson D, Davies P, Petersen RC, Stevens M, de Graaff E, Wauters E, van Baren J, Hillebrand M, Joosse M, Kwon JM, Nowotny P, Che LK, Norton J, Morris JC, Reed LA, Trojanowski J, Basun H, Lannfelt L, Neystat M, Fahn S, Dark F, Tannenberg T, Dodd PR, Hayward N, Kwok JB, Schofield PR, Andreadis A, Snowden J, Craufurd D, Neary D, Owen F, Oostra BA, Hardy J, Goate A, van Swieten J, Mann D, Lynch T, Heutink P. Association of missense and 5'-splice-site mutations in tau with the inherited dementia FTDP-17. Nature 1998;393:702–5.
[119] Iqbal K, Liu F, Gong CX. Tau and neurodegenerative disease: the story so far. Nat Rev Neurol 2016;12:15–27.
[120] Polymeropoulos MH, Lavedan C, Leroy E, Ide SE, Dehejia A, Dutra A, Pike B, Root H, Rubenstein J, Boyer R, Stenroos ES, Chandrasekharappa S, Athanassiadou A, Papapetropoulos T, Johnson WG, Lazzarini AM, Duvoisin RC, Di Iorio G, Golbe LI, Nussbaum RL. Mutation in the alpha-synuclein gene identified in families with Parkinson's disease. Science 1997;276:2045–7.
[121] Spillantini MG, Schmidt ML, Lee VM, Trojanowski JQ, Jakes R, Goedert M. Alpha-synuclein in Lewy bodies. Nature 1997;388:839–40.
[122] Bates GP, Dorsey R, Gusella JF, Hayden MR, Kay C, Leavitt BR, Nance M, Ross CA, Scahill RI, Wetzel R, Wild EJ, Tabrizi SJ. Huntington disease. Nat Rev Dis Prim 2015;1:15005.
[123] Vidal R, Revesz T, Rostagno A, Kim E, Holton JL, Bek T, Bojsen-Moller M, Braendgaard H, Plant G, Ghiso J, Frangione B. A decamer duplication in the 3' region of the BRI gene originates an amyloid peptide that is associated with dementia in a Danish kindred. Proc Natl Acad Sci USA 2000;97:4920–5.
[124] Johnson RT. Prion diseases. Lancet Neurol 2005;4:635–42.
[125] Strittmatter WJ, Saunders AM, Schmechel D, Pericak-Vance M, Enghild J, Salvesen GS, Roses AD. Apolipoprotein E: high-avidity binding to beta-amyloid and increased frequency of type 4 allele in late-onset familial Alzheimer disease. Proc Natl Acad Sci USA 1993;90:1977–81.
[126] Liu CC, Liu CC, Kanekiyo T, Xu H, Bu G. Apolipoprotein E and Alzheimer disease: risk, mechanisms and therapy. Nat Rev Neurol 2013;9:106–18.
[127] Farrer LA, Cupples LA, Haines JL, Hyman B, Kukull WA, Mayeux R, Myers RH, Pericak-Vance MA, Risch N, van Duijn CM. Effects of age, sex, and ethnicity on the association between apolipoprotein E genotype and Alzheimer disease. A meta-analysis. APOE and Alzheimer Disease Meta Analysis Consortium. JAMA 1997;278:1349–56.
[128] Castellano JM, Kim J, Stewart FR, Jiang H, DeMattos RB, Patterson BW, Fagan AM, Morris JC, Mawuenyega KG, Cruchaga C, Goate AM, Bales KR, Paul SM, Bateman RJ, Holtzman DM. Human apoE isoforms differentially regulate brain amyloid-beta peptide clearance. Sci Transl Med 2011;3:89ra57.
[129] Osenkowski P, Ye W, Wang R, Wolfe MS, Selkoe DJ. Direct and potent regulation of gamma-secretase by its lipid microenvironment. J Biol Chem 2008;283:22529–40.

[130] Guerreiro R, Wojtas A, Bras J, Carrasquillo M, Rogaeva E, Majounie E, Cruchaga C, Sassi C, Kauwe JS, Younkin S, Hazrati L, Collinge J, Pocock J, Lashley T, Williams J, Lambert JC, Amouyel P, Goate A, Rademakers R, Morgan K, Powell J, St George-Hyslop P, Singleton A, Hardy J, Alzheimer Genetic Analysis G. TREM2 variants in Alzheimer's disease. N Engl J Med 2013;368:117–27.

[131] Jonsson T, Stefansson H, Steinberg S, Jonsdottir I, Jonsson PV, Snaedal J, Bjornsson S, Huttenlocher J, Levey AI, Lah JJ, Rujescu D, Hampel H, Giegling I, Andreassen OA, Engedal K, Ulstein I, Djurovic S, Ibrahim-Verbaas C, Hofman A, Ikram MA, van Duijn CM, Thorsteinsdottir U, Kong A, Stefansson K. Variant of TREM2 associated with the risk of Alzheimer's disease. N Engl J Med 2013;368:107–16.

[132] Sims R, van der Lee SJ, Naj AC, Bellenguez C, Badarinarayan N, Jakobsdottir J, Kunkle BW, Boland A, Raybould R, Bis JC, Martin ER, Grenier-Boley B, Heilmann-Heimbach S, Chouraki V, Kuzma AB, Sleegers K, Vronskaya M, Ruiz A, Graham RR, Olaso R, Hoffmann P, Grove ML, Vardarajan BN, Hiltunen M, Nothen MM, White CC, Hamilton-Nelson KL, Epelbaum J, Maier W, Choi SH, Beecham GW, Dulary C, Herms S, Smith AV, Funk CC, Derbois C, Forstner AJ, Ahmad S, Li H, Bacq D, Harold D, Satizabal CL, Valladares O, Squassina A, Thomas R, Brody JA, Qu L, Sanchez-Juan P, Morgan T, Wolters FJ, Zhao Y, Garcia FS, Denning N, Fornage M, Malamon J, Naranjo MCD, Majounie E, Mosley TH, Dombroski B, Wallon D, Lupton MK, Dupuis J, Whitehead P, Fratiglioni L, Medway C, Jian X, Mukherjee S, Keller L, Brown K, Lin H, Cantwell LB, Panza F, McGuinness B, Moreno-Grau S, Burgess JD, Solfrizzi V, Proitsi P, Adams HH, Allen M, Seripa D, Pastor P, Cupples LA, Price ND, Hannequin D, Frank-Garcia A, Levy D, Chakrabarty P, Caffarra P, Giegling I, Beiser AS, Giedraitis V, Hampel H, Garcia ME, Wang X, Lannfelt L, Mecocci P, Eiriksdottir G, Crane PK, Pasquier F, Boccardi V, Henandez I, Barber RC, Scherer M, Tarraga L, Adams PM, Leber M, Chen Y, Albert MS, Riedel-Heller S, Emilsson V, Beekly D, Braae A, Schmidt R, Blacker D, Masullo C, Schmidt H, Doody RS, Spalletta G, Longstreth Jr WT, Fairchild TJ, Bossu P, Lopez OL, Frosch MP, Sacchinelli E, Ghetti B, Yang Q, Huebinger RM, Jessen F, Li S, Kamboh MI, Morris J, Sotolongo-Grau O, Katz MJ, Corcoran C, Dunstan M, Braddel A, Thomas C, Meggy A, Marshall R, Gerrish A, Chapman J, Aguilar M, Taylor S, Hill M, Fairen MD, Hodges A, Vellas B, Soininen H, Kloszewska I, Daniilidou M, Uphill J, Patel Y, Hughes JT, Lord J, Turton J, Hartmann AM, Cecchetti R, Fenoglio C, Serpente M, Arcaro M, Caltagirone C, Orfei MD, Ciaramella A, Pichler S, Mayhaus M, Gu W, Lleo A, Fortea J, Blesa R, Barber IS, Brookes K, Cupidi C, Maletta RG, Carrell D, Sorbi S, Moebus S, Urbano M, Pilotto A, Kornhuber J, Bosco P, Todd S, Craig D, Johnston J, Gill M, Lawlor B, Lynch A, Fox NC, Hardy J, Consortium A, Albin RL, Apostolova LG, Arnold SE, Asthana S, Atwood CS, Baldwin CT, Barnes LL, Barral S, Beach TG, Becker JT, Bigio EH, Bird TD, Boeve BF, Bowen JD, Boxer A, Burke JR, Burns JM, Buxbaum JD, Cairns NJ, Cao C, Carlson CS, Carlsson CM, Carney RM, Carrasquillo MM, Carroll SL, Diaz CC, Chui HC, Clark DG, Cribbs DH, Crocco EA, DeCarli C, Dick M, Duara R, Evans DA, Faber KM, Fallon KB, Fardo DW, Farlow MR, Ferris S, Foroud TM, Galasko DR, Gearing M, Geschwind DH, Gilbert JR, Graff-Radford NR, Green RC, Growdon JH, Hamilton RL, Harrell LE, Honig LS, Huentelman MJ, Hulette CM, Hyman BT, Jarvik GP, Abner E, Jin LW, Jun G, Karydas A, Kaye JA, Kim R, Kowall NW, Kramer JH, LaFerla FM, Lah JJ, Leverenz JB, Levey AI, Li G, Lieberman AP, Lunetta KL, Lyketsos CG, Marson DC, Martiniuk F, Mash DC, Masliah E, McCormick WC, McCurry SM, McDavid AN, McKee AC, Mesulam M, Miller BL, Miller CA, Miller JW, Morris JC, Murrell JR, Myers AJ, O'Bryant S, Olichney JM, Pankratz VS, Parisi JE, Paulson HL, Perry W, Peskind E, Pierce A, Poon WW, Potter H, Quinn JF, Raj A, Raskind M, Reisberg B, Reitz C, Ringman JM, Roberson ED, Rogaeva E, Rosen HJ, Rosenberg RN, Sager MA, Saykin AJ, Schneider JA, Schneider LS, Seeley WW, Smith AG, Sonnen JA, Spina S, Stern RA, Swerdlow RH, Tanzi RE, Thornton-Wells TA, Trojanowski JQ, Troncoso JC, Van Deerlin VM, Van Eldik LJ, Vinters HV, Vonsattel JP, Weintraub S, Welsh-Bohmer KA, Wilhelmsen KC, Williamson J, Wingo TS, Woltjer RL, Wright CB, Yu CE, Yu L, Garzia F, Golamaully F, Septier G, Engelborghs S, Vandenberghe R, De Deyn PP, Fernadez CM, Benito YA, Thonberg H, Forsell C, Lilius L, Kinhult-Stahlbom A, Kilander L, Brundin R, Concari L, Helisalmi S, Koivisto AM, Haapasalo A, Dermecourt V, Fievet N, Hanon O,

Dufouil C, Brice A, Ritchie K, Dubois B, Himali JJ, Keene CD, Tschanz J, Fitzpatrick AL, Kukull WA, Norton M, Aspelund T, Larson EB, Munger R, Rotter JI, Lipton RB, Bullido MJ, Hofman A, Montine TJ, Coto E, Boerwinkle E, Petersen RC, Alvarez V, Rivadeneira F, Reiman EM, Gallo M, O'Donnell CJ, Reisch JS, Bruni AC, Royall DR, Dichgans M, Sano M, Galimberti D, St George-Hyslop P, Scarpini E, Tsuang DW, Mancuso M, Bonuccelli U, Winslow AR, Daniele A, Wu CK, Gerad/Perades CAE, Peters O, Nacmias B, Riemenschneider M, Heun R, Brayne C, Rubinsztein DC, Bras J, Guerreiro R, Al-Chalabi A, Shaw CE, Collinge J, Mann D, Tsolaki M, Clarimon J, Sussams R, Lovestone S, O'Donovan MC, Owen MJ, Behrens TW, Mead S, Goate AM, Uitterlinden AG, Holmes C, Cruchaga C, Ingelsson M, Bennett DA, Powell J, Golde TE, Graff C, De Jager PL, Morgan K, Ertekin-Taner N, Combarros O, Psaty BM, Passmore P, Younkin SG, Berr C, Gudnason V, Rujescu D, Dickson DW, Dartigues JF, DeStefano AL, Ortega-Cubero S, Hakonarson H, Campion D, Boada M, Kauwe JK, Farrer LA, Van Broeckhoven C, Ikram MA, Jones L, Haines JL, Tzourio C, Launer LJ, Escott-Price V, Mayeux R, Deleuze JF, Amin N, Holmans PA, Pericak-Vance MA, Amouyel P, van Duijn CM, Ramirez A, Wang LS, Lambert JC, Seshadri S, Williams J, Schellenberg GD. Rare coding variants in PLCG2, ABI3, and TREM2 implicate microglial-mediated innate immunity in Alzheimer's disease. Nat Genet 2017;49:1373–84.

[133] Visscher PM, Wray NR, Zhang Q, Sklar P, McCarthy MI, Brown MA, Yang J. 10 Years of GWAS discovery: biology, function, and translation. Am J Hum Genet 2017;101:5–22.

[134] Lambert JC, Ibrahim-Verbaas CA, Harold D, Naj AC, Sims R, Bellenguez C, DeStafano AL, Bis JC, Beecham GW, Grenier-Boley B, Russo G, Thorton-Wells TA, Jones N, Smith AV, Chouraki V, Thomas C, Ikram MA, Zelenika D, Vardarajan BN, Kamatani Y, Lin CF, Gerrish A, Schmidt H, Kunkle B, Dunstan ML, Ruiz A, Bihoreau MT, Choi SH, Reitz C, Pasquier F, Cruchaga C, Craig D, Amin N, Berr C, Lopez OL, De Jager PL, Deramecourt V, Johnston JA, Evans D, Lovestone S, Letenneur L, Moron FJ, Rubinsztein DC, Eiriksdottir G, Sleegers K, Goate AM, Fievet N, Huentelman MW, Gill M, Brown K, Kamboh MI, Keller L, Barberger-Gateau P, McGuiness B, Larson EB, Green R, Myers AJ, Dufouil C, Todd S, Wallon D, Love S, Rogaeva E, Gallacher J, St George-Hyslop P, Clarimon J, Lleo A, Bayer A, Tsuang DW, Yu L, Tsolaki M, Bossu P, Spalletta G, Proitsi P, Collinge J, Sorbi S, Sanchez-Garcia F, Fox NC, Hardy J, Deniz Naranjo MC, Bosco P, Clarke R, Brayne C, Galimberti D, Mancuso M, Matthews F, European Alzheimer's Disease I, Genetic, Environmental Risk in Alzheimer's D, Alzheimer's Disease Genetic C, Cohorts for H, Aging Research in Genomic E, Moebus S, Mecocci P, Del Zompo M, Maier W, Hampel H, Pilotto A, Bullido M, Panza F, Caffarra P, Nacmias B, Gilbert JR, Mayhaus M, Lannefelt L, Hakonarson H, Pichler S, Carrasquillo MM, Ingelsson M, Beekly D, Alvarez V, Zou F, Valladares O, Younkin SG, Coto E, Hamilton-Nelson KL, Gu W, Razquin C, Pastor P, Mateo I, Owen MJ, Faber KM, Jonsson PV, Combarros O, O'Donovan MC, Cantwell LB, Soininen H, Blacker D, Mead S, Mosley Jr TH, Bennett DA, Harris TB, Fratiglioni L, Holmes C, de Bruijn RF, Passmore P, Montine TJ, Bettens K, Rotter JI, Brice A, Morgan K, Foroud TM, Kukull WA, Hannequin D, Powell JF, Nalls MA, Ritchie K, Lunetta KL, Kauwe JS, Boerwinkle E, Riemenschneider M, Boada M, Hiltuenen M, Martin ER, Schmidt R, Rujescu D, Wang LS, Dartigues JF, Mayeux R, Tzourio C, Hofman A, Nothen MM, Graff C, Psaty BM, Jones L, Haines JL, Holmans PA, Lathrop M, Pericak-Vance MA, Launer LJ, Farrer LA, van Duijn CM, Van Broeckhoven C, Moskvina V, Seshadri S, Williams J, Schellenberg GD, Amouyel P. Meta-analysis of 74,046 individuals identifies 11 new susceptibility loci for Alzheimer's disease. Nat Genet 2013;45:1452–8.

[135] Rogaeva E, Meng Y, Lee JH, Gu Y, Kawarai T, Zou F, Katayama T, Baldwin CT, Cheng R, Hasegawa H, Chen F, Shibata N, Lunetta KL, Pardossi-Piquard R, Bohm C, Wakutani Y, Cupples LA, Cuenco KT, Green RC, Pinessi L, Rainero I, Sorbi S, Bruni A, Duara R, Friedland RP, Inzelberg R, Hampe W, Bujo H, Song YQ, Andersen OM, Willnow TE, Graff-Radford N, Petersen RC, Dickson D, Der SD, Fraser PE, Schmitt-Ulms G, Younkin S, Mayeux R, Farrer LA, St George-Hyslop P. The neuronal sortilin-related receptor SORL1 is genetically associated with Alzheimer disease. Nat Genet 2007;39:168–77.

[136] Shen L, Jia J. An overview of genome-wide association studies in Alzheimer's disease. Neurosci Bull 2016;32:183–90.

[137] Guerreiro R, Ross OA, Kun-Rodrigues C, Hernandez DG, Orme T, Eicher JD, Shepherd CE, Parkkinen L, Darwent L, Heckman MG, Scholz SW, Troncoso JC, Pletnikova O, Ansorge O, Clarimon J, Lleo A, Morenas-Rodriguez E, Clark L, Honig LS, Marder K, Lemstra A, Rogaeva E, St George-Hyslop P, Londos E, Zetterberg H, Barber I, Braae A, Brown K, Morgan K, Troakes C, Al-Sarraj S, Lashley T, Holton J, Compta Y, Van Deerlin V, Serrano GE, Beach TG, Lesage S, Galasko D, Masliah E, Santana I, Pastor P, Diez-Fairen M, Aguilar M, Tienari PJ, Myllykangas L, Oinas M, Revesz T, Lees A, Boeve BF, Petersen RC, Ferman TJ, Escott-Price V, Graff-Radford N, Cairns NJ, Morris JC, Pickering-Brown S, Mann D, Halliday GM, Hardy J, Trojanowski JQ, Dickson DW, Singleton A, Stone DJ, Bras J. Investigating the genetic architecture of dementia with Lewy bodies: a two-stage genome-wide association study. Lancet Neurol 2018;17:64–74.

[138] Ferrari R, Hernandez DG, Nalls MA, Rohrer JD, Ramasamy A, Kwok JB, Dobson-Stone C, Brooks WS, Schofield PR, Halliday GM, Hodges JR, Piguet O, Bartley L, Thompson E, Haan E, Hernandez I, Ruiz A, Boada M, Borroni B, Padovani A, Cruchaga C, Cairns NJ, Benussi L, Binetti G, Ghidoni R, Forloni G, Galimberti D, Fenoglio C, Serpente M, Scarpini E, Clarimon J, Lleo A, Blesa R, Waldo ML, Nilsson K, Nilsson C, Mackenzie IR, Hsiung GY, Mann DM, Grafman J, Morris CM, Attems J, Griffiths TD, McKeith IG, Thomas AJ, Pietrini P, Huey ED, Wassermann EM, Baborie A, Jaros E, Tierney MC, Pastor P, Razquin C, Ortega-Cubero S, Alonso E, Perneczky R, Diehl-Schmid J, Alexopoulos P, Kurz A, Rainero I, Rubino E, Pinessi L, Rogaeva E, St George-Hyslop P, Rossi G, Tagliavini F, Giaccone G, Rowe JB, Schlachetzki JC, Uphill J, Collinge J, Mead S, Danek A, Van Deerlin VM, Grossman M, Trojanowski JQ, van der Zee J, Deschamps W, Van Langenhove T, Cruts M, Van Broeckhoven C, Cappa SF, Le Ber I, Hannequin D, Golfier V, Vercelletto M, Brice A, Nacmias B, Sorbi S, Bagnoli S, Piaceri I, Nielsen JE, Hjermind LE, Riemenschneider M, Mayhaus M, Ibach B, Gasparoni G, Pichler S, Gu W, Rossor MN, Fox NC, Warren JD, Spillantini MG, Morris HR, Rizzu P, Heutink P, Snowden JS, Rollinson S, Richardson A, Gerhard A, Bruni AC, Maletta R, Frangipane F, Cupidi C, Bernardi L, Anfossi M, Gallo M, Conidi ME, Smirne N, Rademakers R, Baker M, Dickson DW, Graff-Radford NR, Petersen RC, Knopman D, Josephs KA, Boeve BF, Parisi JE, Seeley WW, Miller BL, Karydas AM, Rosen H, van Swieten JC, Dopper EG, Seelaar H, Pijnenburg YA, Scheltens P, Logroscino G, Capozzo R, Novelli V, Puca AA, Franceschi M, Postiglione A, Milan G, Sorrentino P, Kristiansen M, Chiang HH, Graff C, Pasquier F, Rollin A, Deramecourt V, Lebert F, Kapogiannis D, Ferrucci L, Pickering-Brown S, Singleton AB, Hardy J, Momeni P. Frontotemporal dementia and its subtypes: a genome-wide association study. Lancet Neurol 2014;13:686–99.

[139] Manzoni C, Kia DA, Vandrovcova J, Hardy J, Wood NW, Lewis PA, Ferrari R. Genome, transcriptome and proteome: the rise of omics data and their integration in biomedical sciences. Brief Bioinform 2016;19(2):286–302.

[140] Karch CM, Goate AM. Alzheimer's disease risk genes and mechanisms of disease pathogenesis. Biol Psychiatry 2015;77:43–51.

[141] Sasaguri H, Nilsson P, Hashimoto S, Nagata K, Saito T, De Strooper B, Hardy J, Vassar R, Winblad B, Saido TC. APP mouse models for Alzheimer's disease preclinical studies. EMBO J 2017;36:2473–87.

[142] Chen G, Chen KS, Knox J, Inglis J, Bernard A, Martin SJ, Justice A, McConlogue L, Games D, Freedman SB, Morris RG. A learning deficit related to age and beta-amyloid plaques in a mouse model of Alzheimer's disease. Nature 2000;408:975–9.

[143] Oakley H, Cole SL, Logan S, Maus E, Shao P, Craft J, Guillozet-Bongaarts A, Ohno M, Disterhoft J, Van Eldik L, Berry R, Vassar R. Intraneuronal beta-amyloid aggregates, neurodegeneration, and neuron loss in transgenic mice with five familial Alzheimer's disease mutations: potential factors in amyloid plaque formation. J Neurosci 2006;26:10129–40.

[144] McGowan E, Pickford F, Kim J, Onstead L, Eriksen J, Yu C, Skipper L, Murphy MP, Beard J, Das P, Jansen K, DeLucia M, Lin WL, Dolios G, Wang R, Eckman CB, Dickson DW, Hutton M, Hardy J, Golde T. Abeta42 is essential for parenchymal and vascular amyloid deposition in mice. Neuron 2005;47:191–9.

[145] Lewis J, McGowan E, Rockwood J, Melrose H, Nacharaju P, Van Slegtenhorst M, Gwinn-Hardy K, Paul Murphy M, Baker M, Yu X, Duff K, Hardy J, Corral A, Lin WL, Yen SH, Dickson DW, Davies P, Hutton M. Neurofibrillary tangles, amyotrophy and progressive motor disturbance in mice expressing mutant (P301L) tau protein. Nat Genet 2000;25:402–5.
[146] Oddo S, Caccamo A, Shepherd JD, Murphy MP, Golde TE, Kayed R, Metherate R, Mattson MP, Akbari Y, LaFerla FM. Triple-transgenic model of Alzheimer's disease with plaques and tangles: intracellular Abeta and synaptic dysfunction. Neuron 2003;39:409–21.
[147] Drummond E, Wisniewski T. Alzheimer's disease: experimental models and reality. Acta Neuropathol 2017;133:155–75.
[148] Koprich JB, Kalia LV, Brotchie JM. Animal models of alpha-synucleinopathy for Parkinson disease drug development. Nat Rev Neurosci 2017;18:515–29.
[149] Lavedan C. The synuclein family. Genome Res 1998;8:871–80.
[150] Griffin EF, Caldwell KA, Caldwell GA. Genetic and pharmacological discovery for Alzheimer's disease using *Caenorhabditis elegans*. ACS Chem Neurosci 2017;8:2596–606.
[151] Bharadwaj P, Martins R, Macreadie I. Yeast as a model for studying Alzheimer's disease. FEMS Yeast Res 2010;10:961–9.
[152] Wittmann CW, Wszolek MF, Shulman JM, Salvaterra PM, Lewis J, Hutton M, Feany MB. Tauopathy in Drosophila: neurodegeneration without neurofibrillary tangles. Science 2001;293:711–4.
[153] Takahashi K, Tanabe K, Ohnuki M, Narita M, Ichisaka T, Tomoda K, Yamanaka S. Induction of pluripotent stem cells from adult human fibroblasts by defined factors. Cell 2007;131:861–72.
[154] Israel MA, Yuan SH, Bardy C, Reyna SM, Mu Y, Herrera C, Hefferan MP, Van Gorp S, Nazor KL, Boscolo FS, Carson CT, Laurent LC, Marsala M, Gage FH, Remes AM, Koo EH, Goldstein LS. Probing sporadic and familial Alzheimer's disease using induced pluripotent stem cells. Nature 2012;482:216–20.
[155] Kondo T, Asai M, Tsukita K, Kutoku Y, Ohsawa Y, Sunada Y, Imamura K, Egawa N, Yahata N, Okita K, Takahashi K, Asaka I, Aoi T, Watanabe A, Watanabe K, Kadoya C, Nakano R, Watanabe D, Maruyama K, Hori O, Hibino S, Choshi T, Nakahata T, Hioki H, Kaneko T, Naitoh M, Yoshikawa K, Yamawaki S, Suzuki S, Hata R, Ueno S, Seki T, Kobayashi K, Toda T, Murakami K, Irie K, Klein WL, Mori H, Asada T, Takahashi R, Iwata N, Yamanaka S, Inoue H. Modeling Alzheimer's disease with iPSCs reveals stress phenotypes associated with intracellular Abeta and differential drug responsiveness. Cell Stem Cell 2013;12:487–96.
[156] Espuny-Camacho I, Arranz AM, Fiers M, Snellinx A, Ando K, Munck S, Bonnefont J, Lambot L, Corthout N, Omodho L, Vanden Eynden E, Radaelli E, Tesseur I, Wray S, Ebneth A, Hardy J, Leroy K, Brion JP, Vanderhaeghen P, De Strooper B. Hallmarks of Alzheimer's disease in stem-cell-derived human neurons transplanted into mouse brain. Neuron 2017;93:1066–81 e8.
[157] Hardy JA, Higgins GA. Alzheimer's disease: the amyloid cascade hypothesis. Science 1992;256:184–5.
[158] Hardy J, Selkoe DJ. The amyloid hypothesis of Alzheimer's disease: progress and problems on the road to therapeutics. Science 2002;297:353–6.
[159] Musiek ES, Holtzman DM. Three dimensions of the amyloid hypothesis: time, space and 'wingmen'. Nat Neurosci 2015;18:800–6.
[160] Reitz C. Alzheimer's disease and the amyloid cascade hypothesis: a critical review. Int J Alzheimers Dis 2012;2012:369808.
[161] Canevelli M, Bruno G, Cesari M. The sterile controversy on the amyloid cascade hypothesis. Neurosci Biobehav Rev 2017;83:472–3.
[162] Morris GP, Clark IA, Vissel B. Inconsistencies and controversies surrounding the amyloid hypothesis of Alzheimer's disease. Acta Neuropathol Commun 2014;2:135.
[163] Karran E, Mercken M, De Strooper B. The amyloid cascade hypothesis for Alzheimer's disease: an appraisal for the development of therapeutics. Nat Rev Drug Discov 2011;10:698–712.

[164] Scheuner D, Eckman C, Jensen M, Song X, Citron M, Suzuki N, Bird TD, Hardy J, Hutton M, Kukull W, Larson E, Levy-Lahad E, Viitanen M, Peskind E, Poorkaj P, Schellenberg G, Tanzi R, Wasco W, Lannfelt L, Selkoe D, Younkin S. Secreted amyloid beta-protein similar to that in the senile plaques of Alzheimer's disease is increased in vivo by the presenilin 1 and 2 and APP mutations linked to familial Alzheimer's disease. Nat Med 1996;2:864–70. Epub 1996/08/01.

[165] Mann DM, Iwatsubo T, Cairns NJ, Lantos PL, Nochlin D, Sumi SM, Bird TD, Poorkaj P, Hardy J, Hutton M, Prihar G, Crook R, Rossor MN, Haltia M. Amyloid beta protein (Abeta) deposition in chromosome 14-linked Alzheimer's disease: predominance of Abeta42(43). Ann Neurol 1996;40:149–56.

[166] Spillantini MG, Goedert M, Jakes R, Klug A. Different configurational states of beta-amyloid and their distributions relative to plaques and tangles in Alzheimer disease. Proc Natl Acad Sci USA 1990;87:3947–51.

[167] Chiti F, Dobson CM. Protein misfolding, functional amyloid, and human disease. Annu Rev Biochem 2006;75:333–66.

[168] Jarrett JT, Berger EP, Lansbury Jr PT. The carboxy terminus of the beta amyloid protein is critical for the seeding of amyloid formation: implications for the pathogenesis of Alzheimer's disease. Biochemistry 1993;32:4693–7.

[169] Nilsberth C, Westlind-Danielsson A, Eckman CB, Condron MM, Axelman K, Forsell C, Stenh C, Luthman J, Teplow DB, Younkin SG, Naslund J, Lannfelt L. The 'Arctic' APP mutation (E693G) causes Alzheimer's disease by enhanced Abeta protofibril formation. Nat Neurosci 2001;4:887–93.

[170] Vassar R, Bennett BD, Babu-Khan S, Kahn S, Mendiaz EA, Denis P, Teplow DB, Ross S, Amarante P, Loeloff R, Luo Y, Fisher S, Fuller J, Edenson S, Lile J, Jarosinski MA, Biere AL, Curran E, Burgess T, Louis JC, Collins F, Treanor J, Rogers G, Citron M. Beta-secretase cleavage of Alzheimer's amyloid precursor protein by the transmembrane aspartic protease BACE. Science 1999;286:735–41.

[171] Sinha S, Anderson JP, Barbour R, Basi GS, Caccavello R, Davis D, Doan M, Dovey HF, Frigon N, Hong J, Jacobson-Croak K, Jewett N, Keim P, Knops J, Lieberburg I, Power M, Tan H, Tatsuno G, Tung J, Schenk D, Seubert P, Suomensaari SM, Wang S, Walker D, Zhao J, McConlogue L, John V. Purification and cloning of amyloid precursor protein beta-secretase from human brain. Nature 1999;402:537–40.

[172] Yan R, Bienkowski MJ, Shuck ME, Miao H, Tory MC, Pauley AM, Brashier JR, Stratman NC, Mathews WR, Buhl AE, Carter DB, Tomasselli AG, Parodi LA, Heinrikson RL, Gurney ME. Membrane-anchored aspartyl protease with Alzheimer's disease beta-secretase activity. Nature 1999;402:533–7.

[173] Hong L, Koelsch G, Lin X, Wu S, Terzyan S, Ghosh AK, Zhang XC, Tang J. Structure of the protease domain of memapsin 2 (beta-secretase) complexed with inhibitor. Science 2000;290:150–3.

[174] De Strooper B, Iwatsubo T, Wolfe MS. Presenilins and gamma-secretase: structure, function, and role in Alzheimer Disease. Cold Spring Harb Perspect Med 2012;2:a006304.

[175] Bai XC, Yan C, Yang G, Lu P, Ma D, Sun L, Zhou R, Scheres SHW, Shi Y. An atomic structure of human gamma-secretase. Nature 2015;525:212–7.

[176] Allinson TM, Parkin ET, Turner AJ, Hooper NM. ADAMs family members as amyloid precursor protein alpha-secretases. J Neurosci Res 2003;74:342–52.

[177] Kuhn PH, Wang H, Dislich B, Colombo A, Zeitschel U, Ellwart JW, Kremmer E, Rossner S, Lichtenthaler SF. ADAM10 is the physiologically relevant, constitutive alpha-secretase of the amyloid precursor protein in primary neurons. EMBO J 2010;29:3020–32.

[178] St George-Hyslop P, Fraser PE. Assembly of the presenilin gamma-/epsilon-secretase complex. J Neurochem 2012;120(Suppl. 1):84–8.

[179] Andrew RJ, Kellett KA, Thinakaran G, Hooper NM. A Greek tragedy: the growing complexity of Alzheimer amyloid precursor protein proteolysis. J Biol Chem 2016;291:19235–44.

[180] Bates KA, Verdile G, Li QX, Ames D, Hudson P, Masters CL, Martins RN. Clearance mechanisms of Alzheimer's amyloid-beta peptide: implications for therapeutic design and diagnostic tests. Mol Psychiatry 2009;14:469–86.

[181] Crehan H, Holton P, Wray S, Pocock J, Guerreiro R, Hardy J. Complement receptor 1 (CR1) and Alzheimer's disease. Immunobiology 2012;217:244–50.

[182] Qiu WQ, Folstein MF. Insulin, insulin-degrading enzyme and amyloid-beta peptide in Alzheimer's disease: review and hypothesis. Neurobiol Aging 2006;27:190–8.

[183] Oddo S. The ubiquitin-proteasome system in Alzheimer's disease. J Cell Mol Med 2008;12:363–73.

[184] Nixon RA, Yang DS. Autophagy failure in Alzheimer's disease–locating the primary defect. Neurobiol Dis 2011;43:38–45.

[185] Fiala M, Cribbs DH, Rosenthal M, Bernard G. Phagocytosis of amyloid-beta and inflammation: two faces of innate immunity in Alzheimer's disease. J Alzheimers Dis 2007;11:457–63.

[186] Morawe T, Hiebel C, Kern A, Behl C. Protein homeostasis, aging and Alzheimer's disease. Mol Neurobiol 2012;46:41–54.

[187] Menzies FM, Fleming A, Caricasole A, Bento CF, Andrews SP, Ashkenazi A, Fullgrabe J, Jackson A, Jimenez Sanchez M, Karabiyik C, Licitra F, Lopez Ramirez A, Pavel M, Puri C, Renna M, Ricketts T, Schlotawa L, Vicinanza M, Won H, Zhu Y, Skidmore J, Rubinsztein DC. Autophagy and neurodegeneration: pathogenic mechanisms and therapeutic opportunities. Neuron 2017;93:1015–34.

[188] Grundke-Iqbal I, Iqbal K, Quinlan M, Tung YC, Zaidi MS, Wisniewski HM. Microtubule-associated protein tau. A component of Alzheimer paired helical filaments. J Biol Chem 1986;261:6084–9.

[189] Grundke-Iqbal I, Iqbal K, Tung YC, Quinlan M, Wisniewski HM, Binder LI. Abnormal phosphorylation of the microtubule-associated protein tau (tau) in Alzheimer cytoskeletal pathology. Proc Natl Acad Sci USA 1986;83:4913–7.

[190] Hasegawa M, Morishima-Kawashima M, Takio K, Suzuki M, Titani K, Ihara Y. Protein sequence and mass spectrometric analyses of tau in the Alzheimer's disease brain. J Biol Chem 1992;267:17047–54.

[191] Fitzpatrick AWP, Falcon B, He S, Murzin AG, Murshudov G, Garringer HJ, Crowther RA, Ghetti B, Goedert M, Scheres SHW. Cryo-EM structures of tau filaments from Alzheimer's disease. Nature 2017;547:185–90.

[192] Ueda K, Fukushima H, Masliah E, Xia Y, Iwai A, Yoshimoto M, Otero DA, Kondo J, Ihara Y, Saitoh T. Molecular cloning of cDNA encoding an unrecognized component of amyloid in Alzheimer disease. Proc Natl Acad Sci USA 1993;90:11282–6. Epub 1993/12/01.

[193] Iwai A, Masliah E, Yoshimoto M, Ge N, Flanagan L, de Silva HA, Kittel A, Saitoh T. The precursor protein of non-A beta component of Alzheimer's disease amyloid is a presynaptic protein of the central nervous system. Neuron 1995;14:467–75.

[194] Lashuel HA, Overk CR, Oueslati A, Masliah E. The many faces of alpha-synuclein: from structure and toxicity to therapeutic target. Nat Rev Neurosci 2013;14:38–48. Epub 2012/12/21.

[195] Singleton A, Myers A, Hardy J. The law of mass action applied to neurodegenerative disease: a hypothesis concerning the etiology and pathogenesis of complex diseases. Hum Mol Genet 2004;13(1):R123–6.

[196] Knowles TP, Vendruscolo M, Dobson CM. The amyloid state and its association with protein misfolding diseases. Nat Rev Mol Cell Biol 2014;15:384–96.

[197] Iversen LL, Mortishire-Smith RJ, Pollack SJ, Shearman MS. The toxicity in vitro of beta-amyloid protein. Biochem J 1995;311(Pt 1):1–16.

[198] Iqbal K, Liu F, Gong CX, Alonso Adel C, Grundke-Iqbal I. Mechanisms of tau-induced neurodegeneration. Acta Neuropathol 2009;118:53–69.

[199] Cremades N, Cohen SI, Deas E, Abramov AY, Chen AY, Orte A, Sandal M, Clarke RW, Dunne P, Aprile FA, Bertoncini CW, Wood NW, Knowles TP, Dobson CM, Klenerman D. Direct observation of the interconversion of normal and toxic forms of alpha-synuclein. Cell 2012;149:1048–59. Epub 2012/05/29.

[200] Bucciantini M, Giannoni E, Chiti F, Baroni F, Formigli L, Zurdo J, Taddei N, Ramponi G, Dobson CM, Stefani M. Inherent toxicity of aggregates implies a common mechanism for protein misfolding diseases. Nature 2002;416:507–11.

[201] Fink AL. Protein aggregation: folding aggregates, inclusion bodies and amyloid. Fold Des 1998;3:R9–23.

[202] Volles MJ, Lansbury Jr PT. Vesicle permeabilization by protofibrillar alpha-synuclein is sensitive to Parkinson's disease-linked mutations and occurs by a pore-like mechanism. Biochemistry 2002;41:4595–602.

[203] Shankar GM, Li S, Mehta TH, Garcia-Munoz A, Shepardson NE, Smith I, Brett FM, Farrell MA, Rowan MJ, Lemere CA, Regan CM, Walsh DM, Sabatini BL, Selkoe DJ. Amyloid-beta protein dimers isolated directly from Alzheimer's brains impair synaptic plasticity and memory. Nat Med 2008;14:837–42.

[204] Ait-Bouziad N, Lv G, Mahul-Mellier AL, Xiao S, Zorludemir G, Eliezer D, Walz T, Lashuel HA. Discovery and characterization of stable and toxic Tau/phospholipid oligomeric complexes. Nat Commun 2017;8:1678.

[205] Gremer L, Scholzel D, Schenk C, Reinartz E, Labahn J, Ravelli RBG, Tusche M, Lopez-Iglesias C, Hoyer W, Heise H, Willbold D, Schroder GF. Fibril structure of amyloid-beta(1-42) by cryo-electron microscopy. Science 2017;358:116–9.

[206] Fandrich M. Oligomeric intermediates in amyloid formation: structure determination and mechanisms of toxicity. J Mol Biol 2012;421:427–40.

[207] Lindwall G, Cole RD. Phosphorylation affects the ability of tau protein to promote microtubule assembly. J Biol Chem 1984;259:5301–5.

[208] Bramblett GT, Goedert M, Jakes R, Merrick SE, Trojanowski JQ, Lee VM. Abnormal tau phosphorylation at Ser396 in Alzheimer's disease recapitulates development and contributes to reduced microtubule binding. Neuron 1993;10:1089–99.

[209] Hanger DP, Hughes K, Woodgett JR, Brion JP, Anderton BH. Glycogen synthase kinase-3 induces Alzheimer's disease-like phosphorylation of tau: generation of paired helical filament epitopes and neuronal localisation of the kinase. Neurosci Lett 1992;147:58–62.

[210] Fujiwara H, Hasegawa M, Dohmae N, Kawashima A, Masliah E, Goldberg MS, Shen J, Takio K, Iwatsubo T. alpha-Synuclein is phosphorylated in synucleinopathy lesions. Nat Cell Biol 2002;4:160–4.

[211] Ittner LM, Gotz J. Amyloid-beta and tau–a toxic pas de deux in Alzheimer's disease. Nat Rev Neurosci 2011;12:65–72. Epub 2011/01/05.

[212] Clinton LK, Blurton-Jones M, Myczek K, Trojanowski JQ, LaFerla FM. Synergistic Interactions between Abeta, tau, and alpha-synuclein: acceleration of neuropathology and cognitive decline. J Neurosci 2010;30:7281–9.

[213] Giasson BI, Forman MS, Higuchi M, Golbe LI, Graves CL, Kotzbauer PT, Trojanowski JQ, Lee VM. Initiation and synergistic fibrillization of tau and alpha-synuclein. Science 2003;300:636–40.

[214] DaRocha-Souto B, Scotton TC, Coma M, Serrano-Pozo A, Hashimoto T, Sereno L, Rodriguez M, Sanchez B, Hyman BT, Gomez-Isla T. Brain oligomeric beta-amyloid but not total amyloid plaque burden correlates with neuronal loss and astrocyte inflammatory response in amyloid precursor protein/tau transgenic mice. J Neuropathol Exp Neurol 2011;70:360–76.

[215] Arrasate M, Mitra S, Schweitzer ES, Segal MR, Finkbeiner S. Inclusion body formation reduces levels of mutant huntingtin and the risk of neuronal death. Nature 2004;431:805–10.

[216] Treusch S, Cyr DM, Lindquist S. Amyloid deposits: protection against toxic protein species? Cell Cycle 2009;8:1668–74.

[217] Prusiner SB. Cell biology. A unifying role for prions in neurodegenerative diseases. Science 2012;336:1511–3.

[218] Collinge J. Molecular neurology of prion disease. J Neurol Neurosurg Psychiatry 2005;76:906–19.

[219] Godec MS, Asher DM, Masters CL, Kozachuk WE, Friedland RP, Gibbs Jr CJ, Gajdusek DC, Rapoport SI, Schapiro MB. Evidence against the transmissibility of Alzheimer's disease. Neurology 1991;41:1320.

[220] Baker HF, Ridley RM, Duchen LW, Crow TJ, Bruton CJ. Evidence for the experimental transmission of cerebral beta-amyloidosis to primates. Int J Exp Pathol 1993;74:441–54.

[221] Ridley RM, Baker HF, Windle CP, Cummings RM. Very long term studies of the seeding of beta-amyloidosis in primates. J Neural Transm (Vienna) 2006;113:1243–51.

[222] Collinge J, Clarke AR. A general model of prion strains and their pathogenicity. Science 2007;318:930–6. Epub 2007/11/10.

[223] Li JY, Englund E, Holton JL, Soulet D, Hagell P, Lees AJ, Lashley T, Quinn NP, Rehncrona S, Bjorklund A, Widner H, Revesz T, Lindvall O, Brundin P. Lewy bodies in grafted neurons in subjects with Parkinson's disease suggest host-to-graft disease propagation. Nat Med 2008;14:501–3.

[224] Kordower JH, Chu Y, Hauser RA, Freeman TB, Olanow CW. Lewy body-like pathology in long-term embryonic nigral transplants in Parkinson's disease. Nat Med 2008;14:504–6.

[225] Meyer-Luehmann M, Coomaraswamy J, Bolmont T, Kaeser S, Schaefer C, Kilger E, Neuenschwander A, Abramowski D, Frey P, Jaton AL, Vigouret JM, Paganetti P, Walsh DM, Mathews PM, Ghiso J, Staufenbiel M, Walker LC, Jucker M. Exogenous induction of cerebral beta-amyloidogenesis is governed by agent and host. Science 2006;313:1781–4.

[226] Eisele YS, Obermuller U, Heilbronner G, Baumann F, Kaeser SA, Wolburg H, Walker LC, Staufenbiel M, Heikenwalder M, Jucker M. Peripherally applied Abeta-containing inoculates induce cerebral beta-amyloidosis. Science 2010;330:980–2.

[227] Watts JC, Condello C, Stohr J, Oehler A, Lee J, DeArmond SJ, Lannfelt L, Ingelsson M, Giles K, Prusiner SB. Serial propagation of distinct strains of Abeta prions from Alzheimer's disease patients. Proc Natl Acad Sci USA 2014;111:10323–8.

[228] Clavaguera F, Bolmont T, Crowther RA, Abramowski D, Frank S, Probst A, Fraser G, Stalder AK, Beibel M, Staufenbiel M, Jucker M, Goedert M, Tolnay M. Transmission and spreading of tauopathy in transgenic mouse brain. Nat Cell Biol 2009;11:909–13.

[229] Goedert M, Eisenberg DS, Crowther RA. Propagation of tau aggregates and neurodegeneration. Annu Rev Neurosci 2017;40:189–210.

[230] Woerman AL, Aoyagi A, Patel S, Kazmi SA, Lobach I, Grinberg LT, McKee AC, Seeley WW, Olson SH, Prusiner SB. Tau prions from Alzheimer's disease and chronic traumatic encephalopathy patients propagate in cultured cells. Proc Natl Acad Sci USA 2016;113:E8187–96.

[231] Kfoury N, Holmes BB, Jiang H, Holtzman DM, Diamond MI. Trans-cellular propagation of Tau aggregation by fibrillar species. J Biol Chem 2012;287:19440–51.

[232] Kaufman SK, Sanders DW, Thomas TL, Ruchinskas AJ, Vaquer-Alicea J, Sharma AM, Miller TM, Diamond MI. Tau prion strains dictate patterns of cell pathology, progression rate, and regional vulnerability in vivo. Neuron 2016;92:796–812.

[233] Luk KC, Kehm V, Carroll J, Zhang B, O'Brien P, Trojanowski JQ, Lee VM. Pathological alpha-synuclein transmission initiates Parkinson-like neurodegeneration in nontransgenic mice. Science 2012;338:949–53.

[234] Watts JC, Giles K, Oehler A, Middleton L, Dexter DT, Gentleman SM, DeArmond SJ, Prusiner SB. Transmission of multiple system atrophy prions to transgenic mice. Proc Natl Acad Sci USA 2013;110:19555–60.

[235] Woerman AL, Stohr J, Aoyagi A, Rampersaud R, Krejciova Z, Watts JC, Ohyama T, Patel S, Widjaja K, Oehler A, Sanders DW, Diamond MI, Seeley WW, Middleton LT, Gentleman SM, Mordes DA, Sudhof TC, Giles K, Prusiner SB. Propagation of prions causing synucleinopathies in cultured cells. Proc Natl Acad Sci USA 2015;112:E4949–58.

[236] Olanow CW, Prusiner SB. Is Parkinson's disease a prion disorder? Proc Natl Acad Sci USA 2009;106:12571–2.

[237] Jarrett JT, Lansbury Jr PT. Seeding "one-dimensional crystallization" of amyloid: a pathogenic mechanism in Alzheimer's disease and scrapie? Cell 1993;73:1055–8.

[238] Rudge P, Jaunmuktane Z, Adlard P, Bjurstrom N, Caine D, Lowe J, Norsworthy P, Hummerich H, Druyeh R, Wadsworth JD, Brandner S, Hyare H, Mead S, Collinge J. Iatrogenic CJD due to pituitary-derived growth hormone with genetically determined incubation times of up to 40 years. Brain J Neurol 2015;138:3386–99.

[239] Hamaguchi T, Noguchi-Shinohara M, Nozaki I, Nakamura Y, Sato T, Kitamoto T, Mizusawa H, Yamada M. The risk of iatrogenic Creutzfeldt-Jakob disease through medical and surgical procedures. Neuropathology 2009;29:625–31.

[240] Jaunmuktane Z, Mead S, Ellis M, Wadsworth JD, Nicoll AJ, Kenny J, Launchbury F, Linehan J, Richard-Loendt A, Walker AS, Rudge P, Collinge J, Brandner S. Evidence for human transmission of amyloid-beta pathology and cerebral amyloid angiopathy. Nature 2015;525:247–50.

[241] Ritchie DL, Adlard P, Peden AH, Lowrie S, Le Grice M, Burns K, Jackson RJ, Yull H, Keogh MJ, Wei W, Chinnery PF, Head MW, Ironside JW. Amyloid-beta accumulation in the CNS in human growth hormone recipients in the UK. Acta Neuropathol 2017;134:221–40.

[242] Duyckaerts C, Sazdovitch V, Ando K, Seilhean D, Privat N, Yilmaz Z, Peckeu L, Amar E, Comoy E, Maceski A, Lehmann S, Brion JP, Brandel JP, Haik S. Neuropathology of iatrogenic Creutzfeldt-Jakob disease and immunoassay of French cadaver-sourced growth hormone batches suggest possible transmission of tauopathy and long incubation periods for the transmission of Abeta pathology. Acta Neuropathol 2017;135(2):201–12.

[243] Heneka MT, Carson MJ, El Khoury J, Landreth GE, Brosseron F, Feinstein DL, Jacobs AH, Wyss-Coray T, Vitorica J, Ransohoff RM, Herrup K, Frautschy SA, Finsen B, Brown GC, Verkhratsky A, Yamanaka K, Koistinaho J, Latz E, Halle A, Petzold GC, Town T, Morgan D, Shinohara ML, Perry VH, Holmes C, Bazan NG, Brooks DJ, Hunot S, Joseph B, Deigendesch N, Garaschuk O, Boddeke E, Dinarello CA, Breitner JC, Cole GM, Golenbock DT, Kummer MP. Neuroinflammation in Alzheimer's disease. Lancet Neurol 2015;14:388–405.

[244] Ulrich JD, Ulland TK, Colonna M, Holtzman DM. Elucidating the role of TREM2 in Alzheimer's disease. Neuron 2017;94:237–48.

[245] Zhao Y, Wu X, Li X, Jiang LL, Gui X, Liu Y, Sun Y, Zhu B, Pina-Crespo JC, Zhang M, Zhang N, Chen X, Bu G, An Z, Huang TY, Xu H. TREM2 is a receptor for beta-amyloid that mediates microglial function. Neuron 2018;97:1023–31 e7.

[246] Leyns CEG, Ulrich JD, Finn MB, Stewart FR, Koscal LJ. Remolina Serrano J, Robinson GO, Anderson E, Colonna M, Holtzman DM. TREM2 deficiency attenuates neuroinflammation and protects against neurodegeneration in a mouse model of tauopathy. Proc Natl Acad Sci USA 2017;114:11524–9.

[247] Lee CYD, Daggett A, Gu X, Jiang LL, Langfelder P, Li X, Wang N, Zhao Y, Park CS, Cooper Y, Ferando I, Mody I, Coppola G, Xu H, Yang XW. Elevated TREM2 gene dosage reprograms microglia responsivity and ameliorates pathological phenotypes in Alzheimer's disease models. Neuron 2018;97:1032–48 e5.

[248] Haigis MC, Yankner BA. The aging stress response. Mol Cell 2010;40:333–44.

[249] Taylor RC, Dillin A. Aging as an event of proteostasis collapse. Cold Spring Harb Perspect Biol 2011:3.

[250] Kapogiannis D, Mattson MP. Disrupted energy metabolism and neuronal circuit dysfunction in cognitive impairment and Alzheimer's disease. Lancet Neurol 2011;10:187–98.

[251] Birks J. Cholinesterase inhibitors for Alzheimer's disease. Cochrane Database Syst Rev 2006:CD005593.

[252] Birks J, Harvey RJ. Donepezil for dementia due to Alzheimer's disease. Cochrane Database Syst Rev 2006:CD001190.

[253] McShane R, Areosa Sastre A, Minakaran N. Memantine for dementia. Cochrane Database Syst Rev 2006:CD003154.

[254] Rolinski M, Fox C, Maidment I, McShane R. Cholinesterase inhibitors for dementia with Lewy bodies, Parkinson's disease dementia and cognitive impairment in Parkinson's disease. Cochrane Database Syst Rev 2012:CD006504.

[255] Emre M, Tsolaki M, Bonuccelli U, Destee A, Tolosa E, Kutzelnigg A, Ceballos-Baumann A, Zdravkovic S, Bladstrom A, Jones R, Study I. Memantine for patients with Parkinson's disease dementia or dementia with Lewy bodies: a randomised, double-blind, placebo-controlled trial. Lancet Neurol 2010;9:969–77.

[256] Cummings J, Lee G, Mortsdorf T, Ritter A, Zhong K. Alzheimer's disease drug development pipeline. Alzheimers Dement (NY) 2017;2017(3):367–84.

[257] MacLeod R, Hillert EK, Cameron RT, Baillie GS. The role and therapeutic targeting of alpha-, beta- and gamma-secretase in Alzheimer's disease. Future Sci OA 2015;1:FSO11.

[258] May PC, Willis BA, Lowe SL, Dean RA, Monk SA, Cocke PJ, Audia JE, Boggs LN, Borders AR, Brier RA, Calligaro DO, Day TA, Ereshefsky L, Erickson JA, Gevorkyan H, Gonzales CR, James DE, Jhee SS, Komjathy SF, Li L, Lindstrom TD, Mathes BM, Martenyi F, Sheehan SM, Stout SL, Timm DE, Vaught GM, Watson BM, Winneroski LL, Yang Z, Mergott DJ. The potent BACE1 inhibitor LY2886721 elicits robust central Abeta pharmacodynamic responses in mice, dogs, and humans. J Neurosci 2015;35:1199–210.

[259] Kennedy ME, Stamford AW, Chen X, Cox K, Cumming JN, Dockendorf MF, Egan M, Ereshefsky L, Hodgson RA, Hyde LA, Jhee S, Kleijn HJ, Kuvelkar R, Li W, Mattson BA, Mei H, Palcza J, Scott JD, Tanen M, Troyer MD, Tseng JL, Stone JA, Parker EM, Forman MS. The BACE1 inhibitor verubecestat (MK-8931) reduces CNS beta-amyloid in animal models and in Alzheimer's disease patients. Sci Transl Med 2016;8:363ra150.

[260] Mullard A. BACE inhibitor bust in Alzheimer trial. Nat Rev Drug Discov 2017;16:155.

[261] Doody RS, Raman R, Farlow M, Iwatsubo T, Vellas B, Joffe S, Kieburtz K, He F, Sun X, Thomas RG, Aisen PS, Alzheimer's Disease Cooperative Study Steering C, Siemers E, Sethuraman G, Mohs R, Semagacestat Study G. A phase 3 trial of semagacestat for treatment of Alzheimer's disease. N Engl J Med 2013;369:341–50. Epub 2013/07/26.

[262] De Strooper B. Lessons from a failed gamma-secretase Alzheimer trial. Cell 2014;159:721–6.

[263] Postina R. Activation of alpha-secretase cleavage. J Neurochem 2012;120(Suppl. 1):46–54.

[264] Lichtenthaler SF. alpha-secretase in Alzheimer's disease: molecular identity, regulation and therapeutic potential. J Neurochem 2011;116:10–21.

[265] Marcello E, Borroni B, Pelucchi S, Gardoni F, Di Luca M. ADAM10 as a therapeutic target for brain diseases: from developmental disorders to Alzheimer's disease. Expert Opin Ther Targets 2017;21:1017–26.

[266] Chu CS, Tseng PT, Stubbs B, Chen TY, Tang CH, Li DJ, Yang WC, Chen YW, Wu CK, Veronese N, Carvalho AF, Fernandes BS, Herrmann N, Lin PY. Use of statins and the risk of dementia and mild cognitive impairment: a systematic review and meta-analysis. Sci Rep 2018;8:5804.

[267] Schenk D, Barbour R, Dunn W, Gordon G, Grajeda H, Guido T, Hu K, Huang J, Johnson-Wood K, Khan K, Kholodenko D, Lee M, Liao Z, Lieberburg I, Motter R, Mutter L, Soriano F, Shopp G, Vasquez N, Vandevert C, Walker S, Wogulis M, Yednock T, Games D, Seubert P. Immunization with amyloid-beta attenuates Alzheimer-disease-like pathology in the PDAPP mouse. Nature 1999;400:173–7.

[268] Bard F, Cannon C, Barbour R, Burke RL, Games D, Grajeda H, Guido T, Hu K, Huang J, Johnson-Wood K, Khan K, Kholodenko D, Lee M, Lieberburg I, Motter R, Nguyen M, Soriano F, Vasquez N, Weiss K, Welch B, Seubert P, Schenk D, Yednock T. Peripherally administered antibodies against amyloid beta-peptide enter the central nervous system and reduce pathology in a mouse model of Alzheimer disease. Nat Med 2000;6:916–9.

[269] Orgogozo JM, Gilman S, Dartigues JF, Laurent B, Puel M, Kirby LC, Jouanny P, Dubois B, Eisner L, Flitman S, Michel BF, Boada M, Frank A, Hock C. Subacute meningoencephalitis in a subset of patients with AD after Abeta42 immunization. Neurology 2003;61:46–54.

[270] Nicoll JA, Wilkinson D, Holmes C, Steart P, Markham H, Weller RO. Neuropathology of human Alzheimer disease after immunization with amyloid-beta peptide: a case report. Nat Med 2003;9:448–52.

[271] Mellman I, Coukos G, Dranoff G. Cancer immunotherapy comes of age. Nature 2011;480:480–9.

[272] Weiner HL, Frenkel D. Immunology and immunotherapy of Alzheimer's disease. Nat Rev Immunol 2006;6:404–16.

[273] Honig LS, Vellas B, Woodward M, Boada M, Bullock R, Borrie M, Hager K, Andreasen N, Scarpini E, Liu-Seifert H, Case M, Dean RA, Hake A, Sundell K, Poole Hoffmann V, Carlson C, Khanna R, Mintun M, DeMattos R, Selzler KJ, Siemers E. Trial of Solanezumab for mild dementia due to Alzheimer's disease. N Engl J Med 2018;378:321–30.

[274] Salloway S, Sperling R, Fox NC, Blennow K, Klunk W, Raskind M, Sabbagh M, Honig LS, Porsteinsson AP, Ferris S, Reichert M, Ketter N, Nejadnik B, Guenzler V, Miloslavsky M, Wang D, Lu Y, Lull J, Tudor IC, Liu E, Grundman M, Yuen E, Black R, Brashear HR. Bapineuzumab, Clinical Trial I. Two phase 3 trials of bapineuzumab in mild-to-moderate Alzheimer's disease. N Engl J Med 2014;370:322–33.

[275] Cummings JL, Cohen S, van Dyck CH, Brody M, Curtis C, Cho W, Ward M, Friesenhahn M, Rabe C, Brunstein F, Quartino A, Honigberg LA, Fuji RN, Clayton D, Mortensen D, Ho C, Paul R. ABBY: a phase 2 randomized trial of crenezumab in mild to moderate Alzheimer disease. Neurology 2018;90(21):e1889–e1897.

[276] Panza F, Solfrizzi V, Seripa D, Imbimbo BP, Lozupone M, Santamato A, Tortelli R, Galizia I, Prete C, Daniele A, Pilotto A, Greco A, Logroscino G. Tau-based therapeutics for Alzheimer's disease: active and passive immunotherapy. Immunotherapy 2016;8:1119–34.

[277] Braczynski AK, Schulz JB, Bach JP. Vaccination strategies in tauopathies and synucleinopathies. J Neurochem 2017;143:467–88.

[278] Novak P, Schmidt R, Kontsekova E, Zilka N, Kovacech B, Skrabana R, Vince-Kazmerova Z, Katina S, Fialova L, Prcina M, Parrak V, Dal-Bianco P, Brunner M, Staffen W, Rainer M, Ondrus M, Ropele S, Smisek M, Sivak R, Winblad B, Novak M. Safety and immunogenicity of the tau vaccine AADvac1 in patients with Alzheimer's disease: a randomised, double-blind, placebo-controlled, phase 1 trial. Lancet Neurol 2017;16:123–34.

[279] Yanamandra K, Patel TK, Jiang H, Schindler S, Ulrich JD, Boxer AL, Miller BL, Kerwin DR, Gallardo G, Stewart F, Finn MB, Cairns NJ, Verghese PB, Fogelman I, West T, Braunstein J, Robinson G, Keyser J, Roh J, Knapik SS, Hu Y, Holtzman DM. Anti-tau antibody administration increases plasma tau in transgenic mice and patients with tauopathy. Sci Transl Med 2017:9.

[280] Mullard A. Pharma pumps up anti-tau Alzheimer pipeline despite first Phase III failure. Nat Rev Drug Discov 2016;15:591–2.

[281] Koros C, Stamelou M. Interventions in progressive supranuclear palsy. Park Relat Disord 2016;22(Suppl. 1):S93–5.

[282] Gauthier S, Feldman HH, Schneider LS, Wilcock GK, Frisoni GB, Hardlund JH, Moebius HJ, Bentham P, Kook KA, Wischik DJ, Schelter BO, Davis CS, Staff RT, Bracoud L, Shamsi K, Storey JM, Harrington CR, Wischik CM. Efficacy and safety of tau-aggregation inhibitor therapy in patients with mild or moderate Alzheimer's disease: a randomised, controlled, double-blind, parallel-arm, phase 3 trial. Lancet 2016;388:2873–84.

[283] Lovestone S, Boada M, Dubois B, Hull M, Rinne JO, Huppertz HJ, Calero M, Andres MV, Gomez-Carrillo B, Leon T, del Ser T, Investigators A. A phase II trial of tideglusib in Alzheimer's disease. J Alzheimers Dis 2015;45:75–88.

[284] Tolosa E, Litvan I, Hoglinger GU, Burn D, Lees A, Andres MV, Gomez-Carrillo B, Leon T, Del Ser T, Investigators T. A phase 2 trial of the GSK-3 inhibitor tideglusib in progressive supranuclear palsy. Mov Disord 2014;29:470–8.

[285] van Roon-Mom WMC, Roos RAC, de Bot ST. Dose-dependent lowering of mutant huntingtin using antisense oligonucleotides in Huntington disease patients. Nucleic Acid Ther 2018;28:59–62.

[286] Schenk DB, Koller M, Ness DK, Griffith SG, Grundman M, Zago W, Soto J, Atiee G, Ostrowitzki S, Kinney GG. First-in-human assessment of PRX002, an anti-alpha-synuclein monoclonal antibody, in healthy volunteers. Mov Disord 2017;32:211–8.

[287] Galimberti D, Scarpini E. Pioglitazone for the treatment of Alzheimer's disease. Expert Opin Investig Drugs 2017;26:97–101.

[288] Mattson MP. Emerging neuroprotective strategies for Alzheimer's disease: dietary restriction, telomerase activation, and stem cell therapy. Exp Gerontol 2000;35:489–502.

[289] Amanatkar HR, Papagiannopoulos B, Grossberg GT. Analysis of recent failures of disease modifying therapies in Alzheimer's disease suggesting a new methodology for future studies. Expert Rev Neurother 2017;17:7–16.

[290] Golde TE. Overcoming translational barriers impeding development of Alzheimer's disease modifying therapies. J Neurochem 2016;139(Suppl. 2):224–36.

[291] Collins FS, Varmus H. A new initiative on precision medicine. N Engl J Med 2015;372:793–5.

[292] Dugger SA, Platt A, Goldstein DB. Drug development in the era of precision medicine. Nat Rev Drug Discov 2018;17:183–96.

[293] Cholerton B, Larson EB, Quinn JF, Zabetian CP, Mata IF, Keene CD, Flanagan M, Crane PK, Grabowski TJ, Montine KS, Montine TJ. Precision medicine: clarity for the complexity of dementia. Am J Pathol 2016;186:500–6.

[294] Kumar-Sinha C, Chinnaiyan AM. Precision oncology in the age of integrative genomics. Nat Biotechnol 2018;36:46–60.

[295] van der Flier WM, Barkhof F, Scheltens P. Shifting paradigms in dementia: toward stratification of diagnosis and treatment using MRI. Ann NY Acad Sci 2007;1097:215–24.

[296] Ahmed RM, Paterson RW, Warren JD, Zetterberg H, O'Brien JT, Fox NC, Halliday GM, Schott JM. Biomarkers in dementia: clinical utility and new directions. J Neurol Neurosurg Psychiatry 2014;85:1426–34.

[297] Carroll CM, Li YM. Physiological and pathological roles of the gamma-secretase complex. Brain Res Bull 2016;126:199–206.

[298] Beel AJ, Sanders CR. Substrate specificity of gamma-secretase and other intramembrane proteases. Cell Mol Life Sci 2008;65:1311–34.

[299] Saftig P, Lichtenthaler SF. The alpha secretase ADAM10: a metalloprotease with multiple functions in the brain. Prog Neurobiol 2015;135:1–20.

[300] Schweisguth F. Regulation of notch signaling activity. Curr Biol 2004;14:R129–38.

[301] Selkoe DJ, Hardy J. The amyloid hypothesis of Alzheimer's disease at 25 years. EMBO Mol Med 2016;8:595–608.

[302] Karran E, De Strooper B. The amyloid cascade hypothesis: are we poised for success or failure? J Neurochem 2016;139(Suppl. 2):237–52.

[303] Voytyuk I, De Strooper B, Chavez-Gutierrez L. Modulation of gamma- and beta-secretases as early prevention against Alzheimer's disease. Biol Psychiatry 2018;83:320–7.

CHAPTER

PARKINSON'S DISEASE

3

CHAPTER OUTLINE

The Molecular and Clinical Pathology of Neurodegenerative Disease. https://doi.org/10.1016/B978-0-12-811069-0.00003-3

3.1 INTRODUCTION

Parkinson's disease, named to recognise the contribution of James Parkinson (the 19th-century physician who first codified the symptoms that define the disorder [1]), is the second most common neurodegenerative disease afflicting humans [2]. Similar to Alzheimer's-type dementia, which (as noted in Chapter 2) is the most common human neurodegenerative disease, Parkinson's disease is a disorder predominantly of old age, and one that is becoming increasingly common as the global population ages. In stark contrast to Alzheimer's disease, however, Parkinson's disease benefits from robust and reliable drugs to target some of the most prominent and debilitating neurological symptoms resulting from neurodegeneration. In common with Alzheimer's disease, there is as yet no disease-modifying therapy available.

3.2 CLINICAL PRESENTATION

The clinical features of Parkinson's disease are well recognized. It is usually thought of in terms of the motor symptoms that affect hand movements and walking, but it has become increasingly apparent that nonmotor symptoms affect the majority of patients and can have a severe impact on quality of life [2,3].

Like Alzheimer's disease, Parkinson's disease is a condition of age, with a mean age of onset of 60 years. Young-onset Parkinson's disease is recognized and is defined as symptoms presenting under the age of 40 years but accounts for less than 5% of cases [4].

It is important to distinguish between the terms parkinsonism and Parkinson's disease [5]. Parkinsonism is a term used to describe a group of movement symptoms, whereas Parkinson's disease implies a clinically and pathologically defined process as described in the UK Parkinson's disease Society Brain Bank criteria (Table 3.1) [6]. This is a critical distinction: parkinsonism, encompassing a group of symptoms, can have many causes in addition to neurodegeneration, including drugs, trauma, postinfectious, and vascular disease, whereas Parkinson's disease is a distinct disease entity. In many ways, this is analogous to the distinction between dementia and Alzheimer disease and can

Table 3.1 UK Parkinson's Disease Society Brain Bank Criteria

Step 1: Diagnosis of parkinsonian syndrome

- Bradykinesia (slowness of initiation of voluntary movement with progressive reduction in speed and amplitude of repetitive actions)

And at least one of the following:

- Muscular rigidity, 4–6 Hz
- Rest tremor
- Postural instability not caused by primary visual, vestibular, cerebellar, or proprioceptive dysfunction

Step 2: Exclusion criteria for Parkinson disease

- History of repeated strokes with stepwise progression of parkinsonian features
- History of repeated head injury
- History of definite encephalitis
- Oculogyric crises
- Neuroleptic treatment at onset of symptoms
- More than one affected relative
- Sustained remission
- Strictly unilateral features after 3 years
- Supranuclear gaze palsy
- Cerebellar signs
- Early severe autonomic involvement
- Early severe dementia with disturbances of memory, language, and praxis
- Babinski sign
- Presence of cerebral tumour or communicating hydrocephalus on computed tomographic scan
- Negative response to large doses of levodopa (if malabsorption excluded)
- MPTP exposure

Step 3: Supportive prospective positive criteria for Parkinson disease (three or more required for diagnosis of definite Parkinson disease)

- Unilateral onset
- Rest tremor present
- Progressive disorder
- Persistent asymmetry affecting side of onset most
- Excellent response (70%–100%) to levodopa
- Severe levodopa-induced chorea
- Levodopa response for 5 years or more
- Clinical course of 10 years or more

MPTP, *1-methyl-4-phenyl-1,2,3,6-tetrahydropyridine.*

be viewed as a semantic debate. It is, however, a very important semantic difference from a therapeutic viewpoint (a patient with Parkinson's disease will, by definition, react favorably when treated with dopamine pathway drugs, which may not be the case if someone has a broader diagnosis with parkinsonism), and also when considering how we understand and categorize the molecular mechanisms underpinning the disease.

There is, at present, no diagnostic test for Parkinson's disease. Patients with atypical presentations may need investigations to rule out other disorders, but the diagnosis is primarily clinical, resting on the history of the patient's symptoms and the findings on clinical examination, with response to therapy as a supportive phenomenon. Scans that use tracers that bind to the dopamine transporter (DAT) protein in the nigrostriatal nerve endings (DAT scans) show a reduction in binding in Parkinson's disease and can be helpful in diagnosis [7]. However, DAT scan results are also abnormal in other disorders such as the atypical parkinsonian conditions (see later discussion) and are not routinely used in diagnosis.

3.2.1 MOTOR SYMPTOMS

The three cardinal motor symptoms of Parkinson's disease are rigidity, bradykinesia, and tremor (Fig. 3.1) [8]. Gait and postural instability are also prominent motor features. The motor symptoms typically have an asymmetrical onset (starting on one side of the body) with symptoms typically manifesting initially in one upper limb. The contralateral side does eventually become involved but a degree of asymmetry tends to persist throughout the disease course.

Bradykinesia is slowness or poverty of movement and is the most fundamental feature of parkinsonism to the extent that its presence is required for a positive diagnosis of Parkinson's disease to be made. Patients with bradykinesia will have difficulty walking with loss of arm swing and may complain of difficulty turning in bed. They may complain that their handwriting has become smaller (micrographia) and that they have difficulties with manual dexterity. On examination, they will have loss of facial expression with masklike facies (hypomimia), and they may notice that their voice has become quieter and more monotonous (hypophonia).

Besides observing facial expressions and gait, bradykinesia can be examined clinically by asking the patient to perform rapid alternating movements such as tapping the forefinger and the thumb or opening and closing the hand repetitively. With repetition the movements become slower and smaller

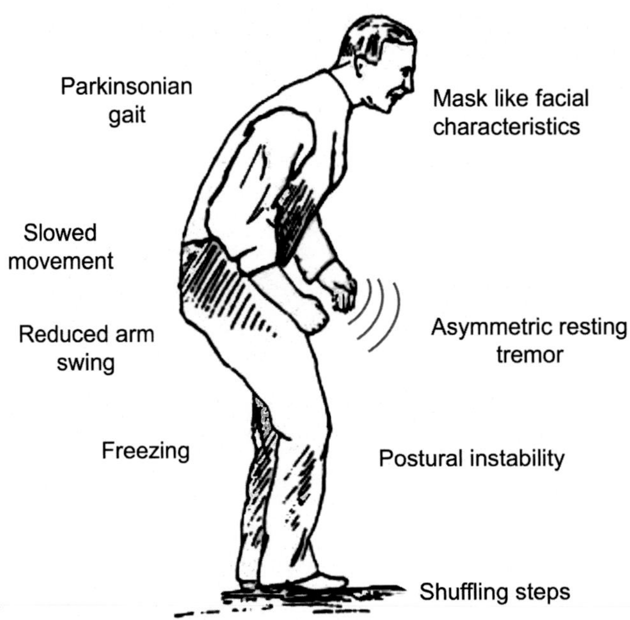

FIGURE 3.1

The movement symptoms associated with Parkinson's disease.

in amplitude; it is this decrementing/fatiguing response that helps distinguish parkinsonism from other conditions that can cause slowness of movement (such as spasticity).

Rigidity is an involuntary increase in muscle tone with a feeling of stiffness or resistance when moving the patient's relaxed limb. Patients will complain of stiffness and will sometimes complain of aching and cramping.

There are two types of rigidity observed in Parkinson's disease:

Lead pipe: Rigidity that is present to the same extent throughout the range of movement.
Cogwheeling rigidity: Clinically this is felt when passively moving the limb as a feeling of stiffness with superimposed jumps.

Rigidity can be exacerbated by certain techniques; for example, asking the patient to move the contralateral arm up and down can increase rigidity in the arm being examined. This phenomenon is known as synkinesis.

Tremor is the involuntary rhythmic oscillatory movement of a body part. Although tremor is the initial presenting symptom in about 70% of patients, 30% of patients have no tremor at all. The typical Parkinson's disease tremor is asymmetrical, most prominent at rest, and increases with distraction. It is a low–frequency, 4–6 Hz, "pill-rolling-type tremor" that is typically a rhythmic movement of the thumb against the index finger. Tremor of body parts, such as the chin and jaw, is frequently seen in Parkinson's disease, but a full head tremor is rare.

Gait/postural instability is a common feature in Parkinson's disease. Patients with Parkinson's disease have a characteristic gait with a narrow base, a hunched posture, and small shuffling steps with reduced arm swing. Patients have difficulty turning and later in the course of the disease can occasionally have the feeling of their feet feeling stuck to the floor, a sensation known as freezing.

Patients with Parkinson's disease may also complain of poor balance and unsteadiness. The "pull test" is used to examine for postural instability. The examiner stands behind the patient and pulls on the patient's shoulders. If the test is positive, patients may require a few steps to correct their balance, and in the later stages, they may fall backward.

3.2.2 NONMOTOR SYMPTOMS

Besides the motor symptoms described earlier, it is increasingly recognized that Parkinson's disease is not just a motor disorder but rather, reflecting the widespread pathology of the disease (see later discussion section 3.3), has a variety of nonmotor symptoms [3]. These symptoms can be as disabling, if not more so, than the motor symptoms and may present years before the motor symptoms become clinically apparent. This has led to increased interest in these symptoms as a way of predicting the onset of Parkinson's disease, which will be crucial if disease-modifying therapies become available in the future.

Nonmotor symptoms can be categorized as in the following sections.

3.2.2.1 Neuropsychiatric

Neuropsychiatric symptoms associated with Parkinson's disease can be varied and are frequently present in the premotor stage of the disease. Depression, anxiety, and apathy are the most prominent neuropsychiatric symptoms in the early stage of the disease, although the mechanisms underlying these presentations are not well understood. Depression, in particular, is a common feature in the case histories of patients with Parkinson's disease, often predating any motor symptoms by a number of years. In the later stages of disease, a high proportion of patients with Parkinson's disease develop some form of cognitive decline, with many reaching the criteria for diagnosis with dementia (see later discussion).

3.2.2.2 Autonomic

The autonomic nervous system, acting to unconsciously control the function of internal organs, is also impacted early in the disease process in Parkinson's disease [9]. Most notably, patients with Parkinson's disease are very likely to suffer from constipation, a symptom that frequently presents many years prior to the onset of motor symptoms and clinical diagnosis of Parkinson's disease. Constipation results from reduced control of the gastrointestinal tract and is likely to reflect a pathological alteration in the brain stem very early on in the disease process. In contrast, incontinence is rarely present in the early stages of Parkinson's disease and is much more associated with a diagnosis of multiple system atrophy (see later discussion).

A number of other autonomic symptoms are observed in patients with Parkinson's disease, including excessive sweating, postural hypotension, erectile dysfunction, and sialorrhea (loss of control of salivation, leading to increased saliva production and drooling). Independently, none of these symptoms are exclusively associated with Parkinson's disease; however, it is now recognized that taken together, they form an important, if not critical, part of the disease.

3.2.2.3 Sleep Disturbance

Another common nonmotor phenomenon, which again forms part of the prodromal symptoms associated with Parkinson's disease, is rapid eye movement (REM) sleep behavior disorder [10]. This is characterized by the loss of the normal muscle atonia during REM sleep causing patients to "act out" their dreams. Data have suggested that REM sleep behavior disorder can precede motor symptoms in almost half of the patients and is present in about half of the patients with Parkinson's disease. Other sleep disturbance symptoms common in PD include daytime somnolence and, in a possible crossover with the movement symptoms of the disease, restless legs and periodic limb movement in sleep.

3.2.2.4 Sensory

Pain is a frequent comorbidity in Parkinson's disease, with a range of presentations [11]. The causes of pain in Parkinson's disease are unclear, and there is a range of sensations that are found in the disease. Neck and shoulder pain, described as coat-hanger pain, has been reported but is also associated with multiple systems atrophy.

3.2.2.5 Anosmia

Anosmia, or loss of sense of smell, is found in 70%–100% of patients with Parkinson's disease [12]. The loss of sense of smell is often the earliest features of Parkinson's disease, retrospectively identified by patients as occurring many years before they notice explicit motor dysfunction. As discussed in Section 3.3, this is likely to reflect degeneration and dysfunction occurring earlier and more severely in the neural circuits responsible for olfaction than in the regions of the brain involved in motor control. Olfactory disturbance is so common that it forms the focus of concerted efforts to develop premotor diagnosis in Parkinson's disease, for example, by the development of scratch and sniff cards that allow a systematic and semiquantitative evaluation of olfactory function.

3.2.3 DEMENTIA AND PARKINSON'S DISEASE

Dementia is common in Parkinson's disease, with a sixfold increased risk compared with the general population. After 15 years, approximately 80% of patients with Parkinson's will have some evidence of cognitive dysfunction. Dementia is diagnosed when at least two cognitive domains are affected and the symptoms are severe enough to impair daily life [13].

Dementia seen in the context of Parkinson's disease can be described as Parkinson's disease dementia or Lewy body dementia (also described as dementia with Lewy bodies), which can be regarded as two ends of a spectrum. The two disorders can be temporally distinguished from each other by the "one-year rule." In Parkinson's disease dementia, motor symptoms develop at least 1 year before the onset of cognitive impairment, whereas in Lewy body dementia, cognitive symptoms develop either before the motor symptoms start or within the first year of onset. Risk factors for Parkinson's disease dementia include older age, more severe parkinsonism, postural instability, and male sex. Different cognitive domains can be affected but impairments in executive and visuospatial function are particularly prominent in Parkinson's disease dementia.

Like Parkinson's disease, dementia with Lewy bodies is an α-synucleinopathy and is characterized by hallucinations, marked fluctuations in cognition, and neuroleptic sensitivity [14]. It is relatively common, accounting for about 20% of all dementia. The management of dementia with Lewy bodies is challenging, not least because the dopaminergic drugs used to treat parkinsonism can exacerbate hallucinations. dementia with Lewy bodies is discussed further in Chapter 2, but it is important to consider this as part of the disease spectrum to which Parkinson's disease belongs; as noted in Section 3.3, there is extensive shared pathology between a number of distinct neurodegenerative disease entities. While we may not fully understand the links between these disorders, this is clear evidence of common disease pathways.

3.2.4 PROGRESSION/PROGNOSIS

Parkinson's disease is progressive and patients will generally experience a decline in their function with time, reflecting the advance of the underlying cell death and pathological condition. There is, however, wide variability in the speed of this progression. The various motor features can affect patients to different extents, with patients described as tremor predominant, akinetic rigid, and mixed. The tremor-predominant form tends to have a better prognosis.

3.2.5 PROGRESSIVE SUPRANUCLEAR PALSY AND MULTIPLE SYSTEM ATROPHY

There are two main forms of atypical degenerative parkinsonism: multiple system atrophy and progressive supranuclear palsy [15,16]. In practice, especially initially, distinguishing between Parkinson's disease and these different disorders can be difficult. It is, however, important to consider them as distinct disorders when possible—they tend to respond less well to treatment and prognosis can be poorer than that in Parkinson's disease.

Red flags clinically for an atypical parkinsonian disorder include early falls, symmetrical onset, axial rigidity, and profound autonomic symptoms.

3.2.5.1 Progressive Supranuclear Palsy

Progressive supranuclear palsy is a degenerative disorder with tau pathology. It does present with parkinsonism, but tends to do so in a more symmetrical fashion than Parkinson's disease, with more severe axial akinesia, rigidity with neck dystonia, and early falls. Tremor is a less prominent feature than that in Parkinson's disease. Patients tend to have cognitive deficits relatively early in the course of the disease, with executive function particularly affected. One of the most characteristic features of progressive supranuclear palsy is the facial expression in which patients have frontalis overactivity that can give a staring appearance. Their eye movements are abnormal, characterized by supranuclear gaze palsy and they may have blepharospasm. Postural instability is often much more severe than that seen in Parkinson's disease, with frequent early falls.

3.2.5.2 Multiple System Atrophy

Multiple system atrophy, in contrast to progressive supranuclear palsy, is an α-synucleinopathy characterized by a combination of autonomic dysfunction with parkinsonism and cerebellar features.

Autonomic symptoms include orthostatic hypotension, erectile dysfunction, and urinary symptoms such as incontinence or incomplete bladder emptying. Cerebellar signs include gait ataxia, dysarthria, and poor coordination. The symptoms of parkinsonism include bradykinesia as well as postural instability, rigidity, and tremor. If the presentation is predominantly parkinsonian, patients are labeled as MSA-P, whereas the term MSA-C is used for those with mainly cerebellar features. Patients often have other features such as dysarthria (strangulated voice), stridor, jerky finger movements (polyminimyoclonus), dystonia, and Raynaud phenomenon and can have a dusky discoloration of the skin. In contrast to progressive supranuclear palsy, dementia is uncommon.

3.3 PATHOLOGY

PD is classically defined as being characterized by two key pathological conditions: loss of neurons in the basal ganglia, specifically in the *substantia nigra pars compacta*, and the presence of intracellular inclusions called Lewy bodies (named for Frederick Lewy, who first described them in the early years of the 20th century) [17,18]. The presence of both phenomena are required for the categorical diagnosis of PD as defined by the Queen Square UK Parkinson's Disease Society Brain Bank criteria (see Table 3.1) [6]. This has several important implications for how people are diagnosed with PD. First, an individual can have all the clinical symptoms consistent with a diagnosis of PD, but if they do not have Lewy bodies in their brain, then they do not have PD (a point that will be returned to in Section 3.4, relating to the genetics of PD). A second implication is that for a categorical diagnosis, an individual has to undergo neuropathological examination post mortem, something that occurs only in a small minority of PD cases.

3.3.1 NIGRAL DEGENERATION

The loss of dopaminergic neurons from the *substantia nigra pars compacta* is the central cellular change associated with the movement symptoms observed in Parkinson's disease. The *substantia nigra*, derived from the Latin for dark substance, is characterized by the presence of deeply pigmented dopaminergic cell bodies containing high levels of melanin and is clearly visible to the naked eye in dissected postmortem samples. In the Parkinson's disease-affected brain, almost all this pigmentation is absent, the result of massive cell loss in this region of the basal ganglia (Fig. 3.2). In contrast to Alzheimer's disease, where there is extensive and widespread cell loss, the pathology in Parkinson's

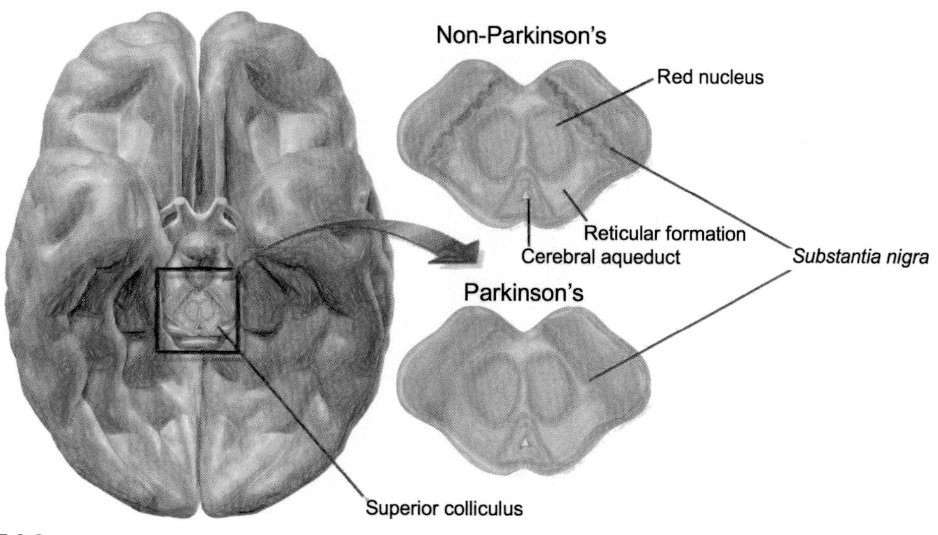

FIGURE 3.2

Degeneration of the *substantia nigra* in Parkinson's disease.

Adapted from: Blausen.com staff (2014). "Medical gallery of Blausen Medical 2014". Wikiversity Journal of Medicine 1 (2): 10. https://doi.org/10.15347/wjm/2014.010 using a creative commons licence

disease is relatively concentrated in discrete regions of the brain; for example, regions of the basal ganglia outside the *substantia nigra* are relatively spared by the disease process. The reason for such an obvious differential vulnerability to disease in Parkinson's disease is unclear (as is the case for the regional and cell-type specificity in most neurodegenerative disorders). It is important to note that cellular pathology in the *substantia nigra*, and indeed regions outside the nigra, is not limited to neuronal cell death. As for the brain in Alzheimer's disease, the loss of neurons is accompanied by a complex array of alterations in other cell types, including the activation of astrocytes and the proliferation of microglia in response to damage. This aspect of neurodegeneration in Parkinson's disease is only just beginning to be understood and expands the view of neurodegeneration in this disorder from one that is purely centered on the loss of neurons to a more nuanced model where neurodegeneration occurs in concert with reactive responses from glial cells and the immune system in the brain, a process that may be beneficial in some cases and at certain points in the disease but may act to accelerate the disease in others.

3.3.2 LEWY BODIES

Lewy bodies are intracellular inclusions found in a range of neurodegenerative disorders but are most closely associated with the pathology of Parkinson's disease and dementia with Lewy bodies (see Chapter 2). They are predominantly made up of the protein α-synuclein, a 140-amino-acid protein of as-yet undetermined function. Lewy bodies were first described by Frederick Lewy during the early 20th century and are dense circular structures made up of concentric rings of aggregated protein (Fig. 3.3) [17,19].

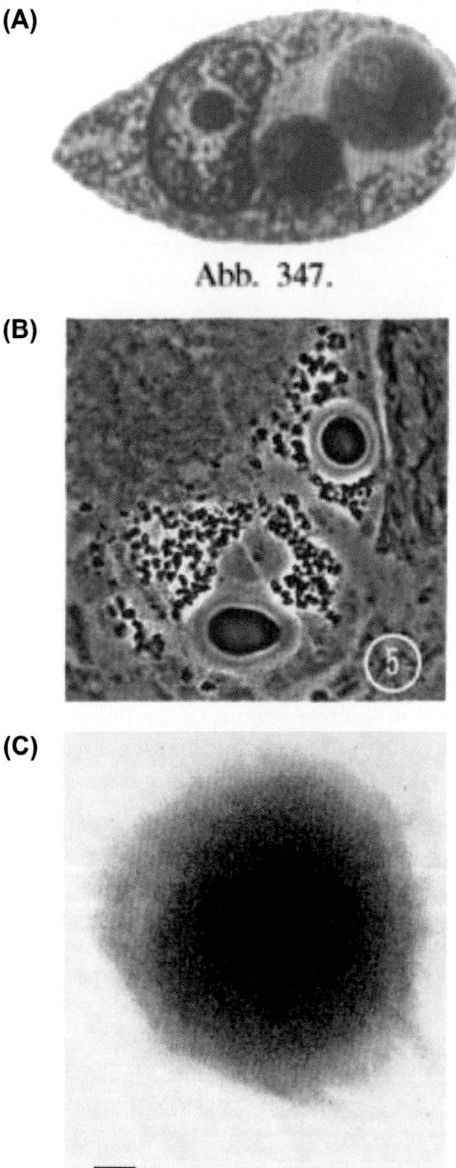

FIGURE 3.3

The structure of a Lewy body through the ages. (A) A drawing by Frederick Lewy of the pathology that now bears his name. (B) An electron micrograph of the structure of a Lewy body in the human brain. (C) Immunostaining for α-synuclein in the *substantia nigra* of a patient with Parkinson's disease.

Images adapted from: A) Lewy FH. Geschichte und Tä tigkeit des Ortslazaretts Haidar Pascha. Berlin: Stritzke Verlag;1920. B) Duffy, Philip E., and Virginia M. Tennyson. "Phase and electron microscopic observations of Lewy bodies and melanin granules in the substantia nigra and locus caeruleus in Parkinson's disease." Journal of Neuropathology & Experimental Neurology 24.3 (1965): 398-414. C) Spillantini, Maria Grazia, et al. "[alpha]-Synuclein in Lewy bodies." Nature 388.6645 (1997): 839.

The precise role of the Lewy body in causing Parkinson's disease (as well as dementia with Lewy bodies and Alzheimer's disease) is obscure. The realization that mutations in the gene that codes for α-synuclein, namely *SNCA*, cause increased protein aggregation and neurodegeneration (see Section 3.4.1), established a direct link between the biology of aggregating proteins and the disease [20]. The causal link between Lewy bodies themselves and neuronal cell death is, however, far less certain. There is provocative research from related disorders (most notably, Huntington's disease, see Chapter 6) that visible organized protein aggregate structures such as Lewy bodies may actually be protective [21]. These data suggest that if you have disruptive protein aggregates multiplying in the cytoplasm of neurons, then these cells respond partly by pulling all these aggregates into one part of the cell to establish, in the case of Parkinson's disease, a Lewy body. In doing so, the damage that these aggregates cause to organelles and cellular processes is limited.

Because the majority of analysis of Lewy bodies comes from postmortem samples, which by definition provide a static measure of pathology, and in the absence of physiologically relevant animal models that recapitulate Lewy body pathology, it is somewhat difficult to test this hypothesis directly. An intriguing insight into the biology of Lewy bodies was provided by the use of fetal cell implants as an experimental treatment for Parkinson's disease during the 1980s and 1990s [22]. Although the clinical results of this cell replacement therapy were mixed (see Section 3.5), a number of patients treated with these cells eventually underwent postmortem neuropathological examination. When the regions that were subjected to cell replacement therapy were examined, two important observations were made. First, a proportion of the fetal cells had survived and had successfully integrated into the basal ganglia. Second, and perhaps surprisingly, there appeared to be Lewy body pathology in these new cells, suggesting that the aggregation of α-synuclein into Lewy bodies could spread from aged diseased cells surrounding the implant into the newly integrated cells [23,24]. This, in turn, supports a model for α-synuclein pathology in the brain where a process of protein aggregate transmission occurs, perhaps analogous to that seen in the prion disorders (see Chapter 4), which is an idea discussed further in Section 3.3.5.

3.3.3 NEUROCHEMICAL DEFICITS

The differential vulnerability of dopaminergic neurons in the parkinsonian brain, and the resulting massive loss of these brain cells, leads to fundamental alteration in the neurochemistry of the central nervous system [25]. Following the identification of dopamine as a neurotransmitter in the 1950s, most notably through the work of Arvid Carlsson who was working at the University of Gothenburg in Sweden (for which he was awarded the Nobel Prize in Physiology or Medicine in 2000), it was soon established that the brains of people with PD had a substantial decrease in the levels of dopamine compared with the healthy brain tissue [26,27].

When acting as a neurotransmitter, dopamine is found in membrane-bound vesicles in the synaptic terminals of dopaminergic cells. When the synaptic terminal is activated by the receipt of an action potential, these vesicles fuse with the synaptic membrane releasing the neurotransmitters into the synaptic cleft where they bind to specialized dopamine receptors on the dendrites of the proximal neuron [28].

Dopamine belongs to the catecholamine family of chemicals and consists of a simple aromatic ring structure with two hydroxyl groups (a catechol) modified with an amine group attached to an ethyl chain (Fig. 3.4) [29]. In the brain, dopamine is derived from a precursor molecule, the amino acid

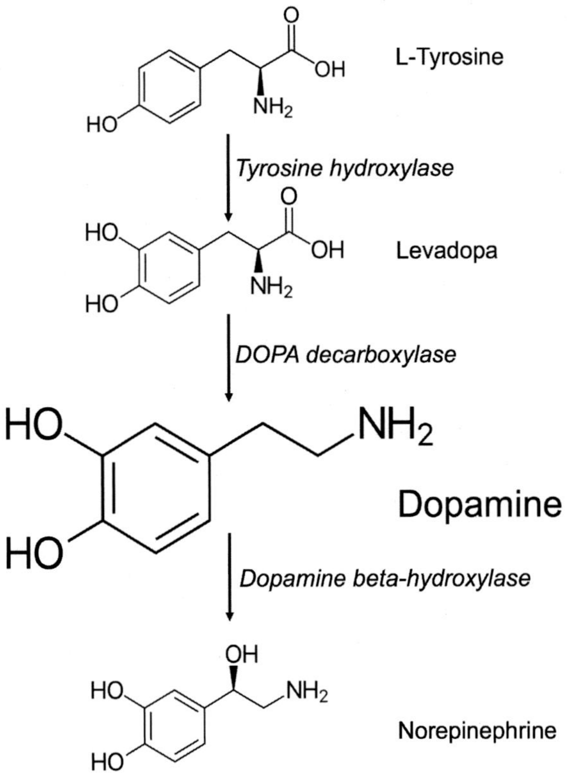

L-Tyrosine

Tyrosine hydroxylase

Levadopa

DOPA decarboxylase

Dopamine

Dopamine beta-hydroxylase

Norepinephrine

FIGURE 3.4

Dopamine metabolism showing the anabolic pathway from L-tyrosine through levodopa to dopamine.

levodopa, which also acts as the precursor for norepinephrine and epinephrine (also known as noradrenaline and adrenaline, respectively). Dopamine is degraded via a number of catabolic enzymes, notably by catechol-*O*-methyltransferase (COMT). Because of the importance of dopamine in PD, the metabolic pathways that generate and degrade this chemical (summarized in Fig. 3.4) have been of critical importance in developing drugs that act to alleviate the symptoms of Parkinson's disease (discussed further in Section 3.5).

3.3.4 DOPAMINERGIC NETWORKS

A key consequence of the death of dopaminergic neurons in the *substantia nigra pars compacta* and the subsequent decrease in the levels of dopamine is the disruption of dopaminergic signaling passing out of the basal ganglia to the rest of the brain [30]. Both excitatory (stimulating downstream action potentials) and inhibitory (suppressing downstream activity) dopaminergic signaling is perturbed in the parkinsonian brain, which has important consequences for therapeutic efforts targeting these networks (see Section 3.5). Axonal projections from the *substantia nigra* extend to the putamen where a series of connections through the globus pallidus and subthalamic nucleus eventually result in activation of

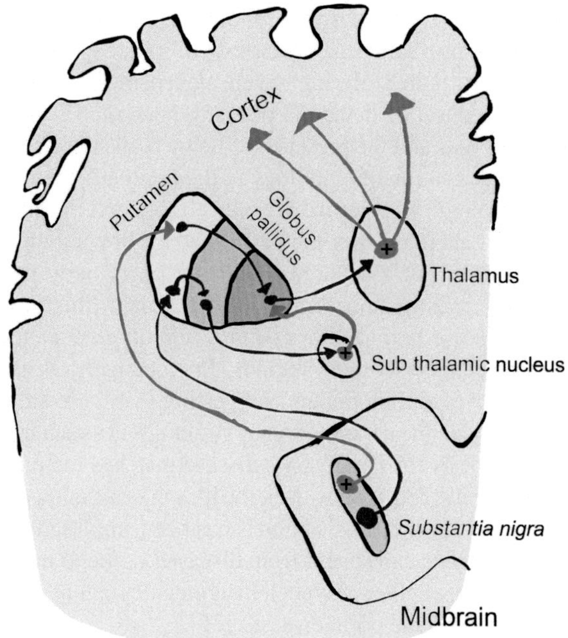

FIGURE 3.5

Dopaminergic enervation downstream from the basal ganglia displaying excitatory and inhibitory neurons extending out from the *substantia nigra*—the main pathways disrupted in Parkinson's disease.

neuronal signals from the thalamus and thence to the motor cortex, eventually connecting outside the central nervous system and acting to initiate and control movement (Fig. 3.5). The loss of both excitatory and inhibitory signals originating in the *substantia nigra pars compacta* eventually reduces the signals exiting the central nervous system and thus impacting the control of movement.

It is important to note that dopamine plays a key role as a neurotransmitter in a number of pathways outside the control of movement, for example, acting as a key component in the neuronal networks involved in reward [31]. Referring back to the differential vulnerability of dopaminergic neurons in the *substantia nigra pars compacta*, it is not entirely clear why dopamine neurons outside this brain region are less susceptible to damage in Parkinson's disease. What is clear is that the role of dopamine apart from movement control and initiation is highly relevant to therapeutic interventions in Parkinson's disease because drugs targeting the dopamine metabolism pathway, particularly drugs that target dopamine receptors, can act to dysregulate global dopamine signaling, leading to significant side effects.

3.3.5 BEYOND THE SUBSTANTIA NIGRA

Over the past 30 years, there has been an increasing interest in Parkinson's disease pathology outside the *substantia nigra*. Returning to the symptoms that are observed in people with Parkinson's disease, there are important deficits and alterations in systems beyond those involved in the control of movement.

A good example of this is loss of sense of smell (anosmia), a symptom that frequently occurs in Parkinson's disease, and often many years before the onset of motor symptoms. This reflects underlying pathology, with the olfactory bulb suffering severe degeneration at an early stage of disease. Likewise, when considering the onset of dementia in many Parkinson's disease cases late on in the disease, there is evidence of cell loss and the presence of Lewy bodies in cortical regions, suggesting a correlation between the physical location of pathology in the brain and the clinical presentation found in a patient. The pattern of pathology, with regard to both cell loss and the presence of Lewy bodies, is one that has garnered significant attention over the past decade following the groundbreaking work of Braak and coworkers [32] to carry out a detailed staging of Lewy body pathology in patients with Parkinson's disease. Similar to investigations in Alzheimer disease, this research suggested that the pathology in the Parkinson's disease brain follows certain stereotypical patterns, starting in the brain stem and then eventually spreading to cortical regions. Provocatively, Braak also reported that the pathology is first detected in nerves around the gut, suggesting that there are elements of the cause of Parkinson's disease that start outside the central nervous system. It is a matter of some debate as to how closely the Braak staging hypothesis for Parkinson's disease matches the pattern of degeneration and pathology in the majority of patients [33,34]; however, the idea that pathology can spread in Parkinson's disease has generated considerable interest. This interest has been amplified by evidence from fetal cell implants suggesting that Lewy bodies can spread from diseased tissue to newly implanted cells, and a number of groups are investigating whether α-synuclein aggregates can be transmitted from cell to cell in a manner analogous to that seen in the prion diseases [35].

3.3.6 BEYOND LEWY BODIES

Finally, it is worth pausing to consider the pathological alterations beyond Lewy bodies. As noted earlier, the broader spectrum of parkinsonism encompasses diverse cellular pathological conditions. This is particularly true of some of the less common disease entities that have a clinical overlap with Parkinson's disease, for example, progressive supranuclear palsy and multiple system atrophy. In progressive supranuclear palsy, intraneuronal neurofibrillary tangles made up of the microtubule-associated protein tau are the predominant pathology, whereas in multiple system atrophy the key protein pathology is found in oligodendrocytes and consists of α-synuclein aggregates termed glial cytoplasmic inclusions [36,37]. To add to this heterogeneity, there are examples of what would otherwise be considered typical Parkinson's disease caused by specific mutations that present with variable pathology, including glial cytoplasmic inclusions and neurofibrillary tangles [38]. What this indicates is that we do not yet fully understand the link between clinical presentation and brain pathology in PD and parkinsonism.

3.4 MOLECULAR MECHANISMS OF DEGENERATION

As noted in Section 3.3, a substantial amount is known about the pathological and neurochemical changes in the brain that underpin the clinical symptoms of Parkinson's disease. Far less progress, at least until relatively recently, has been made in understanding where these pathological and neurochemical changes originate and the molecular events that presage nigral degeneration. Advances in our understanding of the molecular mechanisms of Parkinson's disease have come predominantly from two distinct sources: rare cases of Parkinson's disease (or at least Parkinson's-like syndromes) resulting from specific and identifiable environmental causes and the increasingly appreciated contribution of human genetics to the risk of developing Parkinson's disease.

3.4.1 GENETICS

In many ways, genetics and Parkinson's disease are unexpectedly associated. For decades, Parkinson's disease was considered the prototypical nongenetic neurodegenerative disease, in contrast to, e.g., Huntington's chorea, which is a purely genetic disorder. This changed with the increasing application of molecular genetics to neurological diseases, in particular the identification and careful clinical characterization of families where Parkinson's disease was present in multiple generations [39]. The first such case where a causative mutation was identified was in the Contursi kindred, a family originally from the Contursi region of Italy (close to Salerno), but with members living in the United States and Greece [40]. This family presented with a relatively young-onset form of neurodegeneration (with a mean age of onset of 46 years) and a markedly more aggressive disease course than would normally be expected in Parkinson's disease. Similar to a number of inherited forms of Parkinson's disease, the clinical presentation also included a substantial dementia component, a factor of considerable importance with respect to the implications of the disease genetics of this family. Using a positional cloning approach, geneticists working with the family identified a coding mutation in the *SNCA* gene, coding for α-synuclein. The mutation, exchanging a threonine for an alanine at codon 53 (A53T), segregated with disease and is now clearly established as a rare, but important, cause of Parkinson's disease.

Since then, a large number of mutations have been identified in human genes that cause Parkinson's disease or parkinsonism (summarized in Table 3.2) [41]. These almost universally represent rare causes of disease, with Mendelian causes of Parkinson's disease accounting for less than 5% of cases in most populations. There are exceptions to this; for example, across North Africa, inherited mutations that cause Parkinson's disease can be found in up to 40% of patients in Algeria, Tunisia, and Morocco [42]. The precise reasons for this variation from population to population are unclear. Besides mendelian diseases the explosion in the volume of human genetic data driven by the initial sequencing of the human genome and subsequent genomic research has allowed the identification of common genetic variants that act to predispose individuals to disease in the general population. Genome-wide association studies for Parkinson's disease have produced long lists of potential genetic risk factors, each with a very low (in most cases, fractions of a percentage point) increased lifetime risk of disease associated with them [43]. The task of interpreting and validating the results of genome-wide association studies for Parkinson's disease has only just begun, but a genetic architecture for the disease has already begun to emerge (Fig. 3.6) [44]. These genome-wide significant risk signatures are unlikely to be of any significant utility for diagnosis, as the increase in risk is simply too low to be meaningful in a clinical setting. However, the insight into the pathways leading to degeneration that these studies have yielded, has been of great importance, particularly in terms of providing the rationale for drug development strategies.

Several key themes emerge from these genetic analyses and subsequent functional investigations using cell and animal models to dissect the mechanisms linking genes to the molecular disease process. They are protein aggregation and proteostasis, mitochondrial dysfunction and damage repair mechanisms, and finally, inflammation.

Table 3.2 Mendelian Genes for Parkinson Disease

Locus	Chromosome	Gene	Inheritance
PARK1	4q	*SNCA* (point mutation)	AD
PARK2	6q	*PRKN*	AR
PARK3*	2p	?	AD
PARK4	4q	*SNCA* (multiplication)	AD
PARK5*	4p	*UCHL1*	?
PARK6	1p	*PINK1*	AR
PARK7	1p	*DJ-1*	AR
PARK8	12p-q	*LRRK2*	AD
PARK9	1p	*ATP13A2*	AR
PARK10*	1p	?	?
PARK11*	2q	*GIGYF2*	?
PARK12*	Xq	?	X-linked
PARK13*	2p	*HTRA2*	?
PARK14	22q	*PLA2G6*	?
PARK15	22q	*FBXO7*	AR
PARK16	1q	?	?

(*), uncertain pathogenicity; (?), causative gene has not yet been identified.

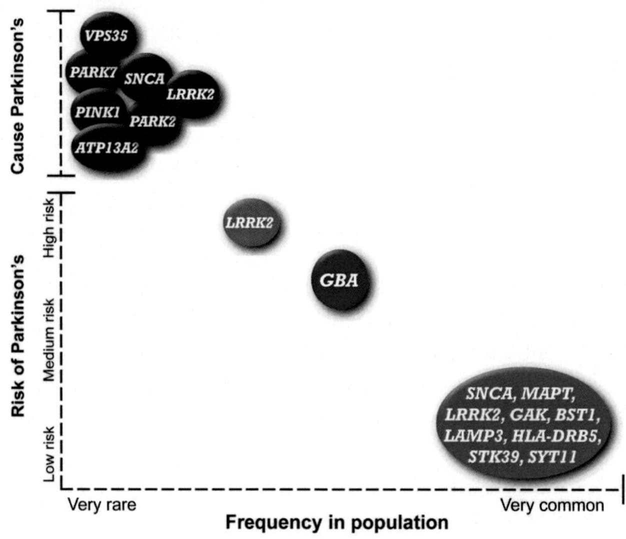

FIGURE 3.6

Genetic architecture of Parkinson's disease showing high-penetrance/low-frequency mutations through to low-effect/common variants.

3.4.1.1 Protein Aggregation/Proteostasis

Following the identification of the A53T mutation in α-synuclein in 1997, it soon became apparent that α-synuclein was the key component of Lewy bodies not just in families with mutations but in the brains of all individuals with Lewy-body-positive pathology [45]. This emphasises the importance of α-synuclein in the disease process leading to Parkinson's disease, and indeed, it is central to how we define the disease. α-Synuclein is a 140-amino-acid protein, and a total of five point mutations, A30P, E46K, G51D, A53T, and A53E, have been identified in it that are robustly associated with Parkinson's disease (Fig. 3.7) [46]. Biochemical analyses of the impact of these mutations have revealed important information about how these coding changes lead to alterations in the properties of α-synuclein and, by implication, to the disease [20]. A consistent impact of mutations is an alteration in the aggregation properties of α-synuclein, increasing the propensity of this protein (which is natively unfolded in solution) to adopt a beta pleated sheet conformation and form multimeric aggregates. The precise nature and impact of these aggregates is a matter of some debate, but it is now firmly established that small multimers (termed oligomers) formed from α-synuclein are cytotoxic and can lead to neuronal cell death [47]. Higher order multimers and much larger fibrils composed of α-synuclein appear to be less toxic to cells and have been proposed to form the molecular species that is eventually incorporated into Lewy bodies. Similar to other protein aggregates associated with neurodegeneration, such as those consisting of amyloid beta and tau, the events that link small oligomers to cellular dysfunction and thence to cell death are not yet clear. The mechanisms that have been proposed include the formation of molecular pores by aggregates that punch holes in membranes, the disruption of organelle function, and the exposure of hydrophobic patches in protein structure that bind to cellular components and lead to their dysfunction [48]. What is indisputable is that the presence of oligomers has serious, and often fatal, consequences for cells in the central nervous system. As noted in Section 3.3, this has led to the suggestion that the accumulation of α-synuclein aggregates in Lewy bodies may be a protective move by neurons in the brain, gathering all the potential cytotoxic species into one place and thereby limiting the damage that they can cause.

An important insight into the mechanisms driving Parkinson's disease linked to α-synuclein was provided in 2003 by the identification of a triplication of the *SNCA* gene in a family from Iowa [49]. The triplication of the gene, increasing the total number of copies of *SNCA* from two to four, was

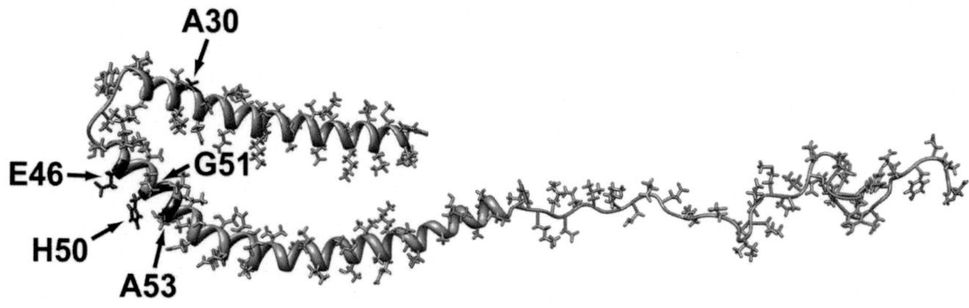

FIGURE 3.7

Coding mutations in α-synuclein are highlighted in the predominantly alpha helical structure that α-synuclein adopts upon binding to lipid membranes.

Structural image derived from PDB 1XQB, image taken from Kara E, Lewis PA, Ling H, Proukakis C, Houlden H, Hardy J. alpha-Synuclein mutations cluster around a putative protein loop. Neurosci Lett 2013;546:67–70. Epub 2013/05/15.

associated with an aggressive, early-onset form of Parkinson's disease with a significant dementia component, again providing an overlap with dementia with Lewy bodies. Intriguingly, the pathology also includes α-synuclein-positive glial cytoplasmic inclusions similar to those found in multiple system atrophy, highlighting the links between this disorder and Parkinson's disease at a cellular as well as clinical level. Subsequent to this, a number of families have been identified where Parkinson's disease is associated with additional copies of the *SNCA* gene, and importantly the level of gene dosage correlates with the severity and onset of the disease [50]. Triplications of *SNCA* are associated with the most severe phenotype, with duplications (resulting in a total of three copies of the gene, increasing α-synuclein protein levels by 50%) result in a less aggressive and later-onset form of Parkinson's disease. This establishes the total level of α-synuclein in the brain as being an important determinant of risk of developing Parkinson's disease, an observation that is further supported by the identification of variation at the *SNCA* locus altering gene expression as being linked to Parkinson's disease in the genome-wide association studies for this disease [51]. There is yet another important parallel here with Alzheimer disease, where additional copies of the *APP* gene found in trisomy 21 and in rare autosomal dominant forms of Alzheimer disease lead to neurodegeneration and dementia. Returning to protein aggregation, it has long been recognized that one of the most important factors determining the kinetics of protein aggregation is the starting concentration of a given protein. Linking together human genetics and protein science, we now have very good evidence that there is a dose-dependent link between α-synuclein, increasing the cellular concentration of the protein and therefore the propensity to aggregate, and the risk for Parkinson's disease. These data have had, and will have in the future, important implications for the efforts to develop new drugs targeting α-synuclein in Parkinson's disease (see Section 3.5).

Again similar to Alzheimer's disease, an important area of research over the past 10 years has been the study of how protein aggregates might be transmitted from one brain cell to the other in a manner analogous to that seen in the prion disorders (the subject of Chapter 4). As noted earlier, there is intriguing evidence from pathological studies in humans that Lewy bodies can spread from cell to cell within the brain. Experimental investigations using cell and animal models have begun to unpick the mechanistic basis for this, with the demonstration that synuclein pathology can be seeded in mouse models by injecting preformed α-synuclein fibrils or by exposure to extracts from the brains of individuals who have died with either Lewy body or glial cytoplasmic inclusions [52,53]. This remains a controversial area of research, and it is unclear how great a contribution protein aggregate templating, the process whereby small aggregates can induce the formation of further aggregation, makes to the disease process in [34].

Importantly, research has started to link other genes associated with Mendelian forms of PD and parkinsonism to the aggregation of proteins or the mechanisms whereby these protein aggregates are disposed of. In 2004, mutations in leucine-rich repeat kinase 2 (LRRK2), a large multidomain kinase, were first identified in families with PD [54,55]. It is now recognized that LRRK2 mutations represent one of the most common genetic causes of PD, with exceptionally high rates in some populations, most notably in individuals of North African Berber or Ashkenazi Jew extraction [56]. The clinical phenotype of *LRRK2*-linked PD is almost indistinguishable from the idiopathic disease; however, there is marked variability in the neuropathology observed in these cases. While the majority of patients with *LRRK2* mutations examined post mortem exhibit Lewy body pathology, a significant proportion do not have Lewy bodies but instead present with neurofibrillary tangles similar to those seen in progressive supranuclear palsy, or with TDP-43 deposition similar to that observed in frontotemporal dementia or amyotrophic lateral sclerosis (see Chapters 2 and 5, respectively). At a molecular level, there is increasing evidence that LRRK2 acts to regulate endosomal trafficking of proteins in the cell and that this has consequences for lysosomal function and the function of the macroautophagy pathway [57]. Both are critical for cellular proteostasis, with macroautophagy gathering up misfolded proteins for degradation and the lysosomes performing a disposal and recycling process. Thus disruption of their functions due to mutations in LRRK2 could disrupt the cellular efforts to dispose of aggregated proteins.

This is also thought to be the case for coding variants in *GBA*. Loss of function of glucocerebrosidase, the protein coded for by the *GBA* gene, was identified over 30 years ago as the cause of an autosomal recessive lysosomal storage disorder called Gaucher's disease [58]. Gaucher's disease is characterized by pathological conditions across a number of organs, with a range of symptoms that can include parkinsonism. In the 1990s, clinicians working with patients with Gaucher's disease observed that there was a higher than expected occurrence of Parkinson's disease in first-degree relatives of patients with Gaucher's disease [59]. Detailed genetic characterization of families with Gaucher's disease revealed that possessing one copy of a mutation that, in the homozygous state, causes a lysosomal storage disorder greatly increases the lifetime risk of developing Parkinson's disease, up to about a 30%, or a one in three, chance of disease. This observation reinforces the importance of lysosomal biology in the cause of Parkinson's disease: glucocerebrosidase is a lysosomal enzyme involved in ceramide metabolism and the loss of *GBA* function explicitly reduces lysosomal function. In the case of heterozygote carriers of *GBA* mutations, there is nigral degeneration and the accumulation of Lewy bodies. Further studies in neurons derived from induced pluripotent stem (iPS) cells from patients with Gaucher's disease have shown that mutations in *GBA* result in the accumulation of α-synuclein, with increased α-synuclein expression (similar to that observed in duplication and triplication mutation carriers) leading to reduced GBA activity [60]. Importantly, analysis of brain tissue from patients with Parkinson's disease with and without mutations in *GBA* confirms that reduced glucocerebrosidase activity and α-synuclein accumulation are common features of neurodegeneration in these cases [61]. These data support a model where α-synuclein and glucocerebrosidase form a bidirectional loop driving disease, where decreased glucocerebrosidase activity results in increased levels of α-synuclein, which then acts to further decrease glucocerebrosidase enzymatic function. The description of mutations in *ATP13a2* (a lysosomal ATPase) causing a complex lysosomal disorder called Kufor Rakeb syndrome, where again parkinsonism and nigral dysfunction play a prominent role, has added to the evidence linking lysosomes to the pathological process in Parkinson's disease [62].

One final example of a Mendelian form of Parkinson's disease linked to the endosomal/lysosomal system and protein aggregate disposal includes mutations in *VPS35* [63]. This gene codes for one of the components of the retromer complex, a multiprotein machine that acts to coordinate protein trafficking in the endosomes. The retromer complex has been relatively well characterized, allowing the single point mutation in *VPS35* reported to date (an autosomal dominant D620N mutation found in a number of independent families) to be studied in some detail. This mutation does not act to completely disrupt the retromer, but instead alters specific trafficking events linked to the WASH complex [64]. This alteration has the consequence of causing changes in the control of macroautophagy, again pointing to catabolic pathways within the cell. Similar to many familial forms of PD/parkinsonism, there is limited information available relating to the brain pathology observed in *VPS35* cases, and it is unclear whether mutations in this gene cause Lewy-body-positive disease. The clinical presentation in these cases, however, is very close to what would be considered typical for Parkinson's disease [65].

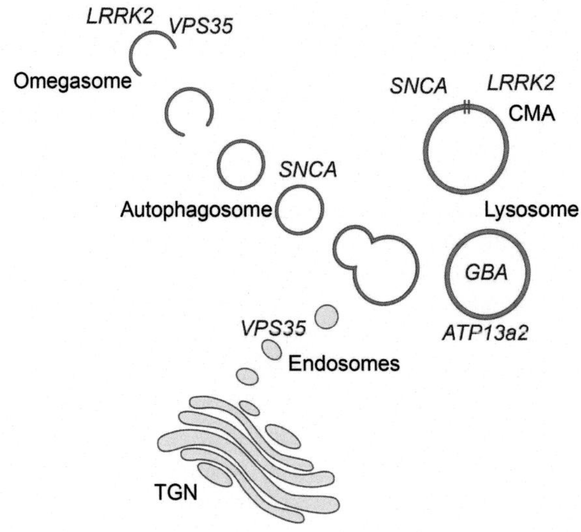

FIGURE 3.8

Molecular pathways linked to protein aggregation and proteostasis, showing genes implicated in Parkinson's disease by genetic analysis of familial diseases.

Together, these genetic studies have started to pinpoint the aggregation of protein (specifically α-synuclein) and the pathways that remove these aggregates as being central to the disease process in Parkinson's disease (summarized in Fig. 3.8). An important caveat to these studies is that, as noted earlier, the mendelian forms of Parkinson's disease account for a small minority of cases. This is where genome-wide association studies have made a major impact on the relevance of rare forms of inherited Parkinson's disease/parkinsonism to the more common idiopathic form of the disease. A number of genes that had previously been identified in autosomal dominant cases of Parkinson's disease have also been identified as being linked to genome-wide significant risk for Parkinson's disease. These include the genes coding for α-synuclein, LRRK2, and glucocerebrosidase, strongly suggesting that understanding the dysfunction of these proteins relating to protein misfolding and catabolism in familial disease will aid in our understanding of the idiopathic disease. Most importantly, this strengthens the case for targeting these proteins as part of the efforts to develop novel therapies for Parkinson's disease.

3.4.1.2 Mitochondrial Dysfunction

A second key theme to come out of genetic studies of Parkinson's disease is that of mitochondrial health and biology, in particular the cellular processes that respond to damaged mitochondria. This area of research emerged out of studies on the second gene to be linked to mendelian forms of parkinsonism caused by mutations in the ubiquitin E3 ligase parkin, the product of *PRKN* gene [66]. Mutations in *PRKN* were originally identified in a series of families from Japan presenting with an early-onset (<40 years old) atypical form of autosomal recessive parkinsonism, and it is now recognized that mutations in *PRKN* are the most common genetic cause of early-onset parkinsonism. Ubiquitin E3 ligases act to catalyze the addition of ubiquitin to proteins, tagging them for processing either through the

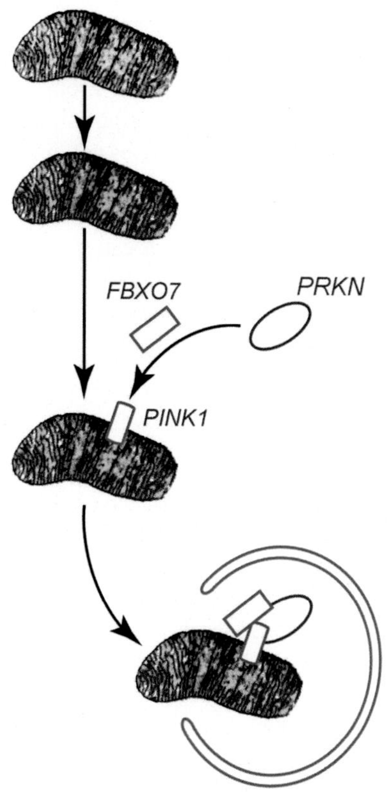

FIGURE 3.9

Mitophagy showing the convergence of familial genes in regulating this process as mitochondria dysfunction and are tagged for degradation by the autophagosome system.

proteasome system or via autophagy. The subsequent identification of mutations in a gene called *PINK1*, coding for a PTEN induced kinase, with a very similar clinical phenotype (young-onset, long-duration, levodopa-responsive parkinsonism) led to a series of seminal studies showcasing the power of combining human genetics and model organisms to understand biological processes [67]. Based on the clinical observation that *PINK1* and *PRKN* mutation carriers shared a similar phenotype, a number of research groups began investigating whether these two genes shared a molecular phenotype by taking advantage of the presence of gene orthologs for both *PINK1* and *PRKN* in the fruit fly, *Drosophila melanogaster*. Knocking out these genes both individually and in unison, it became clear that PINK1 and parkin are critical for mitochondrial quality control in *Drosophila* and that they act in a common pathway to regulate a specialized form of autophagy called mitophagy (Fig. 3.9) [68,69]. Genetic and functional investigations have provided a great deal of insight into how these proteins govern the process of mitophagy, with PINK1 revealed as being able to phosphorylate ubiquitin and an additional member of the pathway *FBXO7* being identified through analysis of a further kindred with juvenile-onset parkinsonism [70–72]. In addition to highlighting mitochondrial health as being critical for the survival of neurons in the *substantia nigra* (echoing evidence from environmental causes of parkinsonism, see Section 3.4.2), these studies stand in testament to the contribution of human disease genetics to our understanding of fundamental biological processes. From the perspective of PD genetics, the roles of PINK1 and parkin, and now FBXO7, in regulating mitophagy have raised some interesting questions. First, the clinical presentation of these cases does not fit with the classical clinical phenotype associated with Parkinson's disease, being younger onset, with a longer disease duration, and with a predominantly movement-associated presentation. Second, the brain pathology in these cases does not clearly match the Lewy body pathology required for a categorical diagnosis of Parkinson's disease [73]. This raises an important, and difficult to address, question: does it therefore follow that mutations in these genes cause a disease that is distinct from Parkinson's disease? This is an issue that impacts on a number of forms of inherited parkinsonism, including that caused by mutations in *ATP13a2* and even *LRRK2*-linked Parkinson's disease.

3.4.1.3 Inflammation

Similar to Alzheimer's disease, the role of inflammation in exacerbating neurodegeneration is an expanding area of research in the field of Parkinson's disease [74]. This comes partly from an increasing appreciation of the pattern of immune response in the pathology of Parkinson's disease, in particular the part played by microglial response, and partly from the identification of risk variants linked to Parkinson's disease in genes with well-characterized immune function. The most notable among these is the *HLA* locus on chromosome 6, which encodes the major histocompatibility complex. A noncoding variant in the *HLA-DRA* region was found to be significantly associated with Parkinson's disease and was predicted to cause alterations in expression of the *HLA-DR and HLA-DQ* variants [75]. The major histocompatibility complex is critical for the immune system to recognize foreign antigens and it governs, as its name suggests, histocompatibility. Intriguingly, microglial cells expressing HLA-DQ had previously been recognized in the pathology of Parkinson's disease [76]. Precisely how this alteration in immune cell function results in causing Parkinson's disease is unclear, although a study suggested that there may be autoimmunity to α-synuclein in a subset of people living with PD [77]. Perhaps another indicator of a link between the immune system and the process leading to neurodegeneration in Parkinson's disease was offered by a study linking changes in the gut microbiome to the severity of pathology in a mouse model of α-synuclein brain deposition [78]. Given the established links between Lewy body pathology in the gut and Parkinson's disease, and a potential role for the gut microbiome in regulating the microglial response in the Parkinson's-affected brain, these are intriguing suggestions that the immune system might be playing a very important role in the events that lead to Parkinson's disease. This has implications for a number of avenues of therapeutic investigation, both in terms of using the immune system to help clear α-synuclein and more broadly in modulating immune function to reduce potential damage to neuronal cells.

3.4.2 ENVIRONMENTAL ROUTES TO PARKINSONISM

As for almost all human traits, the development of Parkinson's disease is thought to be a result of combination of predisposing genetic variation and environmental causes of neurodegeneration. There are significant challenges in identifying the environmental causes of neurodegeneration, partly due to the complex nature of the disease process in Parkinson's disease as well as due to the intrinsic difficulties in moving from a correlation between exposure to a specific insult in the environment to being able to demonstrate that this association is causative. Despite these challenges, there are a number of examples where research has, to a greater or lesser extent, conclusively identified environmental causes of nigral degeneration.

3.4.2.1 1-Methyl-4-Phenyl-1,2,3,6-Tetrahydropyridine

The most famous, and best understood, example of an environmental cause of parkinsonism is provided by the case of 1-methyl-4-phenyl-1,2,3,6-tetrahydropyridine (MPTP). MPTP came to light as an environmental cause of parkinsonism through a combination of detailed clinical detective work and careful experimental investigation. The first evidence linking this chemical to parkinsonism came from a case report describing a student at the University of Maryland in the United States who developed parkinsonism over a very short time in 1976 [79]. The student, Barry Kidston, was addicted to opiates and had attempted to synthesize an opium-like compound called desmethylprodine or MPPP (1-methyl-4-phenyl-4-propionoxypiperidine). During the synthesis, however, he had contaminated the drug with MPTP, which has a related structure and is produced by a similar synthetic pathway (Fig. 3.10A). MPTP, however, has biological properties that are different from those of desmethylprodine. Instead of inducing analgesia and euphoria, it is the precursor to a potent neurotoxin, MPP^+. MPTP is converted to MPP^+ in glial cells within the brain, and MPP^+ is then taken up by the dopamine transporter and concentrated in the dopaminergic neurons of the *substantia nigra pars compacta* (Fig. 3.10B). In these neurons, MPP^+ binds to and inhibits complex I of the mitochondrial respiratory chain, preventing respiration and killing the cells. This sequence of events was unraveled after a second episode of parkinsonism linked to synthetic opioid exposure in the San Francisco bay area in 1982, when six drug addicts developed parkinsonism following MPTP exposure [80]. The causal link between MPTP and nigral degeneration was tested by a neurologist, William Langston, who had worked with the patients in California. Following exposure to MPTP, squirrel monkeys developed rapid nigral degeneration and parkinsonism almost identical to that observed in the Californian drug addicts and Barry Kidston [81].

The identification of MPTP as a chemical that could induce specific nigral degeneration in a range of animal models had a major impact on efforts to mimic the human disease process. Both rodents and nonhuman primates develop severe movement disorders when exposed to MPTP, resulting in an extensive series of investigations by a number of groups to use neuroprotective compounds to ameliorate the damage caused by MPTP [82].

Importantly, and echoing efforts taken during the 1950s to replicate the loss of dopamine associated with movement disorders in humans, the MPTP story resulted in clinicians and pathologists looking in much greater detail at the biology of mitochondria in the Parkinson's disease-affected brain. This revealed a chronic mitochondrial deficit in tissues from idiopathic Parkinson's disease-affected brains [83]. When taken in the context of further studies identifying mitochondrial toxins as being involved with the risk for Parkinson's disease (see later discussion), and the genetic evidence linking mitochondrial quality control to the cellular processes underpinning Parkinson's disease, there is an overwhelming case for the mitochondrion playing an extremely important role in causing this disease.

3.4.2.2 Rotenone and Paraquat

The MPTP story sets a high bar for establishing a link between environmental exposure and parkinsonism. The evidence linking rotenone and paraquat to parkinsonism is more circumstantial and is examined in the context of our understanding of the importance of mitochondrial viability to a healthy *substantia nigra*. The epidemiological link here is much less obvious than that for MPTP, and is also

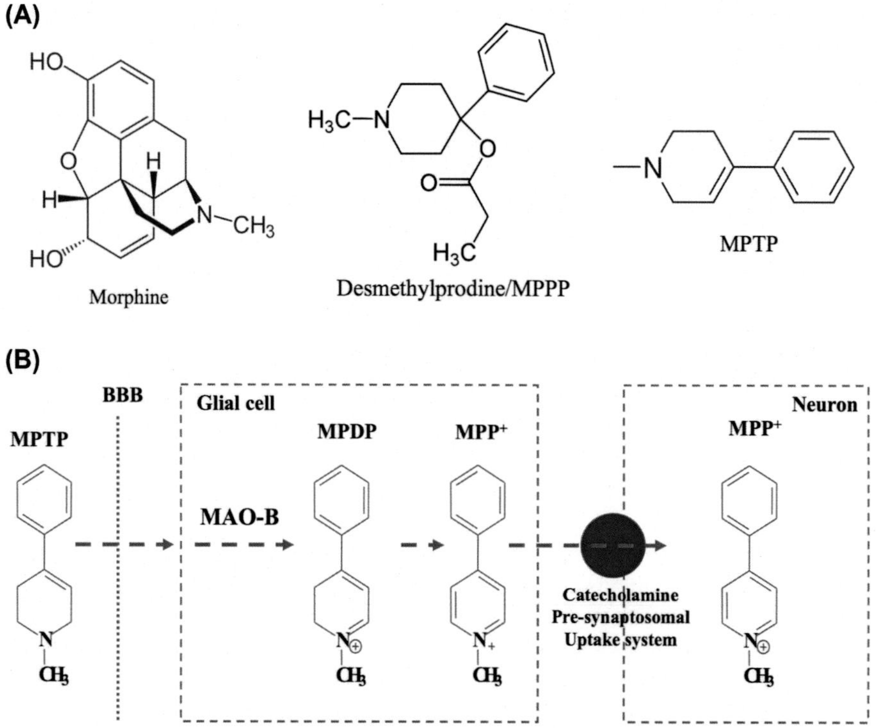

FIGURE 3.10

The structures of (A) morphine, desmethylprodine, and 1-methyl-4-phenyl-1,2,3,6-tetrahydropyridine (MPTP). (B) The uptake and conversion of MPTP to MPP^+. *MPPP*, 1-methyl-4-phenyl-4-propionoxypiperidine.

potentially a link with much wider ramifications. For a number of decades, epidemiological studies have highlighted that there is an increased lifetime risk of developing Parkinson's disease in rural, in particular farming, communities. The reasons underlying this association remain obscure. One possible explanation is provided by a series of experiments investigating the impact of pesticides and herbicides commonly used in farming on the central nervous system [84,85]. Evidence from a number of animal models suggests that exposure to rotenone, a naturally occurring flavonoid, and paraquat, a heterocyclic herbicide (both shown in Fig. 3.11), can lead to the degeneration of dopaminergic neurons in the *substantia nigra pars compacta* [86,87]. This degenerative phenotype, coupled to epidemiological evidence providing a plausible link for these toxins in the environment and increased risk of Parkinson's disease, is strongly suggestive of a causative link. The evidence, however, falls short of the conclusive link that has been demonstrated between MPTP and nigral degeneration.

3.4.2.3 Postencephalitic Parkinsonism

The third example of a potential environmental exposure correlated with parkinsonism is one that raises more questions than answers. It dates back to the first half of the 20th century and the aftermath of the

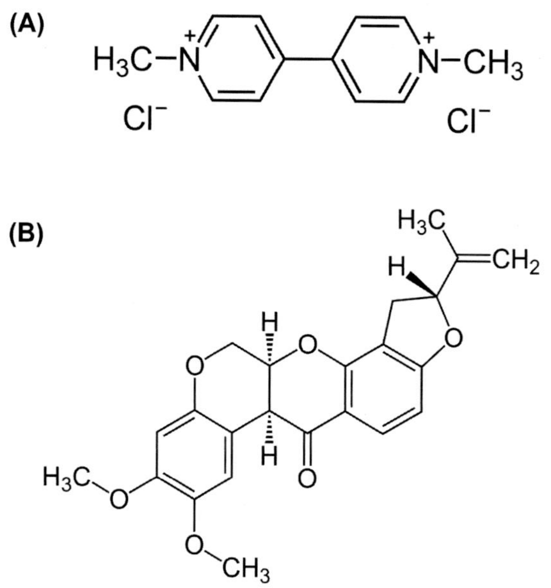

(A)

(B)

FIGURE 3.11

The chemical structures of (A) paraquat and (B) rotenone.

1918 influenza pandemic. This pandemic, caused by the H1N1 influenza virus, was one of the deadliest in human history, infecting over 500 million people worldwide and killing an estimated 50–100 million. Concurrent with this pandemic, a disorder called encephalitis lethargica, or von Economo encephalitis, was reported in a large number of people, possibly up to five million, and with a high mortality rate (approaching 30%) [88]. The symptoms of encephalitis lethargica included lack of energy, high fever, and a severe impact on responsiveness sometimes descending into akinetic mutism and parkinsonism. A proportion of those who survived the disease went on to develop postencephalitic parkinsonism as an ongoing complication [89]. These individuals did not fully recover from the encephalitis and experienced parkinsonian symptoms that, in some cases, resulted in patients remaining in a state of akinetic mutism. Fascinatingly, patients with postencephalitic parkinsonism responded (at least initially) exceptionally well to treatment with levodopa and were one of the earliest groups treated with this drug, which is a story immortalized by Oliver Sacks in his book *Awakenings* [90]. Unfortunately, these patients were also among the first to demonstrate the side effects of sustained and high-dose dopamine therapy (discussed in detail in Section 3.5) [91]. Importantly, the pathological condition observed in cases of postencephalitic parkinsonism is characterised predominantly by neurofibrillary tangles, demonstrating a distinction between this disorder and idiopathic Parkinson's disease. In terms of the precise cause of postencephalitic parkinsonism, the current data is inconclusive. Although there are tentative links with viral infection, it is far from clear that von Economo encephalitis was a direct result of exposure to the 1918 influenza virus, and a direct test of this hypothesis is difficult to carry out due to the challenges of working with this virus.

3.4.2.4 Chronic Traumatic Encephalopathy

The final example of an environmental exposure is one that has overlaps with dementia and motor neuron disease: chronic traumatic encephalopathy. This has been described in some detail in Chapter 2, but it is worth reiterating here, as parkinsonism is very much part of the spectrum of neurodegenerative clinical phenotypes observed in these cases [92]. To recap, head trauma, in particular repeated head trauma and concussion, results in the formation of tau pathology in the human brain and leads to the development of a range of neurodegeneration-related symptoms ranging from cognitive decline to alterations in the control of movement. Whether head trauma has a role in the development of Lewy-body-positive Parkinson's disease remains to be seen, and is the subject of much research.

3.5 THERAPIES

The treatment of Parkinson's disease is a contrasting tale of highly efficacious symptomatic treatment (at least in the early stages of the disease), in the absence of disease-modifying treatments that alter the trajectory of degeneration.

3.5.1 EXISTING TREATMENTS

For several decades, a range of very effective treatments have been used for the motor symptoms associated with the disorder. The realization in the 1950s that the key neurochemical deficit underlying the motor symptoms of Parkinson's disease was loss of the neurotransmitter dopamine swiftly led to trials of dopamine for Parkinson's disease. The trials failed due to the inability of dopamine to cross the blood–brain barrier; however, it was soon realized that the immediate precursor to dopamine, levodopa, has the ability to cross this barrier. Clinical trials ensued, with remarkable results. Treatment with levodopa resulted in an almost complete reversion of the symptoms, restoring normal function in the majority of patients.

3.5.1.1 Dopamine Pathway Drugs

Unfortunately, there are no disease-modifying treatments for Parkinson's disease. However, the symptomatic treatments available are highly efficacious, and in general, the practice is to commence medication when patients are functionally limited by their symptoms. The pharmacological approach to improving symptoms in Parkinson's disease rests on attempting to address the primary pathological finding, that is, the loss of dopaminergic neurons, and thence of dopamine [93]. The different drugs available affect the dopamine pathway in different ways.

3.5.1.1.1 Levodopa

Levodopa is the most effective symptomatic oral drug. It is converted to dopamine by dopa decarboxylase (see Fig. 3.4) and acts on dopamine receptors. In order to prevent breakdown of levodopa to dopamine in the periphery, it must be given with a dopa decarboxylase inhibitor,-i.e., either carbidopa or benserazide. Peripheral dopa decarboxylase inhibitors do not cross the blood–brain barrier, allowing the levodopa to be metabolized to dopamine in the brain.

Levodopa is generally well tolerated. Initially it can cause nausea and other gastrointestinal symptoms but these symptoms usually settle with time and if necessary, antiemetics can be used. Postural hypotension can also occur and sometimes this can limit the dose that can be used.

The long-term use of levodopa has been associated with dyskinesias and motor fluctuations [94]. A "wearing off" phenomenon occurs when the duration of the effect of levodopa decreases and the patient spends more time in the "off stage" with severe bradykinesia and rigidity. Dyskinesias are hyperkinetic movements that typically occur at the peak dose times. They can be severe causing pain, poor balance, and falls. These complications can be addressed by shortening the duration between levodopa doses or by adding a drug to inhibit levodopa or dopamine breakdown (see later discussion). There is a form of levodopa–carbidopa available as a continuous infusion into the proximal jejunum, and it can improve fluctuations.

3.5.1.1.2 Dopamine Agonists

A variety of drugs act as dopamine agonists that stimulate postsynaptic dopamine receptors directly, including pramipexole, ropinirole, cabergoline, and pergolide. They can be used in either monotherapy or adjunctive therapy with levodopa in treating Parkinson's disease. Dopamine agonists are divided into those that have an ergot structure and those that do not. In practice, those with an ergot structure (cabergoline and pergolide) are rarely used nowadays, as they have occasionally been associated with fibrotic reactions affecting the lung, heart, and retroperitoneal space.

Dopamine agonists seem to be less likely associated with the long-term motor fluctuations seen with levodopa and are often used as the first-line therapy in younger patients. As the half-life of dopamine agonists is longer than that of levodopa, they can be more convenient to take and there is also a subcutaneous patch form available (rotigotine) that is useful when patients cannot swallow medications (for example, after surgery). However, it is generally accepted that their antiparkinsonian effect is weaker than that of levodopa.

Dopamine agonists have been associated with neuropsychiatric side effects including impulse control disorders such pathological gambling, hypersexuality, and compulsive shopping/eating [95]. Punding refers to purposeless repetitive stereotyped behaviors and dopamine dysregulation syndrome; compulsive medication overuse has also been associated with the use of dopamine agonists. These behaviors are rarely seen in patients who are on levodopa monotherapy but the incidence is higher in patients taking dopamine agonists. The risk factors for impulse control disorders include male sex, younger age at Parkinson's disease onset, and a personal or family history of substance abuse, impulse control disorders, or bipolar disorder. As these disorders can have a severe impact on the quality of life of the patients and their families, they must be discussed before treatment is initiated. Other side effects of dopamine agonists include gastrointestinal symptoms such as vomiting that may require antiemetic treatment, postural hypotension, leg swelling, and hallucinations.

Apomorphine is the oldest dopamine agonist and must be given parenterally by subcutaneous injection. It can be used as a single injection administered as needed for sudden "off periods" or as a continuous subcutaneous infusion.

3.5.1.1.3 Monoamine Oxidase Inhibitors

Drugs such as selegiline and rasagiline inhibit the monoamine oxidase B enzyme selectively and irreversibly, thus preventing breakdown of dopamine in the synapse [96]. They have a mild effect when given alone and are usually given in combination with levodopa. They are generally well tolerated but can potentiate the side effects of levodopa.

3.5.1.1.4 Catechol-*O*-methyltransferase inhibitors

Entacapone and tolcapone are COMT inhibitors and prevent levodopa metabolism to 3-*O*-methyldopa, thus extending the half-life of levodopa [97]. Entacapone acts only peripherally, whereas tolcapone has

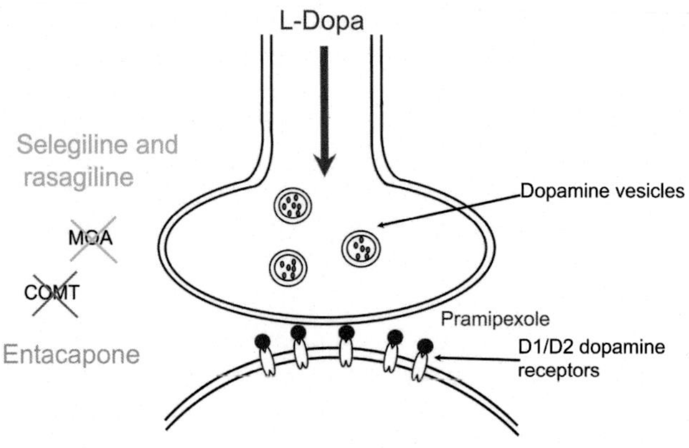

FIGURE 3.12

Pharmacological treatments targeting dopamine metabolism and synaptic function.

a central effect too. The former is used more widely, as the latter can be associated with liver toxicity. COMT inhibitors are given in combination with levodopa, and there is a combined tablet available that contains levodopa, carbidopa, and entacapone (Fig. 3.12).

3.5.1.1.5 Amantadine

Amantadine is a drug with various molecular consequences in the central nervous system. It acts to antagonize NMDA (*N*-methyl-D-aspartate) glutamate receptors, can increase dopamine release, and reduces dopamine reuptake at the synapse; it possesses a combination of actions that has resulted in it being used in the treatment of dyskinesias [98]. It has also been used to specifically treat postural freezing symptoms in Parkinson's disease.

3.5.1.2 Anticholinergics

Anticholinergics were the first type of treatment historically used in Parkinson's disease. They mainly have an effect in reducing tremor but can be associated with worsening of confusion. Owing to their peripheral effects, anticholinergics can also be associated with constipation and urinary retention but can reduce sialorrhea.

3.5.1.3 Deep Brain Stimulation

Occasionally, motor fluctuations, severe tremor, and dyskinesias are not sufficiently managed by medical therapy alone, and in these cases, functional neurosurgery may be considered. In the past, destructive techniques were used but in the recent times these have been superseded by deep brain stimulation [99]. This procedure involves the targeted implantation of electrodes to specific brain nuclei within the subcortical region of the brain together with a pacemaker that is usually implanted in the anterior chest wall (Fig. 3.13). The electrodes and the pacemaker are connected with a subcutaneous wire, allowing the delivery of high-frequency stimuli to the targeted brain regions. The effects of deep brain stimulation can be dramatic and long-lasting in carefully selected patients. In general, only those symptoms that respond to levodopa will be improved by deep brain stimulation. Tremor, which is difficult to treat medically, is an exception to this rule and can be improved by deep brain stimulation.

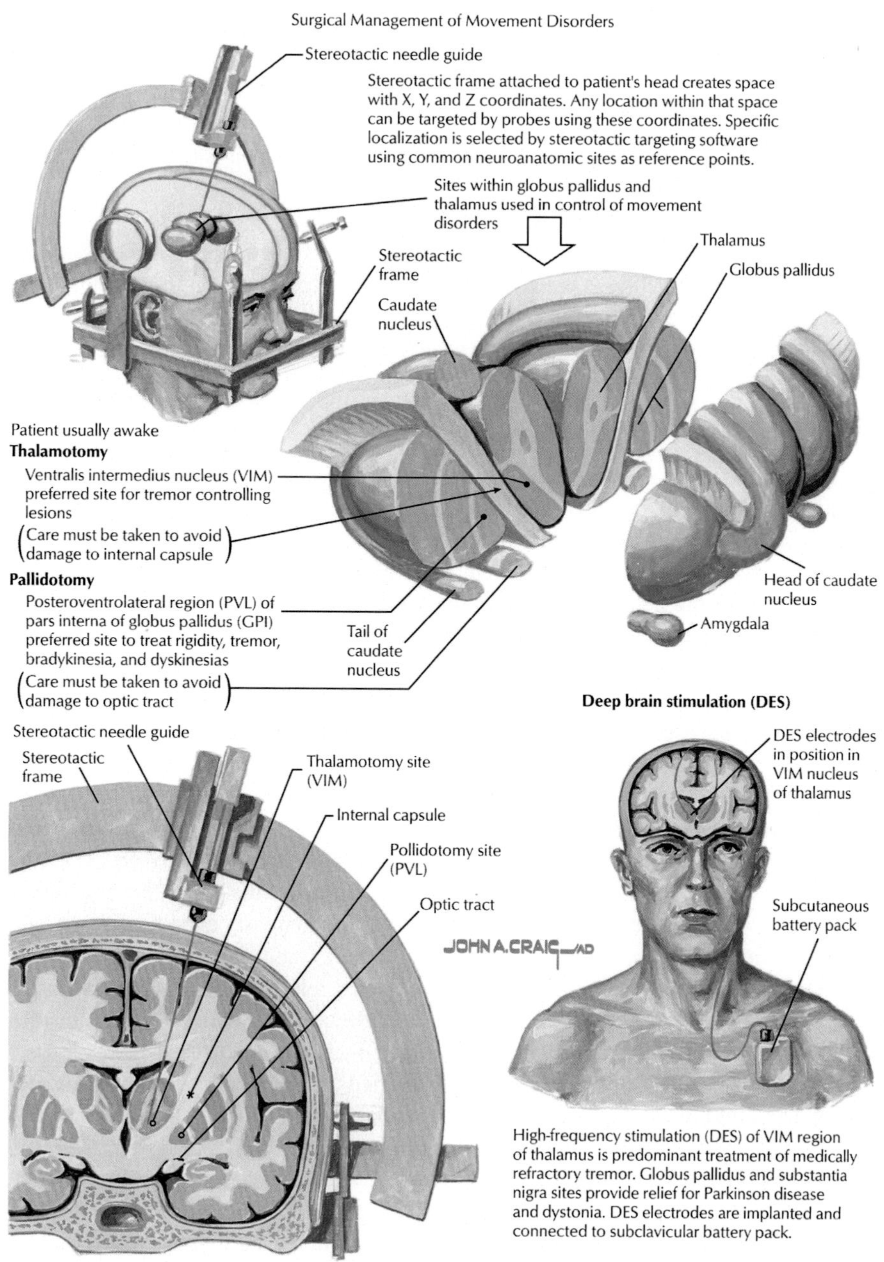

FIGURE 3.13

Surgical management of movement disorders.

Image courtesy of Netter images.

Different brain targets are used depending on which symptoms are the most troublesome and the general health of the patient.

Thalamic stimulation is mainly used for tremor.
Subthalamic nucleus stimulation can improve "off phase" symptoms, dyskinesias, and allow drug doses to be reduced.
Internal pallidum stimulation is mainly used for dyskinesias.

Patient selection for surgery is of paramount importance. Cognitive symptoms and mood can worsen after surgery, and dementia is usually considered a contraindication to deep brain stimulation. Speech can worsen after deep brain stimulation as can gait and postural instability. Other risks include brain hemorrhage, infection of the leads and hardware used, and the risk of the anesthetic itself.

3.5.2 EXPERIMENTAL THERAPIES

As noted in Section 3.1, there is as yet no disease-modifying therapy for Parkinson's disease. As the number of patients afflicted by this disorder increases in line with the aging global population, there is an urgent need for a therapy (or therapies) that either slows down or halts the degenerative process. There is increasing activity in drug development for Parkinson's disease as the fundamental insights into the mechanisms driving neurodegeneration outlined in Section 3.4 reveal more and more targets that could potentially modulate this degeneration. A scan of clinicaltrials.gov, the US registry of clinical trials, reveals that there are over 400 active clinical trials, that is, clinical trials that either are in progress or have been approved and are recruiting participants [100]. Of these, only a relatively small proportion aim to test disease-modifying therapies and some of the most prominent therapies are discussed in the following sections.

3.5.2.1 Cell Replacement Therapy

The concept of cell replacement therapy in Parkinson's disease is an attractive one. The symptoms of Parkinson's disease are driven by the neurochemical deficits resulting from the loss of cells due to neurodegeneration. A simple strategy to treat the disease, and potentially even reverse the degenerative process, would be to replace those cells that are lost. Experimental investigations into whether this strategy would benefit patients with Parkinson's disease have a relatively long history, with cell replacement therapy being carried out several decades ago using fetal brain extracts [101]. In these experiments, which included participants who had developed Parkinson's disease in response to MPTP exposure, purified fetal brain cells extracted from aborted human fetal brain tissue were gathered together and injected into the midbrain, with each transplant requiring cells from a number of aborted fetuses. A proportion of recipients appeared to benefit from this treatment, with evidence that the engrafted cells were able to survive in the human brain, and larger scale trials were initiated based on these positive outcomes [102]. The results of these more comprehensive trials were, however, disappointing with no significant improvements [103]. Coupled to evidence that dyskinesia was a frequent side effect of transplants [104], active investigation of fetal transplants as an experimental treatment for Parkinson's disease waned for a number of years. However, a detailed analysis of the patients who participated and the approaches used in these treatments raised the possibility that a subset of patients with younger onset disease and who were earlier on in their disease course may benefit from fetal transplants. This has resulted in a coordinated trial across Europe, which is due to be completed in 2018 at the time of writing [100]. Renewed interest in the concept of cell replacement in Parkinson's disease also came from major advances in the field of stem cell biology, where the ability of pluripotent stem cells to (potentially) differentiate into any cell type in the human body has been harnessed to produce dopaminergic neurons [105]. For many years, these stem cells were derived from embryos (the so-called embryonic stem cells), but the advent of iPS cells has allowed the derivation of pluripotent stem cells from a range of terminally differentiated tissues [106]. These advances have had several important consequences for cell-based therapies for Parkinson's disease. First, the ability of pluripotent stem cells to replicate has reduced the barrier of requiring brain extracts from multiple human fetuses—an issue that raises a number of ethical and moral issues, as well as practical issues, around the availability of this tissue. Second, iPS cell technology in particular has the potential to allow cell transplants to be personalized to an individual, being derived from their own cells. This removes many serious issues around histocompatibility and the possibility of cells being rejected. Hence, a study is being conducted in parallel to the European fetal transplant trial using embryonic stem cell–derived transplants, again due to be reported in the near future. As stem cell technology continues to advance, it is likely that efforts to develop cell transplants for Parkinson's disease will continue for the foreseeable future.

3.5.2.2 Neuroprotective Compounds

As the key event in Parkinson's disease is the death of brain cells, another widespread approach in drug development for this disorder is to try to bolster neuronal survival by the administration of neuroprotective compounds. There is a wide range of potentially neuroprotective drugs, targeting a number of different pathways in the brain to increase the likelihood of neurons being able to survive the disease process. If enough cells survive, then this will reduce or prevent progressive degeneration. Examples of this class of compound that either have been or are being examined in Parkinson's disease include the glial cell line–derived neurotrophic factor (GDNF), an endogenously expressed protein that has neuroprotective properties; exenatide, a glucagon-like peptide 1 agonist; and nicotine (partly based on data suggesting correlation between smoking and decreased risk for Parkinson's disease) [107]. There is an extensive literature reporting the beneficial impact of neuroprotective drugs in preclinical models for Parkinson's disease, with many drugs demonstrating protection against toxin-induced dopaminergic cell death following exposure to MPTP, rotenone, or 6-hydroxydopamine (another nigra-specific toxin). To date, however, there is limited evidence of their efficacy in human trials although a recent phase II trial for exenatide provided some evidence of sustained improvement in symptoms [116]. This may be partly due to the intrinsic difficulties of targeting the central nervous system. GDNF, for example, is unable to cross the blood–brain barrier, so clinical trials in humans require sophisticated methods to administer the protein directly into the brain.

3.5.2.3 Antisynuclein Drugs

α-Synuclein is central to the pathogenic process in Parkinson's disease, as evidenced by the genetic links to the disease and based on its status as the predominant constituent of Lewy bodies. As such, altering the pathobiology of α-synuclein has long been a goal of both academic and pharmaceutical company researchers. Several strategies have been developed to achieve this, focusing on the properties of α-synuclein that are linked to the disease, which have been highlighted by the extensive cellular and biochemical investigations into this protein. First, there is abundant evidence that the greater the concentration of α-synuclein, the more likely to develop a Lewy body disorder. A simple remedy to this would, therefore, be to reduce the amount of α-synuclein. The main ongoing effort in this area uses an approach that has been attempted in Alzheimer's disease—the use of immune therapy to clear aggregation-prone proteins from the brain [108]. Antibodies targeting α-synuclein have been used in animal models, specifically mice expressing a mutant form of α-synuclein, and have been shown to reduce pathology [109]. This mirrors the research carried out in models for Alzheimer's disease (see Chapter 2) and has led to the development of clinical trials of immunotherapy targeting α-synuclein in humans, although these have yet to advance to phase III testing [110]. The second approach is to directly target the aggregation process either by stabilizing a monomeric nonaggregated form of α-synuclein or by inhibiting the templating process that results in the formation of large multimers [111]. There is extensive in vitro and preclinical data showing that the aggregation of α-synuclein can be targeted, but human trials are limited; however, phase I trials of an antiaggregation drugs targeting α-synuclein have been completed, demonstrating that this approach (at least with one candidate compound) is safe in humans. This paves the way for phase II and III trials to examine whether targeting α-synuclein aggregation can have a beneficial outcome in the disease.

3.5.2.4 LRRK2 Inhibitors

The high frequency of mutations in *LRRK2*, coupled with the multiple enzymatic activities of the LRRK2 protein and evidence suggesting that these activities are disrupted by mutations, has led to this gene attracting much attention as a potential drug target in Parkinson's disease. A number of academic and industrial research groups have developed specific kinase inhibitors that directly target the ATP-binding pocket of the LRRK2 kinase domain, including inhibitors that are able to cross the blood–brain barrier and enter the central nervous system [112]. These have gone through advanced preclinical testing in nonhuman primates, revealing that inhibiting LRRK2 can result in pathological alterations in lung tissue [113]. It is not yet clear if these on-target effects will prevent the current generation of LRRK2 inhibitors from entering into full clinical trials, although the existence of a large number of mutation carriers either presenting with Parkinson's disease or at greatly increased risk of the disease provides an ideal cohort of patients to test whether inhibiting LRRK2 kinase activity will be beneficial in terms of slowing degeneration in people with Parkinson's disease. Limited phase I trials of LRRK2 kinase inhibitors by Denali therapeutics have indicated that at least some of these molecules do not have major adverse affects.

3.5.2.5 Glucocerebrosidase Modulators

Targeting glucocerebrosidase in Parkinson's disease provides an interesting challenge. The key event in Parkinson's disease linked to *GBA* mutations is thought to be a partial loss of glucocerebrosidase activity, which is similar to, but not as extreme as, that observed in Gaucher's disease. In Gaucher disease, where an individual is homozygous for mutations that reduce glucocerebrosidase activity, enzyme replacement therapy is used to restore the lost enzymatic activity. The therapy was initially carried out using glucocerebrosidase purified from animal sources, but now recombinant protein is used to treat patients, with considerable success in treating at least some of the symptoms associated with this disease [114]. In principle, this would seem to be a highly relevant approach for Parkinson's disease; however, the protein used in replacement enzyme therapy is not able to cross the blood–brain barrier and so there are considerable challenges in applying this approach to this disorder. A number of research groups have taken a slightly different approach, looking to test interventions that act to stabilize or increase the activity of existing glucocerebrosidase protein within the brain. To this end, a number of clinical trials are in development aiming to use small molecules that are able to enter the central nervous system and are hypothesized to boost glucocerebrosidase activity. An example of this is ambroxol, an FDA-approved compound currently used to treat respiratory diseases. Preclinical studies have indicated that ambroxol can increase glucocerebrosidase activity in human fibroblast cells from patients with both homozygous (i.e., Gaucher's disease) and heterozygous (i.e., Parkinson's disease) *GBA* mutations [115]. Small-scale clinical trials in Canada and the United Kingdom have been approved and are now recruiting. Importantly, and similar to clinical trials targeting *LRRK2*, initial trials can be carried out for ambroxol in a genetically defined population, that is, carriers of mutations in *GBA*. Although the currently approved trials target those who have already displayed motor symptoms, there is a possibility to target individuals who have not yet developed a motor phenotype and are therefore most likely to benefit from a therapy that either slows or halts neurodegeneration.

3.6 CONCLUSIONS

Parkinson's disease is a devastating degenerative neurological disorder, impacting most obviously motor performance and also now recognized to cause severe symptoms across a range of systems. Although major strides have been made to understand the mechanisms leading to neurodegeneration in Parkinson's disease, the goal of developing disease-modifying treatments for this disorder remains, at present, out of reach. The significant efforts now being directed toward clinical trials of compounds or approaches that directly address neurodegeneration in Parkinson's disease provide hope that treatments that slow or halt disease progression will become available over the coming years.

FURTHER READING

Langston JW, Palfreyman J. Case of the frozen addicts. IOS Press; 2014. ISBN 1614993319. A personal account of the research that elucidated the mechanism of MPTP toxicity.

Lees AJ, Hardy J, Revesz T. Parkinson's disease. Lancet 2009;373:2055–66. A comprehensive review of the clinical symptoms and pathological mechanisms in Parkinson's disease.

Lewis PA. James Parkinson: the man behind the shaking palsy. J Park Dis 2012;2:181–7. A short biography of James Parkinson, who first described Parkinson's disease as the shaking palsy in 1817.

Sacks O. Awakenings. Picador; 2012. ISBN 0330523678. Oliver Sacks' retelling of his experiences treating patients with postencephilitic parkinsonism using levodopa.

REFERENCES

[1] Parkinson J. An essay on the shaking palsy. London: Sherwood, Neely and Jones; 1817.

[2] Lees AJ, Hardy J, Revesz T. Parkinson's disease. Lancet 2009;373:2055–66. Epub 2009/06/16.

[3] Schapira AHV, Chaudhuri KR, Jenner P. Non-motor features of Parkinson disease. Nat Rev Neurosci 2017;18:435–50.

[4] Golbe LI. Young-onset Parkinson's disease: a clinical review. Neurology 1991;41:168–73.

[5] Hardy J, Lees AJ. Parkinson's disease: a broken nosology. Mov Disord 2005;20(Suppl. 12):S2–4.

[6] Hughes AJ, Daniel SE, Kilford L, Lees AJ. Accuracy of clinical diagnosis of idiopathic Parkinson's disease: a clinico-pathological study of 100 cases. J Neurol Neurosurg Psychiatry 1992;55:181–4.

[7] Kagi G, Bhatia KP, Tolosa E. The role of DAT-SPECT in movement disorders. J Neurol Neurosurg Psychiatry 2010;81:5–12.
[8] Lees A. The bare essentials: Parkinson's disease. Pract Neurol 2010;10:240–6.
[9] Pfeiffer RF. Autonomic dysfunction in Parkinson's disease. Expert Rev Neurother 2012;12:697–706.
[10] St Louis EK, Boeve AR, Boeve BF. REM sleep behavior disorder in Parkinson's disease and other synucleinopathies. Mov Disord 2017;32:645–58.
[11] Ha AD, Jankovic J. Pain in Parkinson's disease. Mov Disord 2012;27:485–91.
[12] Haehner A, Hummel T, Reichmann H. A clinical approach towards smell loss in Parkinson's disease. J Park Dis 2014;4:189–95.
[13] Aarsland D, Beyer MK, Kurz MW. Dementia in Parkinson's disease. Curr Opin Neurol 2008;21:676–82. Epub 2008/11/08.
[14] Walker Z, Possin KL, Boeve BF, Aarsland D. Lewy body dementias. Lancet 2015;386:1683–97.
[15] Williams DR, Lees AJ. Progressive supranuclear palsy: clinicopathological concepts and diagnostic challenges. Lancet Neurol 2009;8:270–9.
[16] Krismer F, Wenning GK. Multiple system atrophy: insights into a rare and debilitating movement disorder. Nat Rev Neurol 2017;13:232–43.
[17] Lewy F. Zur pathologischen Anatomie der Paralysis agitans. Dtsch Z Nervenheilk 1913;50:50–5.
[18] Forno LS. Neuropathology of Parkinson's disease. J Neuropathol Exp Neurol 1996;55:259–72.
[19] Goedert M, Spillantini MG, Del Tredici K, Braak H. 100 years of Lewy pathology. Nat Rev Neurol 2013;9:13–24.
[20] Lashuel HA, Overk CR, Oueslati A, Masliah E. The many faces of alpha-synuclein: from structure and toxicity to therapeutic target. Nat Rev Neurosci 2013;14:38–48. Epub 2012/12/21.
[21] Arrasate M, Mitra S, Schweitzer ES, Segal MR, Finkbeiner S. Inclusion body formation reduces levels of mutant huntingtin and the risk of neuronal death. Nature 2004;431:805–10.
[22] Isacson O. The production and use of cells as therapeutic agents in neurodegenerative diseases. Lancet Neurol 2003;2:417–24.
[23] Kordower JH, Chu Y, Hauser RA, Freeman TB, Olanow CW. Lewy body-like pathology in long-term embryonic nigral transplants in Parkinson's disease. Nat Med 2008;14:504–6.
[24] Li JY, Englund E, Holton JL, Soulet D, Hagell P, Lees AJ, Lashley T, Quinn NP, Rehncrona S, Bjorklund A, Widner H, Revesz T, Lindvall O, Brundin P. Lewy bodies in grafted neurons in subjects with Parkinson's disease suggest host-to-graft disease propagation. Nat Med 2008;14:501–3.
[25] Young AB, Penney JB. Neurochemical anatomy of movement disorders. Neurol Clin 1984;2:417–33.
[26] Carlsson A, Lindqvist M, Magnusson T, Waldeck B. On the presence of 3-hydroxytyramine in brain. Science 1958;127:471.
[27] Carlsson A, Lindqvist M, Magnusson T. 3,4-Dihydroxyphenylalanine and 5-hydroxytryptophan as reserpine antagonists. Nature 1957;180:1200.
[28] Bjorklund A, Dunnett SB. Dopamine neuron systems in the brain: an update. Trends Neurosci 2007;30:194–202.
[29] Eisenhofer G, Kopin IJ, Goldstein DS. Catecholamine metabolism: a contemporary view with implications for physiology and medicine. Pharmacol Rev 2004;56:331–49.
[30] Beaulieu JM, Gainetdinov RR. The physiology, signaling, and pharmacology of dopamine receptors. Pharmacol Rev 2011;63:182–217.
[31] Yager LM, Garcia AF, Wunsch AM, Ferguson SM. The ins and outs of the striatum: role in drug addiction. Neuroscience 2015;301:529–41.
[32] Braak H, Del Tredici K, Rub U, de Vos RA, Jansen Steur EN, Braak E. Staging of brain pathology related to sporadic Parkinson's disease. Neurobiol Aging 2003;24:197–211.
[33] Burke RE, Dauer WT, Vonsattel JP. A critical evaluation of the Braak staging scheme for Parkinson's disease. Ann Neurol 2008;64:485–91. Epub 2008/12/11.

[34] Walsh DM, Selkoe DJ. A critical appraisal of the pathogenic protein spread hypothesis of neurodegeneration. Nat Rev Neurosci 2016;17:251–60.

[35] Angot E, Steiner JA, Hansen C, Li JY, Brundin P. Are synucleinopathies prion-like disorders?. Lancet Neurol 2010;9:1128–38. Epub 2010/09/18.

[36] Wray S, Lewis PA. A tangled web - tau and sporadic Parkinson's disease. Front Psychiatry/Front Res Found 2010;1:150. Epub 2010/01/01.

[37] Wenning GK, Stefanova N, Jellinger KA, Poewe W, Schlossmacher MG. Multiple system atrophy: a primary oligodendrogliopathy. Ann Neurol 2008;64:239–46.

[38] Kalia LV, Lang AE, Hazrati LN, Fujioka S, Wszolek ZK, Dickson DW, Ross OA, Van Deerlin VM, Trojanowski JQ, Hurtig HI, Alcalay RN, Marder KS, Clark LN, Gaig C, Tolosa E, Ruiz-Martínez J, Marti-Masso JF, Ferrer I, López de Munain A, Goldman SM, Schüle B, Langston JW, Aasly JO, Giordana MT, Bonifati V, Puschmann A, Canesi M, Pezzoli G, Maues De Paula A, Hasegawa K, Duyckaerts C, Brice A, Stoessl AJ, Marras C. Clinical correlations with Lewy body pathology in LRRK2-related Parkinson disease. JAMA Neurol Jan 2015;72(1):100–5.

[39] Singleton A, Hardy J. The evolution of genetics: Alzheimer's and Parkinson's diseases. Neuron 2016;90:1154–63.

[40] Polymeropoulos MH, Lavedan C, Leroy E, Ide SE, Dehejia A, Dutra A, Pike B, Root H, Rubenstein J, Boyer R, Stenroos ES, Chandrasekharappa S, Athanassiadou A, Papapetropoulos T, Johnson WG, Lazzarini AM, Duvoisin RC, Di Iorio G, Golbe LI, Nussbaum RL. Mutation in the alpha-synuclein gene identified in families with Parkinson's disease. Science 1997;276:2045–7.

[41] Volta M, Milnerwood AJ, Farrer MJ. Insights from late-onset familial parkinsonism on the pathogenesis of idiopathic Parkinson's disease. Lancet Neurol 2015;14:1054–64.

[42] Benamer HT, de Silva R. LRRK2 G2019S in the North African population: a review. Eur Neurol 2010;63:321–5.

[43] Nalls MA, Pankratz N, Lill CM, Do CB, Hernandez DG, Saad M, DeStefano AL, Kara E, Bras J, Sharma M, Schulte C, Keller MF, Arepalli S, Letson C, Edsall C, Stefansson H, Liu X, Pliner H, Lee JH, Cheng R, International Parkinson's Disease Genomics Consortium, Parkinson's Study Group Parkinson's Research: The Organized GENetics Initiative, 23andMe, GenePD, NeuroGenetics Research Consortium, Hussman Institute of Human Genomics, Ashkenazi Jewish Dataset Investigator, Cohorts for Health and Aging Research in Genetic Epidemiology, North American Brain Expression Consortium, United Kingdom Brain Expression Consortium, Greek Parkinson's Disease Consortium, Alzheimer Genetic Analysis Group, Ikram MA, Ioannidis JP, Hadjigeorgiou GM, Bis JC, Martinez M, Perlmutter JS, Goate A, Marder K, Fiske B, Sutherland M, Xiromerisiou G, Myers RH, Clark LN, Stefansson K, Hardy JA, Heutink P, Chen H, Wood NW, Houlden H, Payami H, Brice A, Scott WK, Gasser T, Bertram L, Eriksson N, Foroud T, Singleton AB. Large-scale meta-analysis of genome-wide association data identifies six new risk loci for Parkinson's disease. Nat Genet 2014;46:989–93. Epub 2014/07/30.

[44] van der Brug MP, Singleton A, Gasser T, Lewis PA. Parkinson's disease: from human genetics to clinical trials. Sci Transl Med 2015;7:205ps20.

[45] Spillantini MG, Schmidt ML, Lee VM, Trojanowski JQ, Jakes R, Goedert M. Alpha-synuclein in lewy bodies. Nature 1997;388:839–40.

[46] Kara E, Lewis PA, Ling H, Proukakis C, Houlden H, Hardy J. alpha-Synuclein mutations cluster around a putative protein loop. Neurosci Lett 2013;546:67–70. Epub 2013/05/15.

[47] Cremades N, Cohen SI, Deas E, Abramov AY, Chen AY, Orte A, Sandal M, Clarke RW, Dunne P, Aprile FA, Bertoncini CW, Wood NW, Knowles TP, Dobson CM, Klenerman D. Direct observation of the interconversion of normal and toxic forms of alpha-synuclein. Cell 2012;149:1048–59. Epub 2012/05/29.

[48] Lashuel HA, Petre BM, Wall J, Simon M, Nowak RJ, Walz T, Lansbury Jr PT. Alpha-synuclein, especially the Parkinson's disease-associated mutants, forms pore-like annular and tubular protofibrils. J Mol Biol 2002;322:1089–102.

[49] Singleton AB, Farrer M, Johnson J, Singleton A, Hague S, Kachergus J, Hulihan M, Peuralinna T, Dutra A, Nussbaum R, Lincoln S, Crawley A, Hanson M, Maraganore D, Adler C, Cookson MR, Muenter M, Baptista M, Miller D, Blancato J, Hardy J, Gwinn-Hardy K. alpha-Synuclein locus triplication causes Parkinson's disease. Science 2003;302:841.
[50] Devine MJ, Gwinn K, Singleton A, Hardy J. Parkinson's disease and alpha-synuclein expression. Mov Disord 2011. Epub 2011/09/03.
[51] Soldner F, Stelzer Y, Shivalila CS, Abraham BJ, Latourelle JC, Barrasa MI, Goldmann J, Myers RH, Young RA, Jaenisch R. Parkinson-associated risk variant in distal enhancer of alpha-synuclein modulates target gene expression. Nature 2016;533:95–9.
[52] Watts JC, Giles K, Oehler A, Middleton L, Dexter DT, Gentleman SM, DeArmond SJ, Prusiner SB. Transmission of multiple system atrophy prions to transgenic mice. Proc Natl Acad Sci USA 2013;110:19555–60.
[53] Luk KC, Kehm V, Carroll J, Zhang B, O'Brien P, Trojanowski JQ, Lee VM. Pathological alpha-synuclein transmission initiates Parkinson-like neurodegeneration in nontransgenic mice. Science 2012;338:949–53.
[54] Zimprich A, Biskup S, Leitner P, Lichtner P, Farrer M, Lincoln S, Kachergus J, Hulihan M, Uitti RJ, Calne DB, Stoessl AJ, Pfeiffer RF, Patenge N, Carbajal IC, Vieregge P, Asmus F, Muller-Myhsok B, Dickson DW, Meitinger T, Strom TM, Wszolek ZK, Gasser T. Mutations in LRRK2 cause autosomal-dominant parkinsonism with pleomorphic pathology. Neuron 2004;44:601–7. Epub 2004/11/16.
[55] Paisan-Ruiz C, Jain S, Evans EW, Gilks WP, Simon J, van der Brug M, Lopez de Munain A, Aparicio S, Gil AM, Khan N, Johnson J, Martinez JR, Nicholl D, Carrera IM, Pena AS, de Silva R, Lees A, Marti-Masso JF, Perez-Tur J, Wood NW, Singleton AB. Cloning of the gene containing mutations that cause PARK8-linked Parkinson's disease. Neuron 2004;44:595–600.
[56] Paisan-Ruiz C, Lewis P, Singleton A. LRRK2: cause, risk, and mechanism. J Park Dis 2013. https://doi.org/10.3233/JPD-130192. Online first.
[57] Roosen DA, Cookson MR. LRRK2 at the interface of autophagosomes, endosomes and lysosomes. Mol Neurodegener 2016;11:73.
[58] Grabowski GA. Phenotype, diagnosis, and treatment of Gaucher's disease. Lancet 2008;372:1263–71.
[59] Sidransky E, Lopez G. The link between the GBA gene and parkinsonism. Lancet Neurol 2012;11:986–98. Epub 2012/10/20.
[60] Mazzulli JR, Xu YH, Sun Y, Knight AL, McLean PJ, Caldwell GA, Sidransky E, Grabowski GA, Krainc D. Gaucher disease glucocerebrosidase and alpha-synuclein form a bidirectional pathogenic loop in synucleinopathies. Cell 2011;146:37–52. Epub 2011/06/28.
[61] Gegg ME, Burke D, Heales SJ, Cooper JM, Hardy J, Wood NW, Schapira AH. Glucocerebrosidase deficiency in substantia nigra of Parkinson disease brains. Ann Neurol 2012;72:455–63. Epub 2012/10/05.
[62] Ramirez A, Heimbach A, Grundemann J, Stiller B, Hampshire D, Cid LP, Goebel I, Mubaidin AF, Wriekat AL, Roeper J, Al-Din A, Hillmer AM, Karsak M, Liss B, Woods CG, Behrens MI, Kubisch C. Hereditary parkinsonism with dementia is caused by mutations in ATP13A2, encoding a lysosomal type 5 P-type ATPase. Nat Genet 2006;38:1184–91. Epub 2006/09/12.
[63] Vilarino-Guell C, Wider C, Ross OA, Dachsel JC, Kachergus JM, Lincoln SJ, Soto-Ortolaza AI, Cobb SA, Wilhoite GJ, Bacon JA, Behrouz B, Melrose HL, Hentati E, Puschmann A, Evans DM, Conibear E, Wasserman WW, Aasly JO, Burkhard PR, Djaldetti R, Ghika J, Hentati F, Krygowska-Wajs A, Lynch T, Melamed E, Rajput A, Rajput AH, Solida A, Wu RM, Uitti RJ, Wszolek ZK, Vingerhoets F, Farrer MJ. VPS35 mutations in Parkinson disease. Am J Hum Genet 2011;89:162–7. Epub 2011/07/19.
[64] Zavodszky E, Seaman MN, Moreau K, Jimenez-Sanchez M, Breusegem SY, Harbour ME, Rubinsztein DC. Mutation in VPS35 associated with Parkinson's disease impairs WASH complex association and inhibits autophagy. Nat Commun 2014;5:3828. Epub 2014/05/14.

[65] Puschmann A. Monogenic Parkinson's disease and parkinsonism: clinical phenotypes and frequencies of known mutations. Park Relat Disord 2013;19:407–15.
[66] Kitada T, Asakawa S, Hattori N, Matsumine H, Yamamura Y, Minoshima S, Yokochi M, Mizuno Y, Shimizu N. Mutations in the parkin gene cause autosomal recessive juvenile parkinsonism. Nature 1998;392:605–8.
[67] Valente EM, Abou-Sleiman PM, Caputo V, Muqit MM, Harvey K, Gispert S, Ali Z, Del Turco D, Bentivoglio AR, Healy DG, Albanese A, Nussbaum R, Gonzalez-Maldonado R, Deller T, Salvi S, Cortelli P, Gilks WP, Latchman DS, Harvey RJ, Dallapiccola B, Auburger G, Wood NW. Hereditary early-onset Parkinson's disease caused by mutations in PINK1. Science 2004;304:1158–60.
[68] Clark IE, Dodson MW, Jiang C, Cao JH, Huh JR, Seol JH, Yoo SJ, Hay BA, Guo M. Drosophila pink1 is required for mitochondrial function and interacts genetically with parkin. Nature 2006;441(7097): 1162–6.
[69] Park J, Lee SB, Lee S, Kim Y, Song S, Kim S, Bae E, Kim J, Shong M, Kim JM, Chung J. Mitochondrial dysfunction in Drosophila PINK1 mutants is complemented by parkin. Nature 2006;441(7097):1157–61.
[70] Burchell VS, Nelson DE, Sanchez-Martinez A, Delgado-Camprubi M, Ivatt RM, Pogson JH, Randle SJ, Wray S, Lewis PA, Houlden H, Abramov AY, Hardy J, Wood NW, Whitworth AJ, Laman H, Plun-Favreau H. The Parkinson's disease-linked proteins Fbxo7 and Parkin interact to mediate mitophagy. Nat Neurosci 2013;16:1257–65.
[71] McWilliams TG, Muqit MM. PINK1 and Parkin: emerging themes in mitochondrial homeostasis. Curr Opin Cell Biol 2017;45:83–91.
[72] Lazarou M, Sliter DA, Kane LA, Sarraf SA, Wang C, Burman JL, Sideris DP, Fogel AI, Youle RJ. The ubiquitin kinase PINK1 recruits autophagy receptors to induce mitophagy. Nature 2015;524(7565):309–14.
[73] Cookson MR, Hardy J, Lewis PA. Genetic neuropathology of Parkinson's disease. Int J Clin Exp Pathol 2008;1:217–31. Epub 2008/09/12.
[74] Ransohoff RM. How neuroinflammation contributes to neurodegeneration. Science 2016;353:777–83.
[75] Hamza TH, Zabetian CP, Tenesa A, Laederach A, Montimurro J, Yearout D, Kay DM, Doheny KF, Paschall J, Pugh E, Kusel VI, Collura R, Roberts J, Griffith A, Samii A, Scott WK, Nutt J, Factor SA, Payami H. Common genetic variation in the HLA region is associated with late-onset sporadic Parkinson's disease. Nat Genet 2010;42:781–5. Epub 2010/08/17.
[76] McGeer PL, Itagaki S, Boyes BE, McGeer EG. Reactive microglia are positive for HLA-DR in the substantia nigra of Parkinson's and Alzheimer's disease brains. Neurology 1988;38:1285–91.
[77] Sulzer D, Alcalay RN, Garretti F, Cote L, Kanter E, Agin-Liebes J, Liong C, McMurtrey C, Hildebrand WH, Mao X, Dawson VL, Dawson TM, Oseroff C, Pham J, Sidney J, Dillon MB, Carpenter C, Weiskopf D, Phillips E, Mallal S, Peters B, Frazier A, Lindestam Arlehamn CS, Sette A. T cells from patients with Parkinson's disease recognize alpha-synuclein peptides. Nature 2017;546:656–61.
[78] Sampson TR, Debelius JW, Thron T, Janssen S, Shastri GG, Ilhan ZE, Challis C, Schretter CE, Rocha S, Gradinaru V, Chesselet MF, Keshavarzian A, Shannon KM, Krajmalnik-Brown R, Wittung-Stafshede P, Knight R, Mazmanian SK. Gut microbiota regulate motor deficits and neuroinflammation in a model of Parkinson's disease. Cell 2016;167. 1469–80.e12.
[79] Davis GC, Williams AC, Markey SP, Ebert MH, Caine ED, Reichert CM, Kopin IJ. Chronic Parkinsonism secondary to intravenous injection of meperidine analogues. Psychiatry Res 1979;1:249–54.
[80] Langston JW, Ballard P, Tetrud JW, Irwin I. Chronic Parkinsonism in humans due to a product of meperidine-analog synthesis. Science 1983;219:979–80.
[81] Langston JW, Forno LS, Rebert CS, Irwin I. Selective nigral toxicity after systemic administration of 1-methyl-4-phenyl-1,2,5,6-tetrahydropyrine (MPTP) in the squirrel monkey. Brain Res 1984;292:390–4.
[82] Langston JW. The MPTP story. J Park Dis 2017;7:S11–22.
[83] Schapira AH, Cooper JM, Dexter D, Jenner P, Clark JB, Marsden CD. Mitochondrial complex I deficiency in Parkinson's disease. Lancet 1989;1:1269.

[84] McCormack AL, Thiruchelvam M, Manning-Bog AB, Thiffault C, Langston JW, Cory-Slechta DA, Di Monte DA. Environmental risk factors and Parkinson's disease: selective degeneration of nigral dopaminergic neurons caused by the herbicide paraquat. Neurobiol Dis 2002;10:119–27.
[85] Gorell JM, Johnson CC, Rybicki BA, Peterson EL, Richardson RJ. The risk of Parkinson's disease with exposure to pesticides, farming, well water, and rural living. Neurology 1998;50:1346–50.
[86] Betarbet R, Sherer TB, MacKenzie G, Garcia-Osuna M, Panov AV, Greenamyre JT. Chronic systemic pesticide exposure reproduces features of Parkinson's disease. Nat Neurosci 2000;3:1301–6.
[87] Brooks AI, Chadwick CA, Gelbard HA, Cory-Slechta DA, Federoff HJ. Paraquat elicited neurobehavioral syndrome caused by dopaminergic neuron loss. Brain Res 1999;823:1–10.
[88] Reid AH, McCall S, Henry JM, Taubenberger JK. Experimenting on the past: the enigma of von Economo's encephalitis lethargica. J Neuropathol Exp Neurol 2001;60:663–70.
[89] Casals J, Elizan TS, Yahr MD. Postencephalitic parkinsonism–a review. J Neural Transm (Vienna) 1998;105:645–76.
[90] Sacks O. Awakenings. 1st Vintage Books ed.. New York: Vintage Books; 1999.
[91] Calne DB, Stern GM, Laurence DR, Sharkey J, Armitage P. L-dopa in postencephalitic parkinsonism. Lancet 1969;1:744–6.
[92] Hay J, Johnson VE, Smith DH, Stewart W. Chronic traumatic encephalopathy: the neuropathological legacy of traumatic brain injury. Annu Rev Pathol 2016;11:21–45.
[93] Connolly BS, Lang AE. Pharmacological treatment of Parkinson disease: a review. JAMA 2014;311:1670–83.
[94] Huot P, Johnston TH, Koprich JB, Fox SH, Brotchie JM. The pharmacology of L-DOPA-induced dyskinesia in Parkinson's disease. Pharmacol Rev 2013;65:171–222.
[95] Wolters E, van der Werf YD, van den Heuvel OA. Parkinson's disease-related disorders in the impulsive-compulsive spectrum. J Neurol 2008;255(Suppl. 5):48–56.
[96] Robakis D, Fahn S. Defining the role of the monoamine oxidase-B inhibitors for Parkinson's disease. CNS Drugs 2015;29:433–41.
[97] Muller T. Catechol-O-methyltransferase inhibitors in Parkinson's disease. Drugs 2015;75:157–74.
[98] Crosby N, Deane KH, Clarke CE. Amantadine in Parkinson's disease. Cochrane Database Syst Rev 2003:CD003468.
[99] Okun MS. Deep-brain stimulation for Parkinson's disease. N Engl J Med 2012;367:1529–38.
[100] Available from: http://www.clinicaltrials.gov.
[101] Barker RA, Drouin-Ouellet J, Parmar M. Cell-based therapies for Parkinson disease-past insights and future potential. Nat Rev Neurol 2015;11:492–503.
[102] Kordower JH, Freeman TB, Chen EY, Mufson EJ, Sanberg PR, Hauser RA, Snow B, Olanow CW. Fetal nigral grafts survive and mediate clinical benefit in a patient with Parkinson's disease. Mov Disord 1998;13:383–93.
[103] Olanow CW, Goetz CG, Kordower JH, Stoessl AJ, Sossi V, Brin MF, Shannon KM, Nauert GM, Perl DP, Godbold J, Freeman TB. A double-blind controlled trial of bilateral fetal nigral transplantation in Parkinson's disease. Ann Neurol 2003;54:403–14.
[104] Hagell P, Piccini P, Bjorklund A, Brundin P, Rehncrona S, Widner H, Crabb L, Pavese N, Oertel WH, Quinn N, Brooks DJ, Lindvall O. Dyskinesias following neural transplantation in Parkinson's disease. Nat Neurosci 2002;5:627–8.
[105] Kriks S, Shim JW, Piao J, Ganat YM, Wakeman DR, Xie Z, Carrillo-Reid L, Auyeung G, Antonacci C, Buch A, Yang L, Beal MF, Surmeier DJ, Kordower JH, Tabar V, Studer L. Dopamine neurons derived from human ES cells efficiently engraft in animal models of Parkinson's disease. Nature 2011;480:547–51.
[106] Soldner F, Hockemeyer D, Beard C, Gao Q, Bell GW, Cook EG, Hargus G, Blak A, Cooper O, Mitalipova M, Isacson O, Jaenisch R. Parkinson's disease patient-derived induced pluripotent stem cells free of viral reprogramming factors. Cell 2009;136:964–77. Epub 2009/03/10.
[107] Athauda D, Foltynie T. The ongoing pursuit of neuroprotective therapies in Parkinson disease. Nat Rev Neurol 2015;11:25–40.

[108] Lemere CA. Immunotherapy for Alzheimer's disease: hoops and hurdles. Mol Neurodegener 2013;8:36.
[109] Masliah E, Rockenstein E, Adame A, Alford M, Crews L, Hashimoto M, Seubert P, Lee M, Goldstein J, Chilcote T, Games D, Schenk D. Effects of alpha-synuclein immunization in a mouse model of Parkinson's disease. Neuron 2005;46:857–68.
[110] Schenk DB, Koller M, Ness DK, Griffith SG, Grundman M, Zago W, Soto J, Atiee G, Ostrowitzki S, Kinney GG. First-in-human assessment of PRX002, an anti-alpha-synuclein monoclonal antibody, in healthy volunteers. Mov Disord 2017;32:211–8.
[111] Krishnan R, Hefti F, Tsubery H, Lulu M, Proschitsky M, Fisher R. Conformation as the therapeutic target for neurodegenerative diseases. Curr Alzheimer Res 2017;14:393–402.
[112] Galatsis P. Leucine-rich repeat kinase 2 inhibitors: a patent review (2014–2016). Expert opinion on therapeutic patents. 2017. p. 1–10.
[113] Fuji RN, Flagella M, Baca M, Baptista MA, Brodbeck J, Chan BK, Fiske BK, Honigberg L, Jubb AM, Katavolos P, Lee DW, Lewin-Koh SC, Lin T, Liu X, Liu S, Lyssikatos JP, O'Mahony J, Reichelt M, Roose-Girma M, Sheng Z, Sherer T, Smith A, Solon M, Sweeney ZK, Tarrant J, Urkowitz A, Warming S, Yaylaoglu M, Zhang S, Zhu H, Estrada AA, Watts RJ. Effect of selective LRRK2 kinase inhibition on nonhuman primate lung. Sci Transl Med 2015;7:273ra15.
[114] Brady RO, Schiffmann R. Enzyme-replacement therapy for metabolic storage disorders. Lancet Neurol 2004;3:752–6.
[115] McNeill A, Magalhaes J, Shen C, Chau KY, Hughes D, Mehta A, Foltynie T, Cooper JM, Abramov AY, Gegg M, Schapira AH. Ambroxol improves lysosomal biochemistry in glucocerebrosidase mutation-linked Parkinson disease cells. Brain 2014;137:1481–95. Epub 2014/02/28.
[116] Athauda D, Maclagan K, Skene SS, Bajwa-Joseph M, Letchford D, Chowdhury K, Li Y. Exenatide once weekly versus placebo in Parkinson's disease: a randomised, double-blind, placebo-controlled trial. The Lancet 2017;390(10103):1664–75.

CHAPTER

THE PRION DISEASES

4

CHAPTER OUTLINE

4.1 INTRODUCTION

The prion diseases stand distinct from the majority of neurodegenerative disorders, and indeed from almost all other disease that afflict humans. They are characterized by an almost unique tripartite etiology encompassing both genetic and infectious disease, as well as apparently sporadic cases of unknown cause [1]. The causative agent is hypothesized to be a misfolded aggregated form of an endogenous human protein, the prion protein (PrP), which can spread within a host, from cell to cell, and between organisms independent of nucleic acids. The research that has led to our current understanding of prion diseases is one that could grace the pages of an Arthur Conan Doyle detective story,

The Molecular and Clinical Pathology of Neurodegenerative Disease. https://doi.org/10.1016/B978-0-12-811069-0.00004-5

FIGURE 4.1

"It is an old maxim of mine that when you have excluded the impossible, whatever remains, however improbable, must be the truth." Sherlock Holmes [2].

worthy of Sherlock Holmes at his best (Fig. 4.1) [2], and has fundamental implications for our comprehension of infectious disease biology and mechanisms of heritability. Importantly, there is now a growing appreciation that the mechanisms that drive pathology in the prion diseases may have important implications for neurodegenerative diseases such as Alzheimer disease, Parkinson disease, and amyotrophic lateral sclerosis [3].

4.2 CLINICAL PRESENTATION

The human prion diseases are a group of related disorders connected by an almost unique cause. They include Creutzfeldt–Jakob disease, Gerstmann-Sträussler-Scheinker syndrome, fatal familial insomnia, variant Creutzfeldt–Jakob disease, and a neurological disorder found only in the tribes of the highlands of Papua New Guinea called kuru (Table 4.1). The clinical spectrum across these disorders is heterogeneous, spanning cognitive changes and a range of movement symptoms [4]. In contrast to the majority of neurological disorders covered in this book, the prion diseases found in humans are mirrored by a range of naturally occurring disorders found in animals [5]. These include scrapie (affecting sheep),

Table 4.1 Human Prion Diseases

Disease	Clinical Presentation	Pathology	Etiology
Creutzfeldt–Jakob disease	Dementia, myoclonus, ataxia, and psychiatric issues. Rapid disease course (4–6 months). Average age at onset, 65 years.	Spongiform degeneration, astrogliosis, PrP deposition	Inherited autosomal dominant mutations in *PRNP*, transmissible (iatrogenic), sporadic
Variant Creutzfeldt–Jakob disease	Dementia, myoclonus, ataxia, and psychiatric issues (earlier, and more prominent, than standard Creutzfeldt–Jakob disease). Rapid disease course >12 months, younger age of onset compared to standard Creutzfeldt–Jakob disease.	Spongiform degeneration, astrogliosis, PrP deposition, florid plaques	Transmissible zoonosis, evidence for horizontal transmission between humans
Fatal familial insomnia	Altered sleep patterns leading to insomnia, hallucinations, and myoclonus.	Spongiform degeneration, astrogliosis	Inherited autosomal dominant mutations in *PRNP*, some evidence for sporadic disease
Gerstmann-Sträussler-Scheinker syndrome	Cerebellar syndrome with pronounced ataxia, progressing to dementia over years.	Spongiform degeneration, astrogliosis, amyloid deposition	Inherited autosomal dominant mutations in *PRNP*
Kuru	Prominent ataxia and tremor, progressing to akinetic mutism over 1–2 years. Predominantly observed in preadolescent children and females due to transmission.	Spongiform degeneration, astrogliosis, kuru plaques	Transmissible (endocannibalism)

PrP, *prion protein.*

chronic wasting disease (affecting cervids, including a number of deer species), and bovine spongiform encephalopathy (BSE, also known as mad cow disease, affecting cows). These disorders share a number of pathological and symptomatic features with the human disorders, and there is strong evidence that at least one animal prion disease, BSE, has been transmitted to humans as a novel zoonosis [6].

4.2.1 CLINICAL FEATURES OF ANIMAL PRION DISEASES

The three naturally occurring animal prion diseases (scrapie, chronic wasting disease, and BSE) are characterized by progressive deterioration in movement and by gradual wasting due to reduced ability to eat. Scrapie is the prion disease that has the longest documented history, with descriptions dating back to at least the 18th century [7]. As the name suggests, it is notable for a striking compulsion to scrape up against fences and trees resulting in a characteristic unkempt appearance and alopecia [8]. Additional symptoms include trembling (from which is derived a French term for the disorder "le Tremblante Mouton"), gait ataxia, and impaired social behavior [9,10]. Chronic wasting disease was initially reported in captive deer held at a US government research center in Colorado during the 1960s and was recognized as a transmissible spongiform encephalopathy in 1978 [11]. It was subsequently identified in free-range cervids and is characterized by weight loss, progressive deterioration in upkeep, polydipsia, and polyuria, as well as alterations in social behavior (resulting in separation from herds). BSE, first identified in 1986, shares a number of symptoms with scrapie, including gait ataxia [12,13]. Cows afflicted with BSE also present with apprehension and hyperesthesia, as well as aggression [14]. In an early indication that BSE could be transmitted to species other than cows, a related disorder has been reported in domestic cats, exotic felines (such as captive cheetahs), and a range of other ungulates [15]. Prion disease has also been reported in farmed mink (transmissible mink encephalopathy) [16]. All these disorders progress rapidly, with death resulting from malnutrition after a period of months, and share similar neuropathological conditions characterized by vacuolar degeneration of the brain, leading to a characteristic spongiform appearance [17,18]. For BSE and scrapie, a number of countries enforce culling of domestic cattle and sheep, with wild deer being monitored for chronic wasting disease [19,20].

4.2.2 CREUTZFELDT–JAKOB DISEASE

Creutzfeldt–Jakob disease was first described in the 1920s by Hans Gerhard Creutzfeldt and independently by Alfons Jakob [21,22]. The disorder is characterized by a broad spectrum of neurological symptoms, including prominent dementia and myoclonus, as well as extrapyramidal symptoms, ataxia, and psychiatric issues such as depression [23,24]. It is also associated with a wide range of nonspecific symptoms such as fatigue, insomnia, and weight loss. Combined, these make the differential diagnosis of Creutzfeldt–Jakob disease challenging, as there is significant overlap with a number of other neurological disorders, leading to misdiagnosis in the initial stages of the disease. This is compounded by the rarity of the disease, with a global incidence of around 1 per million per annum, reducing the number of neurologists who have direct experience of evaluating and diagnosing the disorder. A key characteristic that helps to distinguish Creutzfeldt–Jakob disease from other dementias is the rapidity of the disease progression: death normally results after the patient has reached a state of akinetic mutism 4–6 months after the initial presentation. Combined with postmortem analysis of brain tissue and bioassay for infectivity, diagnosis can be confirmed after death.

As noted earlier and in Section 4.4, Creutzfeldt–Jakob disease has a complex etiology. The sporadic, and most common, form of the disease has an average age of onset of 65 years, but a wide range of age of onset is observed [25]. In addition to the sporadic form, there are inherited familial forms caused by autosomal dominant mutations in the PrP gene (*PRNP*) and transmitted forms [1].

4.2.3 GERSTMANN-STRÄUSSLER-SCHEINKER SYNDROME

Gerstmann-Sträussler-Scheinker syndrome is caused by autosomal dominant mutations in the *PRNP* gene on chromosome 20. It was first described in an Austrian family by Drs. Gerstmann, Sträussler, and Scheinker in the 1930s; however, the genetic basis for the disease was not revealed until 1989 with the identification of a P102L mutation in the *PRNP* gene in American and British patients with the syndrome [26,27]. The disorder is distinct from Creutzfeldt–Jakob disease, presenting as a cerebellar syndrome with prominent ataxia progressing to dementia. The age of onset is variable, ranging from the fifth to the seventh decade of life, suggesting that other factors (potentially other genetic modifiers) in addition to coding mutations can affect the underlying pathogenesis of this disorder [28]. In contrast to Creutzfeldt–Jakob disease, Gerstmann-Sträussler-Scheinker syndrome can have an extended disease course, lasting up to 6 years. The eventual outcome, however, is inevitable death.

4.2.4 KURU

Kuru (Fig. 4.2) is characterized by a combination of ataxia and prominent tremor, coupled with dysarthria, dysphagia, and incontinence [29]. These symptoms progress to akinetic mutism and death, often from malnutrition, after a disease course of between 1 and 2 years. It was found exclusively in a

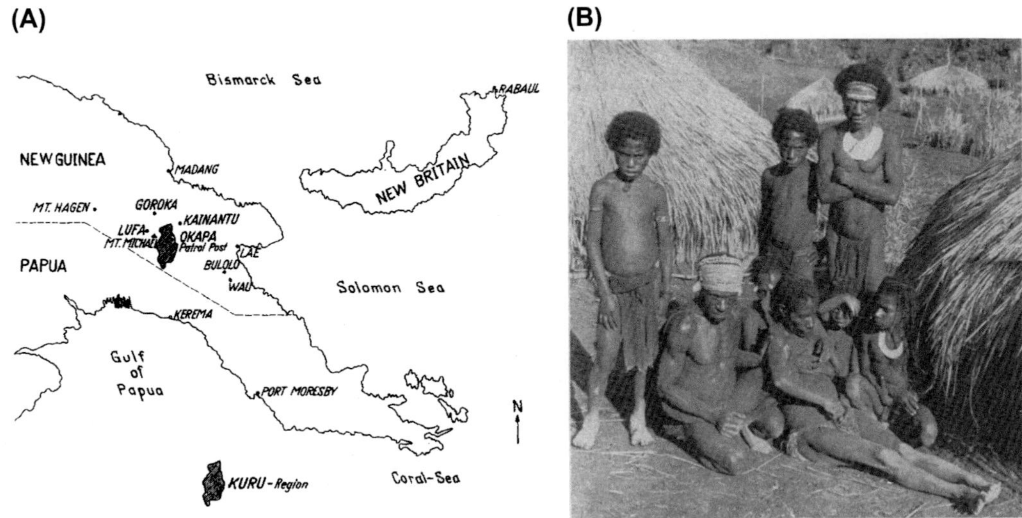

FIGURE 4.2

Kuru: (A) a map of Papua New Guinea showing the Fore region and (B) a woman in the late stages of Kuru, supported by her family.

Images taken from Gajdusek DC, Zigas V. Kuru; clinical, pathological and epidemiological study of an acute progressive degenerative disease of the central nervous system among natives of the Eastern Highlands of New Guinea. Am J Med 1959;26:442–469.

geographically limited region of the highlands of Papua New Guinea inhabited by the Fore people and is caused by exposure to infectious material from individuals with the disease. This exposure occurred during ritual endocannibalistic mortuary feasts that were a traditional part of Fore life [30]. In contrast to Creutzfeldt–Jakob disease, kuru does not present with prominent dementia.

4.2.5 FATAL FAMILIAL INSOMNIA

Fatal familial insomnia provides a key example of the phenotypic variability of the prion diseases. It is a genetic disorder caused by mutations in the *PRNP* gene and presents with altered sleep/wake cycles leading to insomnia, hallucinations, and myoclonus [31,32]. The average age at onset is 49 years, but similar to Gerstmann-Sträussler-Scheinker syndrome, there is a wide range of age of onset. The disease course is, on average, 12 months, but again a wide range is observed [28]. Similar to kuru, there is not a prominent impact on cognition. Intriguingly, the clinical presentation of individuals with the D178N mutation is partially determined by an additional polymorphism in the *PRNP* gene, an M129V variant (see Section 4.4.4).

4.2.6 VARIANT CREUTZFELDT–JAKOB DISEASE

The first cases of variant Creutzfeldt–Jakob disease were described in the early 1990s following the advent of the BSE epidemic in cows during the late 1980s [33]. While sharing a similar clinical presentation with Creutzfeldt–Jakob disease, cases with what came to be known as variant Creutzfeldt–Jakob disease displayed an earlier age at onset (in the third and fourth decades of life), an extended disease course compared to the orthodox form of the disorder (over a year, extending beyond 14 months), and more prominent psychiatric involvement in the early stages of the disorder [34]. As discussed in further detail below (Section 4.4.7), the pathogenesis of variant Creutzfeldt–Jakob disease is closely linked to exposure to BSE [35,36].

4.3 PATHOLOGY

The prion disorders are characterized primarily by the development of widespread spongiform degeneration in the central nervous system, leading to the grouping of these diseases under the term transmissible spongiform encephalopathies following the demonstration that they could be transmitted horizontally from individual to individual (or from organism to organism) [37]. This spongiform degeneration reflects massive cell loss in the brain, leading to the formation of large vacuoles that appear like sponge at a macroscopic level. There is widespread astrogliosis, with activated glial fibrillary acidic protein stain positive astrocytes colocalizing with areas of spongiform degeneration. After the identification of the PrP as a central component of the disease process in prion disorders, immunohistochemical analysis of brain tissue from animals and humans afflicted by prion diseases revealed extensive deposition of aggregated PrP in the brain. In some cases, most notably in individuals with Gerstmann-Sträussler-Scheinker syndrome, this deposition takes the form of extracellular amyloid plaques sharing some characteristics with those observed in Alzheimer-type dementia. The neuropathological findings observed in prion disorders is summarized in Fig. 4.3. Molecular analysis of the pathological forms of PrP found in the brains of prion disease cases has provided a detailed description of a range of different protease-resistant aggregated isoforms of the PrP; forms that can be discriminated by using immunoblot analysis following separation using SDS (sodium dodecyl sulfate) gel electrophoresis (for a more detailed description of strain type analysis see Section 4.4.8) [39]. The development of animal bioassays for infectivity has led to the detailed analysis of the distribution of tissue infectivity. In Creutzfeldt–Jakob disease, this infectivity is concentrated in the central nervous system. Some forms of prion diseases, for example, variant Creutzfeldt–Jakob disease, display a wider distribution of infectivity, with prominent involvement of the lymphoid system (this is also observed in chronic wasting disease) [40]. This has led to the use of tonsil biopsy to identify the presence of protease-resistant PrP, using this as a proxy for the presence of disease. The involvement of tissues outside the central nervous system has raised significant public health concerns over the potential for infection to be passed on by, for example, contaminated blood [41].

4.4 MOLECULAR MECHANISMS OF DEGENERATION

The mechanisms underpinning degeneration in the prion diseases have been an enduring source of fascination and puzzlement for over a century (Fig. 4.4) [42]. Efforts to understand how what we now recognize as prion diseases are caused start with studies of scrapie in the early 20th century. For many years, it had been noted that scrapie was contagious and that it appeared to be transmitted from sheep to sheep. The first experimental demonstration that this could occur by inoculation occurred during the 1920s in a series of experiments carried out by Jean Cuillé and Paul-Louis Chelle in France [43,44].

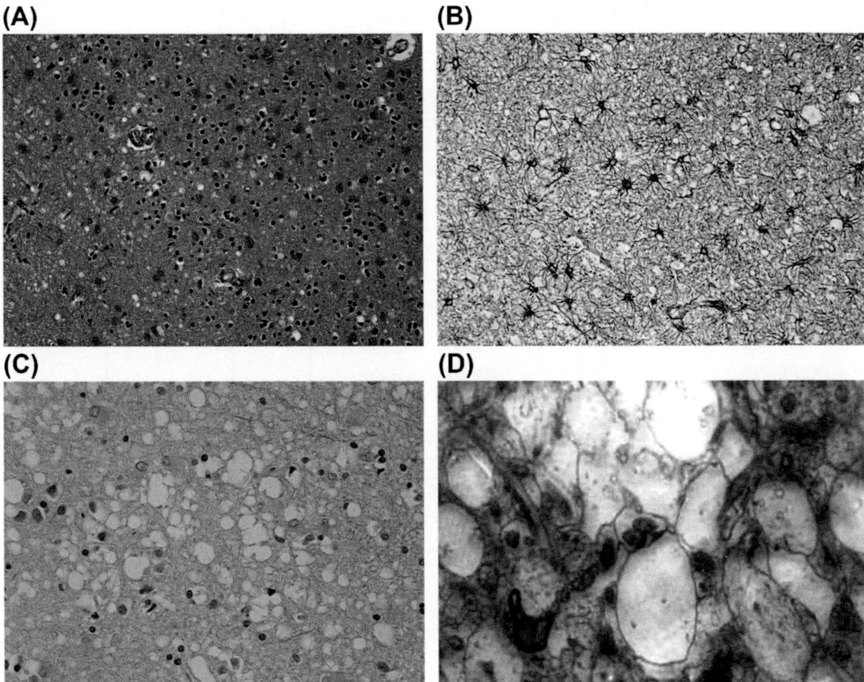

FIGURE 4.3

The pathology of prion disease. (A) Astrocytic gliosis with spongiform vacuolation in the occipital cortex of a sporadic Creutzfeldt–Jakob disease (hematoxylin-eosin [H&E], original magnification ×200). (B) Astrogliosis in the occipital cortical of a patient with sporadic Creutzfeldt–Jakob disease highlighted by glial fibrillary acidic protein (GFAP) immunostaining (original magnification ×200). (C) Severe spongiform change in occipital cortex of a patient with sporadic Creutzfeldt–Jakob disease (H&E, original magnification ×400). (D) Electron micrograph showing spongiform change with vacuolation and swelling of neuritic processes.

Courtesy of *the United Kingdom National CJD Surveillance Unit, Edinburgh, UK. Images taken from du Plessis DG. Prion protein disease and neuropathology of prion disease. Neuroimaging Clin N Am 2008;18: 163–182; ix.*

At this point, however, it had not been noted that the clinical and pathological presentation of scrapie shared key aspects with human diseases such as Creutzfeldt–Jakob disease. This link was not to be made until much later in the century.

4.4.1 KURU

The first key event in unraveling the cause of human prion diseases came from an unlikely source, namely, the mountainous interior of Papua New Guinea. Tribes in this region had been isolated from contact with the outside world for centuries, but the advent of the Second World War opened up the highlands to outsiders. This revealed a fascinating, if somewhat macabre, medical story. In the Fore tribe, living in the steep-sided valleys around the town of Goroka, an unexplained neurological disease had reached epidemic proportions. This disease, called kuru by the Fore (kuru meaning to shake

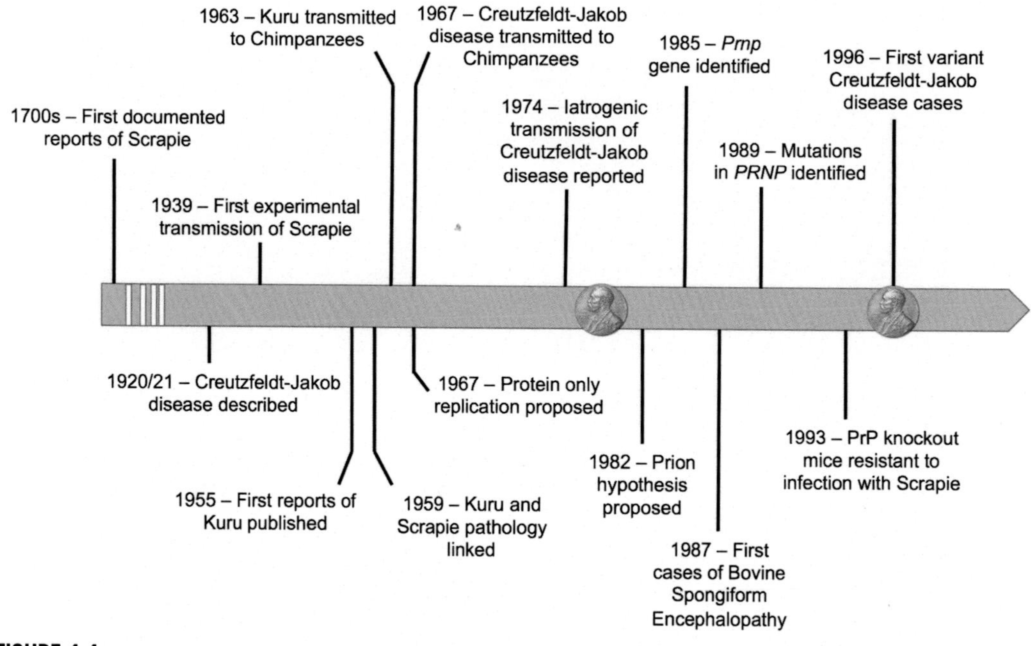

FIGURE 4.4

Timeline for prion disease discovery, including the award of Nobel prizes to Carleton Gajdusek and Stanley Prusiner in 1976 and 1997 respectively. *PrP*, prion protein.

or tremble), had a profound impact on movement and cognition, resulting in incapacitation and akinetic mutism after a disease course of between 12 and 24 months [29]. Death followed, frequently from starvation due to loss of the ability to swallow (for a more detailed description of the symptoms of kuru see Section 4.2.4 and Fig. 4.2). Carleton Gajdusek, a neurologist from the United States, studied the disorder in great detail during field work in Papua New Guinea and published his observations along with those of his colleague Vincent Zigas in a series of seminal papers on the subject. Importantly, he noted that the epidemiology of kuru possessed some significant peculiarities. Those afflicted by the disease were disproportionally made up of females (of all ages) and young males, with very few older males. At this point, there was no understanding of how kuru was spreading through the Fore population, so Gajdusek examined whether there were specific circumstances in the lifestyle of the Fore that might explain this phenomenon. Here, an almost unique aspect of Fore life came into play. The Fore practiced a particular type of endocannibalism, that is, cannibalism practiced on members of your own group or community [30]. Upon the death of a close relative, the females and preadult males and females in the family group would eat the flesh of that relative in a ritualized mortuary feast. Given the pattern of disease incidence focused on these individuals, this suggested to Gajdusek that mortuary feasts could play a key part in the transmission of the causative agent of kuru and he set about testing whether this was the case. Intriguingly, just prior to this, William Hadlow at the UK government Agricultural Research Station in Newbury noted that there were close similarities between the neuropathological findings observed in kuru and those found in the brains of sheep that were afflicted with scrapie, bringing to mind previous experiments demonstrating the transmissibility of the scrapie-causing agent [45]. To test whether kuru could be directly transmitted in a way analogous to that observed in scrapie, Gajdusek carried out a series of inoculation experiments using nonhuman primates in his laboratories at the National Institutes of Health in Bethesda, Maryland. About 18–30 months following injection with kuru-causing material, a subset of chimpanzees developed a disorder matching the clinical and pathological profile of the disease, providing the first experimental demonstration that kuru could be directly transmitted [46]. In a further set of experiments, Gajdusek and his colleagues, in collaboration with clinicians from the Institute of Psychiatry in London, tested whether the Creutzfeldt–Jakob disease (a hitherto underinvestigated disorder that, similar to scrapie, shared some important characteristics with kuru) could also be transmitted. In a study published in Nature in 1968, the team reported that Creutzfeldt–Jakob disease was also experimentally transmissible to chimpanzees, bringing together scrapie, kuru, and Creutzfeldt–Jakob disease for the first time as a family of related disorders termed the transmissible spongiform encephalopathies [47,48].

4.4.2 SLOW VIRUSES AND OTHER THEORIES

Gajdusek's research during the 1950s and 1960s demonstrated that a number of rare neurodegenerative diseases of humans could be transmitted, but it did not reveal the precise mechanisms of transmission or the nature of the causative agent. The stated explanation for the capacity of the transmissible spongiform encephalopathies to be passed from individual to individual was that they were caused by an as yet unidentified virus with a long incubation and/or prodromal disease stage. The term slow virus, first proposed in relation to scrapie, was adopted to describe these disorders [49]. This virus, however, defied identification despite significant efforts by Gajdusek and many other researchers worldwide. Indeed, a number of experimental observations seemed to be inconsistent with a viral origin for the transmissible spongiform encephalopathies. Work carried out by Tikvah Alper and coworkers [50] during the 1960s in London and Newbury, following the development of a mouse adapted form of scrapie that could be transmitted in an experimental setting with a predictable disease course, suggested that the scrapie agent was exceptionally small. In addition to this, it appeared to be resistant to treatments that would inactivate a disease agent containing nucleic acid (including high levels of ultraviolet radiation), an observation that appeared to preclude scrapie being caused by a viral entity [51]. One possibility, proposed in 1967 by John Griffith and entertained by Pattison and Jones following their work on scrapie in the same year, was that scrapie was caused by a self-replicating protein [52,53]. Despite being consistent with the data from Alper and colleagues, experimental proof was lacking and the theory ran counter to the prevailing dogmas surrounding the nature of life.

Although unable to positively identify the causative agent of kuru and Creutzfeldt–Jakob disease, Gajdusek was awarded the Nobel Prize in 1976 for his work on the transmissible spongiform encephalopathies [48] (Box 4.1). For the Fore people, the impact of kuru lessened as the mortuary feasts that had propagated the infection were phased out following increased exposure to the outside world. As a postscript to Carleton Gajdusek's career and contribution to this field, he was convicted of child molestation in 1996 (relating to offences committed upon children that he had brought back to the United States from the Pacific) and sentenced to 18 months in jail [54].

BOX 4.1 NOBEL PRIZES

Prion diseases are notable for being associated with not one but two Nobel Prizes. Carleton Gajdusek was awarded the Nobel Prize for Physiology or Medicine in 1976 for his work to uncover the cause of kuru (and subsequent experiments demonstrating that Creutzfeldt–Jakob disease could be transmitted in a similar fashion). Stan Prusiner received the accolade two decades later for his statement of the prion hypothesis and his work to test its veracity.

Photographs taken from Poser CM. Notes on the history of the prion diseases. Part I. Clin Neurol Neurosurg 2002;104:1–9.

4.4.3 THE PRION HYPOTHESIS

The next major step forward in our understanding of these disorders came about through assiduous experimental studies led by another neurologist, Stan Prusiner, working at the University of California, San Francisco. He was fascinated by the nature of the scrapie agent and worked to improve and optimize existing animal models for the transmissible spongiform encephalopathies. This work focused on Syrian golden hamsters as a model for the transmission of scrapie and provided an extremely sensitive bioassay for the level of infectivity in an isolated tissue sample [56,57]. Prusiner then set out to establish biochemical and biophysical parameters for the scrapie agent by using fractionation of infectious tissue coupled with treatment using a range of chemical and radiological methods. These investigations confirmed and extended the earlier work of Tikvah Alper and prompted Prusiner to put forth a formal statement of a hypothesis for the cause of transmissible spongiform encephalopathies. In an article in Science in 1982, he proposed that these disorders were caused by a protein capable of limited self-replication, echoing Griffith's suggestion from 1967 [58]. The key difference was that Prusiner was able to refer to a substantial body of work supporting this hypothesis (summarized in Fig. 4.5). He also coined a neologism to capture the concept, calling these self-replicating proteins "prions," derived (with a certain degree of vowel juggling) from protein and infectious. Prusiner even provided a guide to pronunciation: "pree-on." This hypothesis, described almost universally as the prion hypothesis, proved controversial. The concept that a protein could, independent of nucleic acids, act as a causative transmissible agent of disease and possess the ability to self-perpetuate was somewhat difficult to reconcile with the central dogma of molecular biology as laid out by Francis Crick a decade or more previously, concerning the flow of information in biological systems (Fig. 4.6A) [59]. This was not the first modification of the central dogma. Data had emerged in the 1960s and 1970s that RNA could self-replicate [60], and in an interesting quirk of science, the central dogma was being challenged by another set of discoveries in the early 1970s and 1980s that centered around the identification of RNA viruses that could reverse-transcribe their genomes into DNA [61]. These retroviruses, including the agents

Properties of the Scrapie Agent

Resistant to treatment with ribonucleases and deoxyribonucleases, UV at 254nm (8 times more resitant than most resistant virus), Zn^{2+} catalysed hydrolysis, psorelen photoadduct formation and NaOH chemical modification

Susceptible to protein denaturants and proteolysis

Stable at 90°C for 30 minutes

Molecular mass < 50,000 daltons

Hydrophobic protein required for infectivity

FIGURE 4.5

Prusiner's arguments for a protein-based mechanism of transmission (derived from reference [58]).

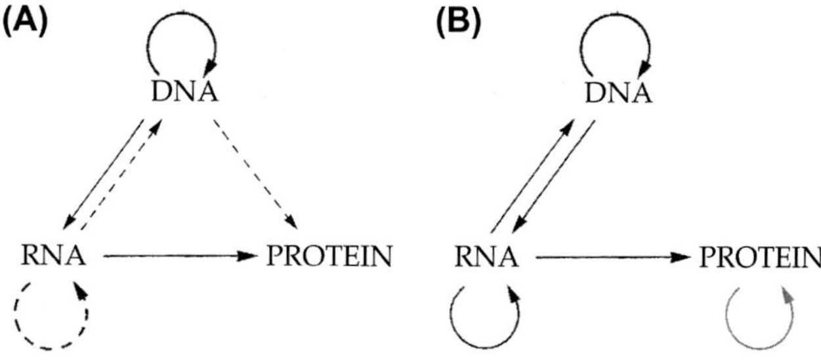

FIGURE 4.6

The central dogma of molecular biology (A) before and (B) after the identification of RNA viruses, retroviruses, and prions.

Images modified from Crick F. Central dogma of molecular biology. Nature 1970;227:561–563.

that cause the acquired immune deficiency syndrome, namely, human immunodeficiency viruses 1 and 2, required the modification of the central dogma, which as originally stated viewed information flowing from DNA to RNA and not in the other direction. The prion hypothesis suggested that another aspect of the central dogma was malleable and that in addition to DNA being able to self-replicate, certain proteins, under the right conditions, could also pass on (or replicate) information relating to structure (Fig. 4.6B). Despite resistance to the hypothesis, Prusiner continued to work in this area and identified the protein in question in 1983 [62,63]. This led to the identification, by a Swiss group led by Charles Weissmann and Bruce Chesebro and colleagues at the NIH Rocky Mountain Laboratories, of first a rodent gene and, subsequently, a human gene (*PRNP*) encoding this protein [64–67]. These findings had significant implications for the prion hypothesis because they demonstrated that the misfolded disease-associated PrP (dubbed PrP^{Sc} for prion protein scrapie form) had a non-disease-associated endogenous counterpart (dubbed PrP^{C} for prion protein cellular form). The possibility thus emerged that the prion diseases could be developing by PrP^{Sc} acting upon PrP^{C} to propagate itself and thus spread from cell to cell and from organism to organism.

4.4.4 INHERITED PRION DISEASE

The identification of a gene encoding the protein involved in the prion diseases led to an important breakthrough in one critical aspect of the human disorders. For many years, it has been recognized that a small number of prion/transmissible spongiform encephalopathy cases occurred in a stereotypical mendelian fashion, passed from one generation to the next. With the identification of a prion gene, mutations in *PRNP* were an obvious candidate for these familial cases, and indeed, mutations in the prion gene were identified in a number of families with Creutzfeldt–Jakob disease, Gerstmann-Sträussler-Scheinker syndrome, and fatal familial insomnia (Fig. 4.7) [27,32,68]. These mutations include point mutations that alter the amino acid sequence of PrP, as well as alterations in an octopeptide repeat sequence found in the N-terminal region of the protein (resulting in an increased number of repeats). The genetics of human prion disorders made a major contribution to the generation of mouse models for inherited forms of these disorders, initially the P102L Gerstmann-Sträussler-Scheinker syndrome–associated mutation but soon followed by a number of other coding variants [69]. The role of polymorphisms in the *PRNP* gene with regard to both susceptibility to prion infection and the precise phenotype of inherited disease has been the subject of great interest. A methionine/valine polymorphism at codon 129 in the human PrP gene plays a key role in both these functions. For variant Creutzfeldt–Jakob disease, the majority of cases have been homozygous for methionine at codon 129, with a limited number of recent exceptions to this [70–72]. The status of codon 129 has been noted to have an impact on the clinical presentation of individuals with mutations at codon 178 (a D178N mutation), in addition to determining the onset of either Creutzfeldt–Jakob disease or fatal familial insomnia [73]. The mechanisms underpinning these phenomena are unclear, although the codon 129 polymorphism has been shown to affect the kinetics of amyloid formation by PrP [74]. Analysis of the Fore population exposed to the kuru-causing agent has revealed an additional polymorphism at codon 127, G127V, that is thought to have provided protection against infection with kuru [75,76].

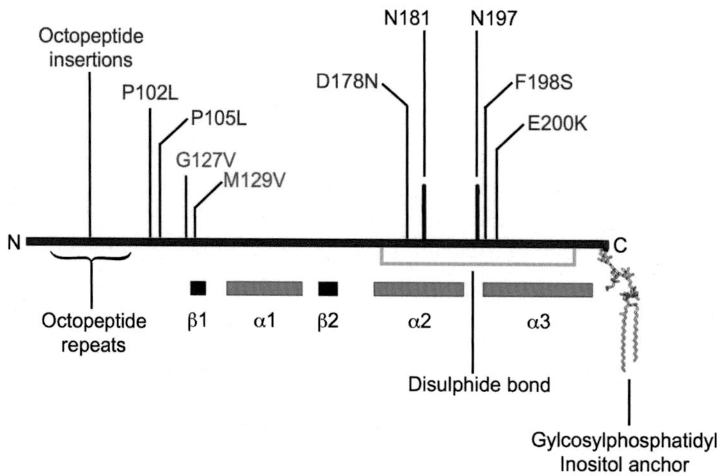

FIGURE 4.7

Secondary structure, posttranslational modification (including glycosylation and addition of a GPI anchor), and genetic variants linked to disease in the prion protein (mutations causative for disease in red, disease modifying polymorphisms in green).

4.4.5 IATROGENIC/TRANSMISSIBLE DISEASES

Further evidence relating to the transmission of prion diseases accumulated throughout the 1980s. In addition to kuru, there have been a number of carefully documented cases where there is clear evidence that human prion diseases have been transmitted horizontally, from person to person iatrogenically (i.e., due to medical procedures). These include cases where individuals who had a prion disease died and donated tissue (e.g., dura mater grafts or corneal material) [77,78] or had human growth hormone extracted post mortem for subsequent use in individuals unable to synthesize their own [79]. In a small number of cases, this resulted in the direct transmission of Creutzfeldt–Jakob disease, highlighting the potential public health risks posed by the prion diseases—a risk further emphasized by the possibility that variant Creutzfeldt–Jakob disease could be transmitted via contaminated blood products [41,80]. Parallels have been drawn between the use of donated human tissue and organs and endocannibalistic practices such as those observed in the Fore people [81]. Returning once again to the underlying genetics of prion diseases, analysis of *PRNP* polymorphisms in patients with iatrogenic Creutzfeldt–Jakob disease has again revealed a key role for the codon 129 variant in susceptibility to the disease [82].

4.4.6 FALSIFYING THE HYPOTHESIS

The use of animal models has been of critical importance throughout the history of investigations into the biology of the prion diseases, and nowhere more so than in testing the prion hypothesis. The identification of a host gene encoding the PrP provided information on the mechanisms underlying the prion diseases. First, transgenic approaches developed in the 1980s made it possible to model the impact of mutations in the prion gene on rodents, with the first transgenic mouse carrying a Gerstmann-Sträussler-Scheinker syndrome–associated point mutation reported in 1989 [69]. Second, knockout gene engineering technology opened up the possibility of completely removing the prion gene [83]. This allowed scientists to test a putative physiological role for the PrP and (equally as importantly) to test whether the host prion gene was a requirement for prion replication and thence disease. In all three areas of investigation, important insights were revealed. First, the expression of a mutant transgene resulted in mice developing neurodegeneration, thus strongly supporting a direct mechanistic link between coding variants in the PrP and the initiation of disease (although not directly proving the prion hypothesis, as it remained possible that the coding variants acted to facilitate disease by an unidentified exogenous infectious agent) [69]. Knockout of the PrP, again in mice, by the Weissmann group resulted in a surprisingly mild phenotype, suggesting that whatever the physiological role of the PrP was it could be dispensed with for most aspects of life [83]. Indeed, the cellular role of the PrP is still not understood. Finally, and taking advantage of the PrP knockout mice, the role of the endogenous protein in the disease process was tested by inoculating these mice with a mouse-adapted scrapie strain. Removing endogenous PrP^C resulted in mice being completely resistant to the disease following inoculation with scrapie [84], and expression of PrP was required for the propagation of infectious prions [85]. This did not prove that the prion hypothesis was correct (there were a number of possibilities as to why PrP-null mice were resistant to infection), but it was an important piece of the puzzle relating to what was happening in the prion diseases. Importantly, had the mice been susceptible to infection, this would have been a serious (although not fatal) blow to the prion hypothesis.

The development and refinement of the prion hypothesis by Stan Prusiner resulted in the award of the second Nobel Prize in relation to the transmissible encephalopathies, which was presented to Professor Prusiner in 1997 (Box 4.1) [86].

4.4.7 BOVINE SPONGIFORM ENCEPHALOPATHY AND VARIANT CREUTZFELDT–JAKOB DISEASE

Up until the end of the 1980s, the transmissible spongiform encephalopathies were a very rare set of human diseases, of greatest note for their exotic properties, and not considered a major public health issue. This began to change with the identification of a novel animal transmissible spongiform encephalopathy afflicting cows, BSE, colloquially known as Mad Cow disease. The first cases of this were identified in cattle in the United Kingdom in 1986, and the disease soon reached epidemic proportions with thousands of cases in a month being identified by the early 1990s (Fig. 4.8) [87]. Although it is still not clear exactly how the BSE epidemic started, a number of possibilities exist. It is possible that scrapie had been transmitted to cows—a possibility that was deemed unlikely given that cows and sheep had coexisted for many centuries with scrapie present in the ovine population, without any evidence for transmission. It is also possible that a mutation appeared in the UK bovine population that generated an initial case and that material from this mutated cow was subsequently transmitted to other cows. A further possibility is that a sporadic case of the disease occurred (similar to what has been observed in the human population) based on a stochastic misfolding event and that this was then transmitted among the UK cattle stock [87,88].

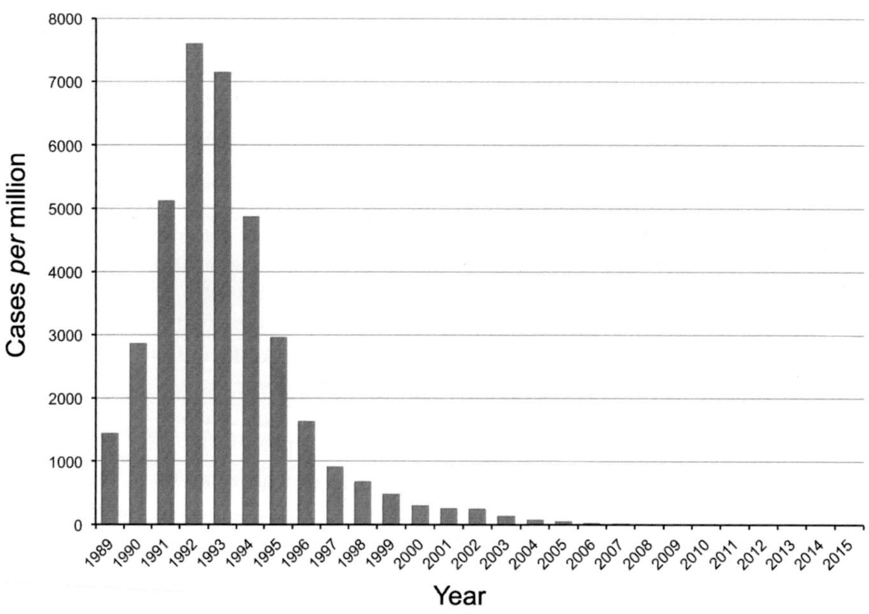

FIGURE 4.8

The incidence of bovine spongiform encephalopathy in the United Kingdom from 1989 to 2015.

Data from the World Organisation for Animal Health (www.oie.int).

What is clear is that farming practices in the 1980s turned a potentially containable problem into a serious crisis for the farming sector. At the time, it was common practice for the parts of cattle slaughtered for their meat deemed not suitable for human consumption to be recycled into meat and bone meal, a protein-rich food source [89]. This was fed back to other cattle to supplement their diet, a practice reminiscent and perhaps analogous to the cannibalistic funeral traditions used by the Fore tribe in the 1950s. It is highly likely that the reuse of bovine material in this fashion resulted in the rapid spread of the BSE during the late 1980s/early 1990s. Efforts to eradicate the disease eventually resulted in millions of cattle being slaughtered, with a single confirmed case in a herd being enough to result in the whole herd being destroyed. This slowed and then decreased the incidence of the disorder, with new cases reducing month on month from 1992 [87]. Although not eradicated, BSE was no longer an existential threat to the UK cattle industry.

The advent of BSE also raised substantial concerns relating to human health, as it was not clear if the disease could be transmitted to humans. Initial advice from the UK government considered the risk to humans as being small, based primarily on the lack of evidence that scrapie could, or ever had been, be transmitted to humans [90]. This changed with the identification of new cases of Creutzfeldt–Jakob disease that did not fit easily with established symptomatic profiles for the disease. This new variant, or variant, form of Creutzfeldt–Jakob disease was first identified in 1996 [33]. As the decade progressed, evidence accumulated that this was a distinct form of Creutzfeldt–Jakob disease and that it was directly linked to exposure to the BSE-causing agent. This evidence took a number of guises, including the identification of a molecular signature (or strain type, see Section 4.4.8 for a more detailed explanation of this phenomenon) of the protein aggregates found in the brains of patients with variant Creutzfeldt–Jakob disease that matched those found in the brains of cows infected with BSE [35,91,92]. Further research with transgenic mice demonstrated that mice expressing the human form of the PrP were susceptible to bovine prions [36,93] and that the pattern and molecular characteristics of pathology were strikingly similar between BSE and variant Creutzfeldt–Jakob disease [94]. The incidence of variant Creutzfeldt–Jakob disease increased from a handful in 1995 to a peak of 28 in 2000, with the number of new cases then decreasing annually to five or fewer from 2005 [95]. The reason for this spike in incidence and subsequent decrease is not clear. It is possible that the similar pattern of incidence between BSE and variant Creutzfeldt–Jakob disease represents the totality of transmission from cows to humans and that the species barrier between the two meant that only a subset of individuals were susceptible to disease. It is also possible that this pattern represents an initial cohort of disease in humans, with the most susceptible individuals succumbing after a short incubation period, and that there are a number of other individuals in the UK population who are incubating the disease but have not yet developed symptoms [96]. Given the large scale of potential exposure to bovine prions in the United Kingdom, this is a matter of great importance for public health; however, it is not possible at present to be certain which scenario is correct. Genetic data highlighting that the majority of cases of new variant Creutzfeldt–Jakob disease shared a similar genotype (homozygous for a common polymorphism in the prion gene, methionine at codon 129) certainly supports the idea that the variant Creutzfeldt–Jakob disease cases to date represent a susceptible subpopulation, and there is data to suggest that there may be a number of people incubating the disorder either asymptomatically or as a prodromal stage of the disease. Whether these individuals will develop variant Creutzfeldt–Jakob disease will only become clear in the fullness of time. Following the realization that it was highly likely that BSE had been transmitted to humans, the UK government set up a stringent reporting system for all forms of human prion disease. This has provided detailed data relating to the incidence of inherited prion disease, sporadic forms, iatrogenic prion disease, and variant Creutzfeldt–Jakob disease (Fig. 4.9) [95]. It is noteworthy that the incidence of

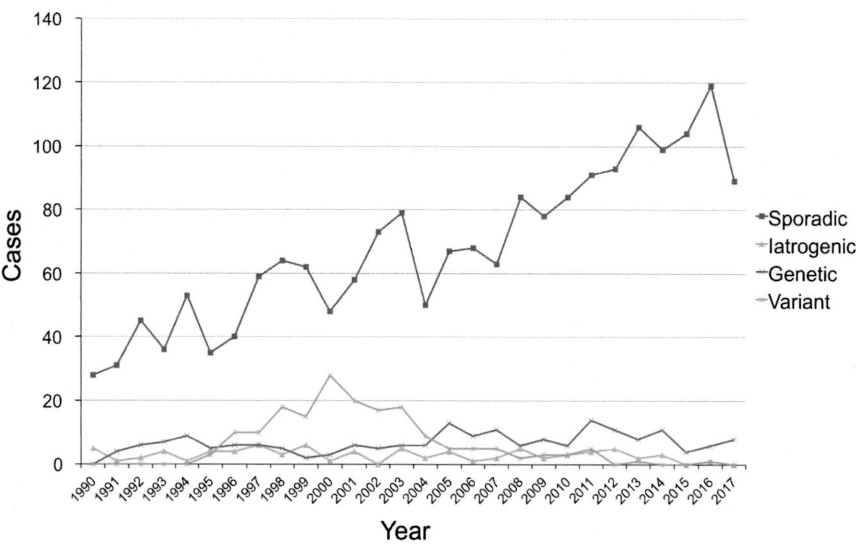

FIGURE 4.9

The occurrence of human prion disease in the United Kingdom from 1990 to 2017.

Data from the CJD surveillance unit (www.cjd.ed.ac.uk).

sporadic Creutzfeldt–Jakob disease has steadily increased from 1990, when the first reliable figures were collected (from 28 cases in 1990 to over 100 in 2013). Again, it is not clear what underlies this increase, although it is likely that improved recognition of Creutzfeldt–Jakob disease under the heightened reporting system explains at least some of this.

One further observation can be made relating to the incidence of variant Creutzfeldt–Jakob disease. The only previous outbreak of a prion disease with which we can compare the transmission of bovine prions to humans is that of the kuru epidemic. In the case of kuru transmission, the last documented exposures due to mortuary feasts occurred in the 1950s prior to the eradication of cannibalism. Despite this, individuals continued to develop disease into the 21st century, suggesting that transmitted human prion disease may have an exceptionally long presymptomatic period [97]. Although the data is much more limited for iatrogenic Creutzfeldt–Jakob disease, there is some evidence that this may also be the case for the human growth hormone–linked disease [98]. We do not yet know if this is also the case for variant Creutzfeldt–Jakob disease.

4.4.8 STRUCTURAL AND MOLECULAR BASES OF PRION TRANSMISSION

The identification of the PrP and the *PRNP* gene provided the opportunity to examine directly the structural and molecular basis of these disorders. In work that paralleled research in a number of other neurological diseases linked to protein aggregation (most notably Alzheimer disease), a number of groups began to investigate the characteristics of the agent responsible for prion disease. The PrP itself is a 253-amino-acid, 26-kDa protein expressed throughout the body (Fig. 4.7 and 4.10A). In its cellular form, PrP^{C}, it is predominantly alpha helical and the subject of a number of posttranslational modifications including glycosylation of asparagine located at codons 181 and 197 and the attachment of a

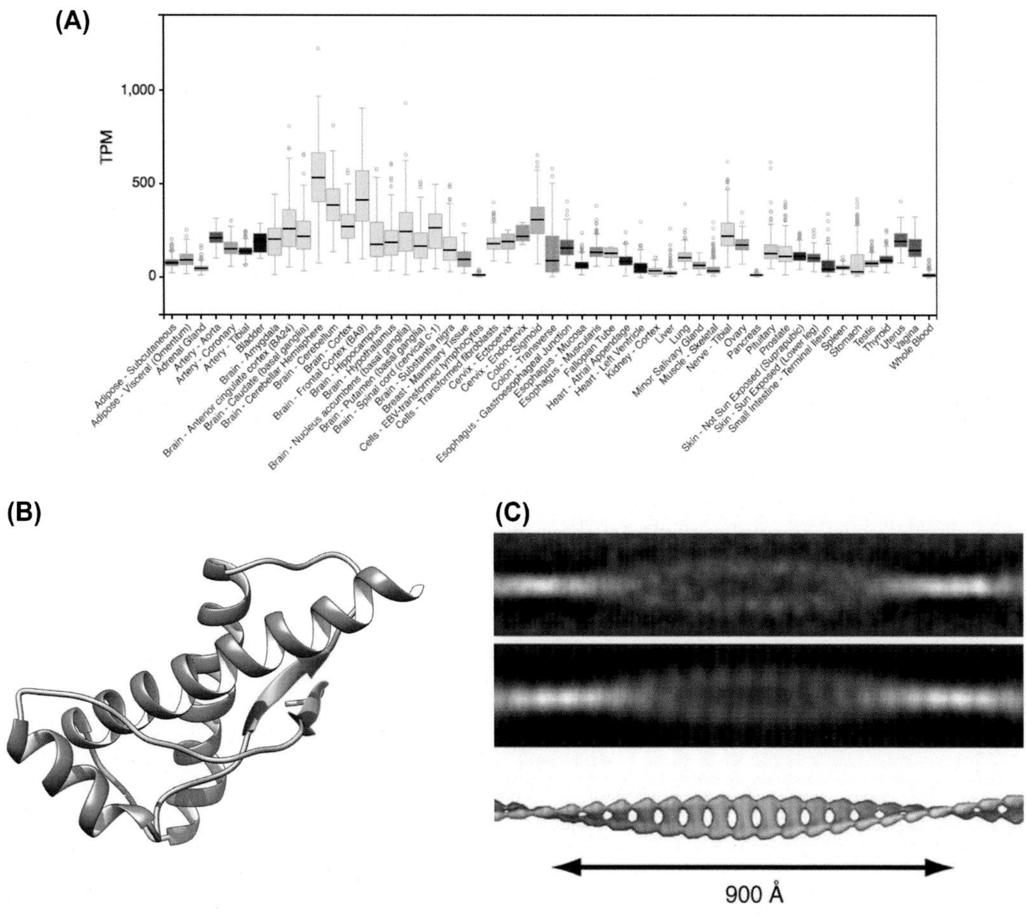

FIGURE 4.10

(A) Prion protein expression in human tissue. (B) Nuclear magnetic resonance structure for the human prion protein. (C) Electron micrograph of fibril structure formed from recombinant prion protein.

(A) Data from GTeX (www.gtexportal.org). (B) Based on coordinates from PDB file 1qlx (reference 98). (C) Image taken from Tattum MH, Cohen-Krausz S, Thumanu K, Wharton CW, Khalili-Shirazi A, Jackson GS, Orlova EV, Collinge J, Clarke AR, Saibil HR. Elongated oligomers assemble into mammalian PrP amyloid fibrils. J Mol Biol 2006;357:975–85.

glycosylphosphatidylinositol anchor (Fig. 4.10B) [99–101]. The latter acts to attach the PrP to the extracellular envelope of the plasma membrane. The cellular function of PrP^{C} remains obscure. As noted earlier, PrP knockout mice are viable and do not present with an obvious phenotype, suggesting that there is a degree of redundancy to PrP function, at least in rodents.

A great deal of effort has gone in to trying to gain structural information relating to the pathogenic PrP^{Sc} form of the PrP. This includes analysis of protein extracted from the brains of individuals and animal models with prion disease, experiments that suggest that the aggregated form of PrP is predominantly in the beta pleated sheet conformation based on diffraction and spectroscopic studies [102–104]. Although Ångstrom resolution data relating to the molecular structure of the disease-associated form of PrP has not yet been forthcoming, data from other disorders as well as lower resolution analysis of fibrils made up of recombinant protein suggest that PrP^{Sc} forms an ordered polymer structure (Fig. 4.10C) [102,105,106].

The ability to produce recombinant PrP opened up the possibility of undertaking a further test of the prion hypothesis. If prion replication was truly protein only, it should be possible to replicate this process with artificially generated recombinant protein. Evidence that this could be achieved was published in 2005 [107] and replicated by an independent group [108], although the challenges of replicating in vitro generated prions and characterizing these remain significant [109].

Insights into the protein biochemistry of PrP, in particular the generation of antibodies specific for PrP, have allowed detailed analysis of protein aggregates from both model systems and clinical specimens [110]. These revealed that the PrP^C and PrP^{Sc} forms of the PrP can be differentiated at a biochemical level by a number of molecular features, including their protease resistance. The most prevalent technique applied to uncover this is treatment with proteinase K, a relatively nonspecific protease that can cleave at multiple points on a given polypeptide [63]. PrP^C is sensitive to proteinase K and is completely degraded, whereas PrP^{Sc} is resistant to proteinase K and will persist as a protease-resistant core (Fig. 4.11A). The glycosylation states of both PrP^C and PrP^{Sc} yield a characteristic triplet banding pattern when analyzed by immunoblot, representing unglycosylated protein,

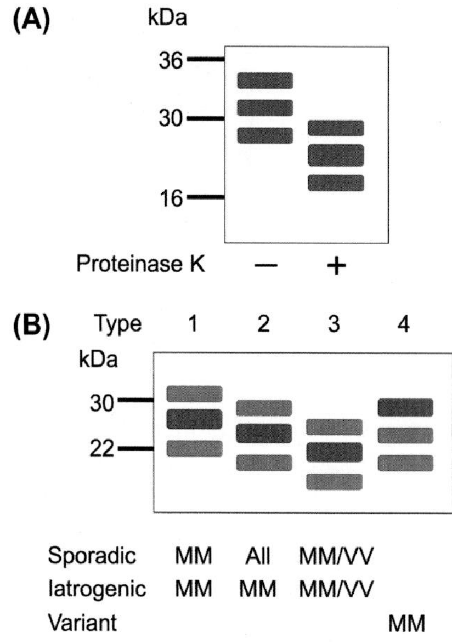

FIGURE 4.11

Prion strain typing by immunoblotting. (A) Cartoon representation of PrP^{Sc} showing the three major glycoforms of PrP (upper band, diglycosylated; middle band, monoglycosylated; and lower band, unglycosylated) and the impact of proteinase K digestion. (B) Human Creutzfeldt–Jakob disease strain types following proteinase K digestion and immunoblot, showing reported codon 129 status.

Image adapted from Hill AF, Collinge J. Subclinical prion infection. Trends Microbiol 2003;11:578–584; Jackson GS, Collinge J. The molecular pathology of CJD: old and new variants. Mol Pathol 2001;54:393–399.

monoglycosylated protein, and diglycosylated protein [113]. A fascinating aspect of prion disease biology, uncovered by carrying out immunoblot analysis of tissue samples from different prion disorders, is that the precise ratio of these glycoforms (and the apparent molecular masses of the protease-resistant fragments remaining after digestion with proteinase K) is surprisingly distinct between different clinical prion disease entities [113,114]. This is exemplified by the different glycoform ratios present in human prion disease (Fig. 4.11B). There are a number of competing classification systems for these different protease-resistant forms, but these differences in biochemical properties of PrP^{Sc} act as a proxy for the underlying differences in the pathogenic properties of the protein aggregates in different forms, or strains, of the prion disorders [115,116]. This has led to a refinement of the prion hypothesis, encompassing conformational strain types as a mechanism for propagation of specific neurological disease states [117]. This provided an explanation for the existence of multiple naturally occurring strains of disease, an observation that dated back to studies of scrapie during the 1950s and 1960s [118]. This propagation of a specific conformation of PrP^{Sc} is underpinned by a number of models that describe precisely how an aggregated protein can self-perpetuate a conformation under the right circumstances (Fig. 4.12). In many ways, these derive from the models proposed in the 1960s to account for the possibility of protein-only transmission, although the impact of several decades of molecular investigations has led to an increasing level of sophistication [120].

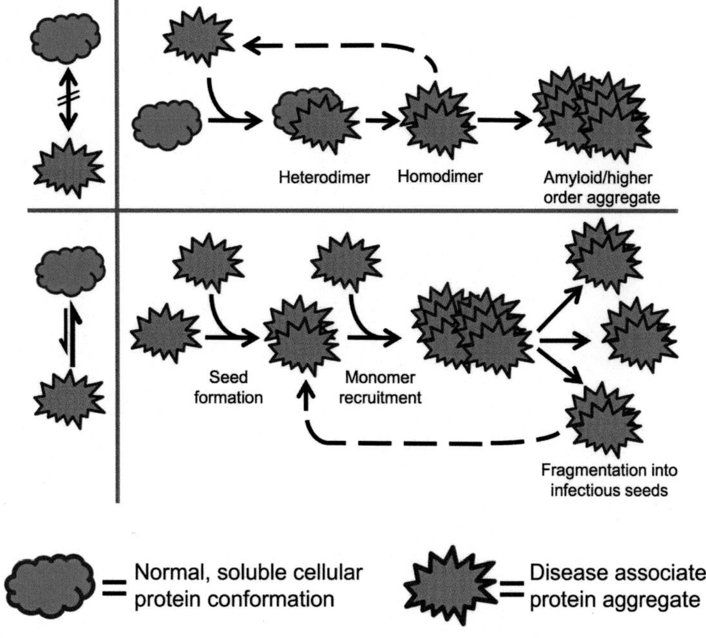

FIGURE 4.12

Models proposed for mechanisms of prion misfolding and propagation, showing the template refolding model (upper panel) and seeding nucleation model (lower panel).

Image adapted from Aguzzi A, Polymenidou M. Mammalian prion biology: one century of evolving concepts. Cell 2004;116:313–327.

The seeded nucleation model postulates that PrP^C and PrP^{Sc} exist in equilibrium within the cell, an equilibrium that is (under normal conditions) heavily weighted toward the PrP^C form [121]. In this scenario, PrP folds and refolds back and forth from the predominantly alpha helical cellular conformation to the beta sheet rich aggregation-prone form. The conversion to PrP^{Sc} is slow, and energetically demanding, whereas the reverse event (from PrP^{Sc} to PrP^C) is rapid and energetically favored. This equilibrium can be tipped in favor of PrP^{Sc} by the presence of aggregated seeds of PrP^{Sc}, which in turn recruit and stabilize other aggregates. As these aggregates grow in size, further seeds can fracture off to perpetuate the process, with these seeds capable of transferring between cells and between organisms. The equilibrium between PrP^C and PrP^{Sc} can be influenced by exposure to exogenous seeds, for example, from dietary sources or medical exposure, or by the presence of coding variants in the amino acid sequence of the PrP. Seeds can also be formed stochastically by the randomly determined coming together of PrP^{Sc} monomers, an event that, although rare, would lead to a low but predictable incidence of sporadic disease.

The template refolding model presupposes that the PrP^{Sc} form possesses intrinsic chaperone-like properties and is capable of binding to and refolding the PrP^C form [119,122]. In this model, mutations and/or exposure to exogenous PrP^{Sc} results in a catalytic process that causes the gradual accumulation of the PrP^{Sc} form, with individual PrP^{Sc} units able to pass from cell to cell spreading the disease process.

Regardless of the precise mechanism whereby PrP^{Sc} replicates, these models can account for the peculiar etiology of the prion disorders: occurring sporadically, as an inherited disease, and as a transmissible infectious entity (Fig. 4.13). The events that connect this cause to cell death and

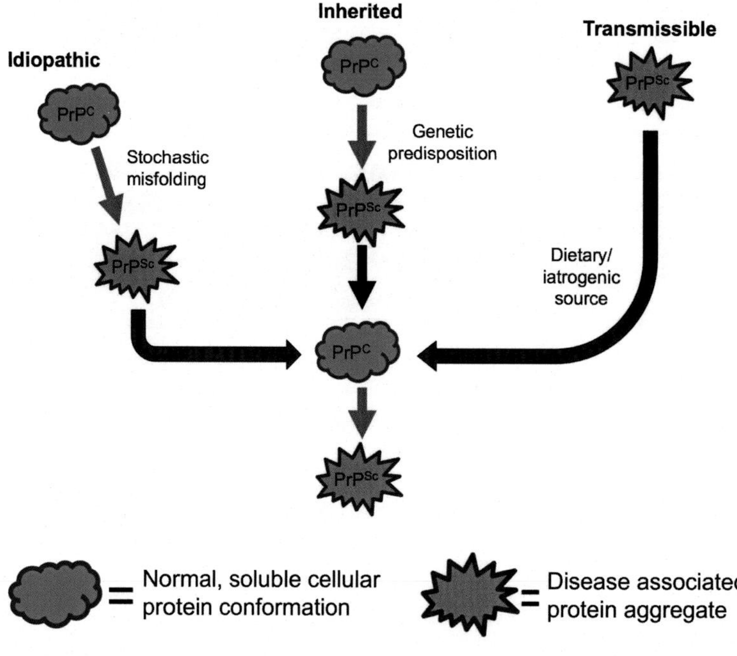

FIGURE 4.13

The tripartite etiology of prion disease. *PrP^C*, prion protein cellular; *PrP^{Sc}*, prion protein scrapie.

neurodegeneration driven by cytotoxicity, however, remain to be conclusively determined [123]. Expression of the PrP^{C} form is required for cytotoxicity [124], and there is some evidence that PrP^{Sc}-forming dimers on the outer face of the plasma membrane can result in a toxic cascade leading to cell death [125], although how this relates to in vivo prion infection is not clear. Data from a range of protein misfolding disorders has highlighted that cytotoxicity is a common property of proteins that misfold into aggregates and that the presence of these aggregates, in particular actively replicating small-order oligomeric species, can lead to intracellular damage and eventually cell death [126].

As noted earlier, animal (and, increasingly, cellular) models for prion propagation have been central to many of the advances that have increased our understanding of the prion disorders. These range from animal bioassays for prion disease, including inoculation of nonhuman primates and rodents, to high-throughput cellular assays for prion infection [119]. Animal models in particular have aided in the precise identification and characterization of the prion agent and have been central to understanding why some animal prion diseases have been transmitted to humans (e.g., BSE) and others have not (e.g., scrapie) [127]. These studies have revealed a significant species barrier to transmission that is partly derived from differences in the primary sequence of the PrP between species, as this barrier can be overcome in rodents by expression of equivalent transgenes. An example of this is the expression of human PrP in mice to enable the transmission of variant Creutzfeldt–Jakob disease. Transgenic models have also been essential to dissect the route taken by prion infection from the periphery, for example, through oral transmission and the gut [128]. These investigations have highlighted the role of specific classes of cells in allowing prions to make their way from the periphery to the brain [129].

Animal models for prion propagation and disease, although resembling much of the pathology and pathogenic course of naturally occurring prion diseases, are characterized by extended incubation periods ranging from weeks to years. The development of cellular models for prion propagation has had a major impact on our ability to tease out the cellular events that underlie prion infection and on more fundamental research into the principles governing prion behavior. For the former, cell models based on propagation of infectious PrP^{Sc} derived from scrapie (such as the mouse-adapted prion Rocky Mountain Laboratory and Chandler strains) in mammalian cells, in particular the N2a mouse neuroblastoma line, have yielded valuable insights [130–132]. The ability to propagate stable strains, differentiated by biochemical and cellular properties, facilitated some experiments investigating the evolution of PrP^{Sc} strains in cells, suggesting that the specific properties of a prion strain can adapt to environmental pressures [133].

An intriguing area of research that has complemented the mammalian prion disease field is that of yeast prions [134]. This research has derived from observations dating back several decades that some heritable yeast phenotypes could not be accounted for by traditional, nucleic acid–based inheritance (see Box 4.2) [136]. Importantly, these investigations have provided a great deal of information relating to the structural properties of yeast prions, which in turn has aided our understanding of the mechanistic basis of mammalian prion propagation [137].

4.4.9 OVERLAP WITH OTHER PROTEIN FOLDING DISORDERS

At an early stage in the investigations into kuru and Creutzfeldt–Jakob disease, it was noted that the protein-rich pathological condition observed in these disorders shared a number of features with other neurological disorders, including amyotrophic lateral sclerosis, Parkinson disease, and Alzheimer disease. This led to early attempts to test whether the pathological conditions in these disorders could also

BOX 4.2 YEAST PRIONS

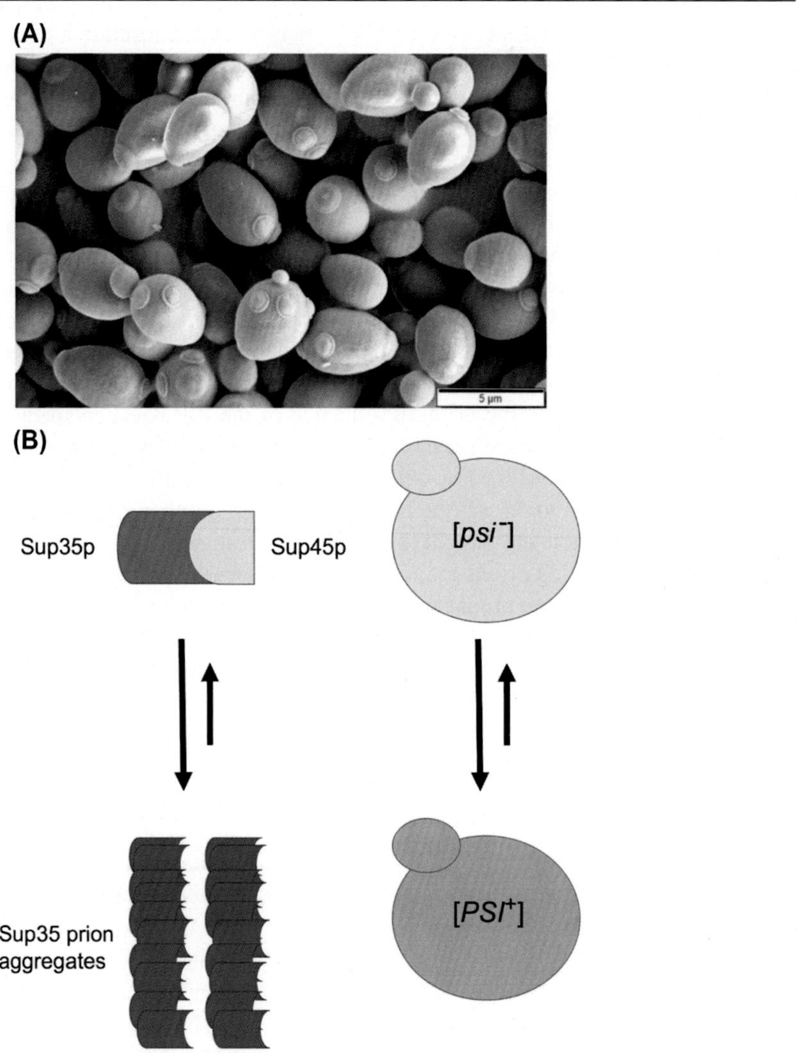

(A) The unicellular eukaryotic organism *Saccharomyces cerevisiae* (commonly known as baker's yeast) has provided a valuable model for understanding physiological prionlike behavior. (B) A number of yeast proteins have been shown to act in a prionlike manner to influence the inheritance of specific traits. In a [*psi*⁻] strain, Sup35 protein has a globular fold and acts as part of a transcriptional complex. Upon exposure to aggregated, prionlike Sup35, the transcriptional activity of this protein is suppressed and the yeast adopts a [*PSI*⁺] phenotype.

Image in panel (B) adapted from True HL. The battle of the fold: chaperones take on prions. Trends Genet 2006;22:110–117.

be transmitted in a manner analogous to kuru and Creutzfeldt–Jakob disease. Using a similar approach to that pioneered with kuru, Carleton Gajdusek and his colleagues inoculated nonhuman primates with brain extracts from individuals with Parkinson disease, a rare form of amyotrophic lateral sclerosis from the pacific island of Guam, and several dementias [138]. None of these inoculations resulted in disease as seen for kuru and Creutzfeldt–Jakob disease, leading Gajdusek to conclude that these disorders were not transmissible in the same manner [139]. Despite this, a number of research groups continued to investigate whether other neurodegenerative diseases could be transmitted, with studies from the 1980s and 1990s providing data that some aspects of the protein aggregate pathology observed in Alzheimer disease could be propagated in animal models [139]. With the establishment of the Braak staging hypotheses for Alzheimer and Parkinson diseases, which proposed that the amyloid-β (abeta), tau, and α-synuclein pathological conditions followed a stereotypical pattern of spread through the brain [140,141], a pathological correlate for propagation of protein aggregates between cells became available. In parallel to this, in vitro studies of recombinant protein have highlighted important structural similarities between the aggregates formed by the PrP and those formed by proteins such as abeta and α-synuclein (deposited in Alzheimer's and Parkinson's diseases) [142].

A major breakthrough occurred in 2008 with data from a number of groups suggesting that the Lewy body pathology found in Parkinson disease could spread from diseased brain tissue to fetal cell implants used to treat the disorder [143,144]. A potential explanation put forward for this observation was that the α-synuclein aggregates that make up Lewy bodies were propagating in a prionlike manner, spreading from cell to cell [145]. This led to a number of laboratories exploring this explanation using cellular and animal models for α-synuclein aggregation, and there is now a wealth of data suggesting that limited cell-to-cell propagation of α-synuclein aggregates can occur within an experimental setting [146–148]; however, how this relates to the human synucleinopathies is unclear and a matter of some controversy [149–151].

Investigations into the propagation of abeta and tau pathology in animal and cellular models have also yielded hints that aggregates of these proteins can propagate in a manner analogous to PrP^{Sc} aggregates in the prion disorders [152–156]. A key observation is that individuals with iatrogenic Creutzfeldt–Jakob disease linked to exposure to contaminated human growth hormone display not only prion disease–linked pathology but also abeta pathology similar to that more normally associated with Alzheimer-type dementia [157]. Similar results have also been reported by two independent analyses [158,159]. This led to pathologists proposing that these cases may represent an example of human transmission of abeta pathology, but this remains controversial [160].

It is important to note that the premise that Alzheimer and Parkinson diseases (as well as other protein misfolding disorders) share some characteristics with the prion disorders, and that their cause might be driven by a prionlike mechanism, is the subject of vigorous debate [151].

An intriguing, and related, area of research has come from investigations highlighting a possible direct functional link between the cellular PrP and proteins involved in Alzheimer and Parkinson diseases. A screen for cell membrane receptors for abeta in 2009 identified PrP^C as a potential binding partner for abeta [161], although similar to prionlike mechanisms in Alzheimer and Parkinson diseases, this has been contested [162]. Of interest, research has suggested that α-synuclein can also bind to PrP^C [163]. These data may provide a cellular link between these diverse proteins involved in neurodegeneration, although more work is required to demonstrate the physiological relevance of these observations.

4.5 THERAPIES

Despite many decades of research, there is to date no validated therapy that ameliorates the symptoms of prion disease, or slows or stops the progress of these disorders [164]. As our understanding of the mechanisms underpinning the prion diseases has progressed, a number of different approaches have undergone evaluation, in some cases progressing through to clinical trials in humans.

Early attempts to intervene in the Creutzfeldt–Jakob disease used antivirals, including amantadine and acyclovir [165,166]. These were applied in isolated cases, making evaluation of their use in terms of increased survival time or amelioration of symptoms difficult, given the underlying variability in disease course. Where more detailed analyses have been carried out, however, these have not indicated any beneficial impact of antivirals in prion disease. Interferon, also used to target viral infections, was tested as a potential therapy for Creutzfeldt–Jakob disease, although again with no beneficial outcomes [166].

Since the development of the prion hypothesis in the 1980s, a number of studies have been carried out in vivo and using cellular models looking to directly intervene in the replication of prions. Using the theoretical models for prion propagation (such as the seeded nucleation model) as a starting point, these have sought to block the formation or propagation of PrP^{Sc} or to accelerate the degradation of the disease-associated form of PrP [167].

A therapeutic strategy for targeting prion disease that parallels efforts in other neurodegenerative disorders is that of using antibodies to clear infectious prion seeds, to intervene in the misfolding/refolding process by blocking propagation, or to reduce the levels of template protein. Both active and passive immunization therapies have been investigated [168–170]. Again, this approach has yielded promising preclinical data in rodent models for prion disease but has yet to be translated into large-scale clinical trials in humans.

An area of recent investigation has been the unfolded protein response in cells, a reaction to endoplasmic reticulum stress that can act to modulate protein translation at a cellular level. Using a range of small-molecule inhibitors to target the signaling pathways that control the unfolded protein response, in particular the protein kinase R-like endoplasmic reticulum kinase (PERK) cascade, several studies have demonstrated benefits in mouse models for prion disease, extending survival in infected animals [171,172]. The utility of this approach in humans with prion disease has yet to be investigated. Similar to antibody therapies, the research groups leading this approach have also examined the utility of PERK pathway modulation in animal models for other neurodegenerative disorders. These studies demonstrated a protective impact in a mouse model for tau dysfunction and aggregation, potentially highlighting the unfolded protein response as a mechanism that may be common to a number of neurodegenerative disorders [173].

The largest human studies undertaken to date involved the use of quinacrine, a drug originally developed as an antimalarial agent and also used in the treatment of giardiasis. These have their origin in cellular studies that suggested that quinacrine was able to reduce prion replication in vitro and in rodent models [174,175]. Following these initially promising results, human trials were initiated to examine whether quinacrine treatment in patients with Creutzfeldt–Jakob disease provided clinical benefits [176]. No clinical benefits were observed in the largest of these trials, the UK-based Prion-1 trial co-ordinated by the National Prion Clinic in London [177]. These human trials for antiprion therapies have highlighted some of the major challenges facing clinical trials targeting these disorders. First, trials are limited by the number of patients available to take part in studies, especially if targeting a specific type of prion disease. This is compounded by an unpredictable and variable disease course, as well as a relatively short potential therapeutic window between diagnosis and death. Combined, these make it exceptionally challenging to achieve the level of statistical power to identify clinical benefits. This is magnified by the clinical heterogeneity exhibited by the prion diseases, making quantitative measures for clinical efficacy difficult to establish [178]. Of course, these challenges are not unique to the prion diseases and are shared by many neurodegenerative diseases, but the unique etiology and rarity of the prion disorders exaggerate these issues.

4.6 CONCLUSIONS

The prion diseases have been a source of fascination for many decades, with their rapid disease course and extraordinary mode of transmission. The impact of research into these disorders on our understanding of neurodegenerative diseases, and of biology, is emphasized by the Nobel Prizes awarded to investigators in this field. Intriguingly, data from the Alzheimer and Parkinson disease studies indicates that, if anything, the relevance of prion biology to human neurological disease has been underestimated. The study of prion disease also provides a sobering lesson in the challenges posed by neurological diseases; despite decades of research, and the availability of animal models for diseases that mimic the human disorders much more closely than those available for many other neurodegenerative disorders, we still await a therapy that makes a clear clinical difference.

FURTHER READING

Gajdusek DC. Unconventional viruses and the origin and disappearance of kuru; 1977, p. 161–215.

Prusiner SB. Prions. Proc Natl Acad Sci 1998;95(23):13363–83.

Shorter J, Lindquist S. Prions as adaptive conduits of memory and inheritance. Nat Rev Genet 2005; 6(6):435–450.

Jucker M, Walker LC. Self-propagation of pathogenic protein aggregates in neurodegenerative diseases. Nature 2013;501(7465):45–51.

REFERENCES

[1] Collinge J. Molecular neurology of prion disease. J Neurol Neurosurg Psychiatry 2005;76:906–19.

[2] Conan-Doyle A. The adventure of the Beryl Coronet. In: Strand magisine, vol. 8. 1892. London.

[3] Jucker M, Walker LC. Self-propagation of pathogenic protein aggregates in neurodegenerative diseases. Nature 2013;501:45–51.

[4] Johnson RT. Prion diseases. Lancet Neurol 2005;4:635–42.

[5] Watts JC, Balachandran A, Westaway D. The expanding universe of prion diseases. PLoS Pathog 2006;2:e26.

[6] Baron T. Identification of inter-species transmission of prion strains. J Neuropathol Exp Neurol 2002;61:377–83.

[7] Leopoldt JG. Nützliche und auf die Erfahrung gegründete Einleitung zu der Landwirthschaft. Berlin: Glogau; 1759.

[8] Jeffrey M, Gonzalez L. Classical sheep transmissible spongiform encephalopathies: pathogenesis, pathological phenotypes and clinical disease. Neuropathol Appl Neurobiol 2007;33:373–94.

[9] Parry HB, Oppenheimer DR. Scrapie disease in sheep: historical, clinical, epidemiological, pathological and practical aspects of the natural disease. London: Academic; 1983.
[10] Schneider K, Fangerau H, Michaelsen B, Raab WH. The early history of the transmissible spongiform encephalopathies exemplified by scrapie. Brain Res Bull 2008;77:343–55.
[11] Sigurdson CJ. A prion disease of cervids: chronic wasting disease. Vet Res 2008;39:41.
[12] Wells GA, Scott AC, Johnson CT, Gunning RF, Hancock RD, Jeffrey M, Dawson M, Bradley R. A novel progressive spongiform encephalopathy in cattle. Vet Rec 1987;121:419–20.
[13] van Keulen LJ, Langeveld JP, Garssen GJ, Jacobs JG, Schreuder BE, Smits MA. Diagnosis of bovine spongiform encephalopathy: a review. Vet Q 2000;22:197–200.
[14] Hope J, Ritchie L, Farquhar C, Somerville R, Hunter N. Bovine spongiform encephalopathy: a scrapie-like disease of British cattle. Prog Clin Biol Res 1989;317:659–67.
[15] Pearson GR, Gruffydd-Jones TJ, Wyatt JM, Hope J, Chong A, Scott AC, Dawson M, Wells GA. Feline spongiform encephalopathy. Vet Rec 1991;128:532.
[16] Marsh RF, Hadlow WJ. Transmissible mink encephalopathy. Rev Sci Tech 1992;11:539–50.
[17] Hope J, Reekie LJ, Hunter N, Multhaup G, Beyreuther K, White H, Scott AC, Stack MJ, Dawson M, Wells GA. Fibrils from brains of cows with new cattle disease contain scrapie-associated protein. Nature 1988;336:390–2.
[18] Wells GA, Wilesmith JW, McGill IS. Bovine spongiform encephalopathy: a neuropathological perspective. Brain Pathol 1991;1:69–78.
[19] Donnelly CA, Ferguson NM, Ghani AC, Woolhouse ME, Watt CJ, Anderson RM. The epidemiology of BSE in cattle herds in Great Britain. I. Epidemiological processes, demography of cattle and approaches to control by culling. Philos Trans R Soc Lond B Biol Sci 1997;352:781–801.
[20] Uehlinger FD, Johnston AC, Bollinger TK, Waldner CL. Systematic review of management strategies to control chronic wasting disease in wild deer populations in North America. BMC Vet Res 2016;12:173.
[21] Jakob A. Über eigenartige Erkrankungen des Zentralnervensystems mit bemerkenswertem anatomischen Befunde. Zeitschrift für die gesamte Neurologie und Psychiatrie 1921;64:147–228.
[22] Creutzfeldt HG. Über eine eigenartige herdförmige Erkrankung des Zentralnervensystems (vorläufige Mitteilung). Zeitschrift für die gesamte Neurologie und Psychiatrie 1920;57:1–18.
[23] Richardson Jr EP, Masters CL. The nosology of Creutzfeldt-Jakob disease and conditions related to the accumulation of PrPCJD in the nervous system. Brain Pathol 1995;5:33–41.
[24] Kim MO, Geschwind MD. Clinical update of Jakob-Creutzfeldt disease. Curr Opin Neurol 2015;28:302–10.
[25] Gambetti P, Kong Q, Zou W, Parchi P, Chen SG. Sporadic and familial CJD: classification and characterisation. Br Med Bull 2003;66:213–39.
[26] Gerstmann J, Sträussler E, Scheinker I. Über eine eigenartige hereditär-familiäre Erkrankung des Zentralnervensystems. Zeitschrift für die gesamte Neurologie und Psychiatrie 1935;154:736–62.
[27] Hsiao K, Baker HF, Crow TJ, Poulter M, Owen F, Terwilliger JD, Westaway D, Ott J, Prusiner SB. Linkage of a prion protein missense variant to Gerstmann-Straussler syndrome. Nature 1989;338:342–5.
[28] Collins S, McLean CA, Masters CL. Gerstmann-Straussler-Scheinker syndrome, fatal familial insomnia, and kuru: a review of these less common human transmissible spongiform encephalopathies. J Clin Neurosci 2001;8:387–97.
[29] Gajdusek DC, Zigas V. Kuru; clinical, pathological and epidemiological study of an acute progressive degenerative disease of the central nervous system among natives of the Eastern Highlands of New Guinea. Am J Med 1959;26:442–69.
[30] Whitfield JT, Pako WH, Collinge J, Alpers MP. Mortuary rites of the South Fore and kuru. Philos Trans R Soc Lond B Biol Sci 2008;363:3721–4.
[31] Lugaresi E, Medori R, Montagna P, Baruzzi A, Cortelli P, Lugaresi A, Tinuper P, Zucconi M, Gambetti P. Fatal familial insomnia and dysautonomia with selective degeneration of thalamic nuclei. N. Engl J Med 1986;315:997–1003.

[32] Medori R, Tritschler HJ, LeBlanc A, Villare F, Manetto V, Chen HY, Xue R, Leal S, Montagna P, Cortelli P, et al. Fatal familial insomnia, a prion disease with a mutation at codon 178 of the prion protein gene. N. Engl J Med 1992;326:444–9.

[33] Will RG, Ironside JW, Zeidler M, Cousens SN, Estibeiro K, Alperovitch A, Poser S, Pocchiari M, Hofman A, Smith PG. A new variant of Creutzfeldt-Jakob disease in the UK. Lancet 1996;347:921–5.

[34] Will RG, Ward HJ. Clinical features of variant Creutzfeldt-Jakob disease. Curr Top Microbiol Immunol 2004;284:121–32.

[35] Collinge J, Sidle KC, Meads J, Ironside J, Hill AF. Molecular analysis of prion strain variation and the aetiology of 'new variant' CJD. Nature 1996;383:685–90.

[36] Bruce ME, Will RG, Ironside JW, McConnell I, Drummond D, Suttie A, McCardle L, Chree A, Hope J, Birkett C, Cousens S, Fraser H, Bostock CJ. Transmissions to mice indicate that 'new variant' CJD is caused by the BSE agent. Nature 1997;389:498–501.

[37] Liberski PP, Ironside JW. An outline of the neuropathology of transmissible spongiform encephalopathies (prion diseases). Folia Neuropathol 2004;42(Suppl. B):39–58.

[38] du Plessis DG. Prion protein disease and neuropathology of prion disease. Neuroimaging Clin N Am 2008;18:163–82. ix.

[39] Hill AF, Joiner S, Wadsworth JD, Sidle KC, Bell JE, Budka H, Ironside JW, Collinge J. Molecular classification of sporadic Creutzfeldt-Jakob disease. Brain J Neurol 2003;126:1333–46.

[40] Wadsworth JD, Joiner S, Hill AF, Campbell TA, Desbruslais M, Luthert PJ, Collinge J. Tissue distribution of protease resistant prion protein in variant Creutzfeldt-Jakob disease using a highly sensitive immunoblotting assay. Lancet 2001;358:171–80.

[41] Wroe SJ, Pal S, Siddique D, Hyare H, Macfarlane R, Joiner S, Linehan JM, Brandner S, Wadsworth JD, Hewitt P, Collinge J. Clinical presentation and pre-mortem diagnosis of variant Creutzfeldt-Jakob disease associated with blood transfusion: a case report. Lancet 2006;368:2061–7.

[42] Liberski PP. Historical overview of prion diseases: a view from Afar. Folia Neuropathol 2012;50:1–12.

[43] Cuille J, Chelle P. La tremblante du mouton est bien inoculable. CR Seances Acad Sci 1938;206:1687–8.

[44] Cuillé J. Pathologie animale-La maladie dite trmblante du mouton est-elle inoculable. CR Acad Sci Paris 1936;203:1552–4.

[45] Hadlow WJ. Scrapie and kuru. Lancet 1959;274:289–90.

[46] Gajdusek DC, Gibbs CJ, Alpers M. Experimental transmission of a Kuru-like syndrome to chimpanzees. Nature 1966;209:794–6.

[47] Gibbs Jr CJ, Gajdusek DC, Asher DM, Alpers MP, Beck E, Daniel PM, Matthews WB. Creutzfeldt-Jakob disease (spongiform encephalopathy): transmission to the chimpanzee. Science 1968;161:388–9.

[48] Gajdusek DC. Unconventional viruses and the origin and disappearance of kuru. Science 1977;197: 943–60.

[49] Sigurdsson B. Rida, a chronic encephalitis of sheep: with general remarks on infections which develop slowly and some of their special characteristics. Br Vet J 1954;110:341–54.

[50] Alper T, Haig DA, Clarke MC. The exceptionally small size of the scrapie agent. Biochem Biophys Res Commun 1966;22:278–84.

[51] Alper T, Cramp WA, Haig DA, Clarke MC. Does the agent of scrapie replicate without nucleic acid? Nature 1967;214:764–6.

[52] Griffith JS. Self-replication and scrapie. Nature 1967;215:1043–4.

[53] Pattison I, Jones KM. The possible nature of the transmissible agent of scrapie. Vet Rec 1967;80:2–9.

[54] Sacks O. D. Carleton Gajdusek, MD (1923-2008). Arch Neurol 2009;66:676–7.

[55] Poser CM. Notes on the history of the prion diseases. Part I. Clin Neurol Neurosurg 2002;104:1–9.

[56] Prusiner SB, Cochran SP, Groth DF, Downey DE, Bowman KA, Martinez HM. Measurement of the scrapie agent using an incubation time interval assay. Ann Neurol 1982;11:353–8.

[57] Kimberlin RH, Walker C. Characteristics of a short incubation model of scrapie in the golden hamster. J Gen Virol 1977;34:295–304.
[58] Prusiner SB. Novel proteinaceous infectious particles cause scrapie. Science 1982;216:136–44.
[59] Crick F. Central dogma of molecular biology. Nature 1970;227:561–3.
[60] Baltimore D, Franklin RM. Preliminary data on a virus-specific enzyme system responsible for the synthesis of viral RNA. Biochem Biophys Res Commun 1962;9:388–92.
[61] Temin HM, Mizutani S. RNA-dependent DNA polymerase in virions of Rous sarcoma virus. Nature 1970;226:1211–3.
[62] Bolton DC, McKinley MP, Prusiner SB. Identification of a protein that purifies with the scrapie prion. Science 1982;218:1309–11.
[63] McKinley MP, Bolton DC, Prusiner SB. A protease-resistant protein is a structural component of the scrapie prion. Cell 1983;35:57–62.
[64] Oesch B, Westaway D, Walchli M, McKinley MP, Kent SB, Aebersold R, Barry RA, Tempst P, Teplow DB, Hood LE, et al. A cellular gene encodes scrapie PrP 27-30 protein. Cell 1985;40:735–46.
[65] Basler K, Oesch B, Scott M, Westaway D, Walchli M, Groth DF, McKinley MP, Prusiner SB, Weissmann C. Scrapie and cellular PrP isoforms are encoded by the same chromosomal gene. Cell 1986;46:417–28.
[66] Kretzschmar HA, Stowring LE, Westaway D, Stubblebine WH, Prusiner SB, Dearmond SJ. Molecular cloning of a human prion protein cDNA. DNA 1986;5:315–24.
[67] Chesebro B, Race R, Wehrly K, Nishio J, Bloom M, Lechner D, Bergstrom S, Robbins K, Mayer L, Keith JM, et al. Identification of scrapie prion protein-specific mRNA in scrapie-infected and uninfected brain. Nature 1985;315:331–3.
[68] Owen F, Poulter M, Lofthouse R, Collinge J, Crow TJ, Risby D, Baker HF, Ridley RM, Hsiao K, Prusiner SB. Insertion in prion protein gene in familial Creutzfeldt-Jakob disease. Lancet 1989;1:51–2.
[69] Hsiao KK, Scott M, Foster D, Groth DF, DeArmond SJ, Prusiner SB. Spontaneous neurodegeneration in transgenic mice with mutant prion protein. Science 1990;250:1587–90.
[70] Zeidler M, Stewart G, Cousens SN, Estibeiro K, Will RG. Codon 129 genotype and new variant CJD.. Lancet 1997;350:668.
[71] Mead S, Joiner S, Desbruslais M, Beck JA, O'Donoghue M, Lantos P, Wadsworth JD, Collinge J. Creutzfeldt-Jakob disease, prion protein gene codon 129VV, and a novel PrPSc type in a young British woman. Arch Neurol 2007;64:1780–4.
[72] Mok T, Jaunmuktane Z, Joiner S, Campbell T, Morgan C, Wakerley B, Golestani F, Rudge P, Mead S, Jager HR, Wadsworth JD, Brandner S, Collinge J. Variant Creutzfeldt-Jakob disease in a patient with heterozygosity at PRNP codon 129. N. Engl J Med 2017;376:292–4.
[73] Goldfarb LG, Petersen RB, Tabaton M, Brown P, LeBlanc AC, Montagna P, Cortelli P, Julien J, Vital C, Pendelbury WW, et al. Fatal familial insomnia and familial Creutzfeldt-Jakob disease: disease phenotype determined by a DNA polymorphism. Science 1992;258:806–8.
[74] Lewis PA, Tattum MH, Jones S, Bhelt D, Batchelor M, Clarke AR, Collinge J, Jackson GS. Codon 129 polymorphism of the human prion protein influences the kinetics of amyloid formation. J Gen Virol 2006;87:2443–9.
[75] Mead S, Whitfield J, Poulter M, Shah P, Uphill J, Campbell T, Al-Dujaily H, Hummerich H, Beck J, Mein CA, Verzilli C, Whittaker J, Alpers MP, Collinge J. A novel protective prion protein variant that colocalizes with kuru exposure. N. Engl J Med 2009;361:2056–65.
[76] Asante EA, Smidak M, Grimshaw A, Houghton R, Tomlinson A, Jeelani A, Jakubcova T, Hamdan S, Richard-Londt A, Linehan JM, Brandner S, Alpers M, Whitfield J, Mead S, Wadsworth JD, Collinge J. A naturally occurring variant of the human prion protein completely prevents prion disease. Nature 2015;522:478–81.
[77] Duffy P, Wolf J, Collins G, DeVoe AG, Streeten B, Cowen D. Letter: possible person-to-person transmission of Creutzfeldt-Jakob disease. N. Engl J Med 1974;290:692–3.

[78] Thadani V, Penar PL, Partington J, Kalb R, Janssen R, Schonberger LB, Rabkin CS, Prichard JW. Creutzfeldt-Jakob disease probably acquired from a cadaveric dura mater graft. Case report. J Neurosurg 1988;69:766–9.
[79] Powell-Jackson J, Weller RO, Kennedy P, Preece MA, Whitcombe EM, Newsom-Davis J. Creutzfeldt-Jakob disease after administration of human growth hormone. Lancet 1985;2:244–6.
[80] Llewelyn CA, Hewitt PE, Knight RS, Amar K, Cousens S, Mackenzie J, Will RG. Possible transmission of variant Creutzfeldt-Jakob disease by blood transfusion. Lancet 2004;363:417–21.
[81] Sharp LA. The commodification of the body and its parts. Annu Rev Anthropol 2000;29:287–328.
[82] Collinge J, Palmer MS, Dryden AJ. Genetic predisposition to iatrogenic Creutzfeldt-Jakob disease. Lancet 1991;337:1441–2.
[83] Bueler H, Fischer M, Lang Y, Bluethmann H, Lipp HP, DeArmond SJ, Prusiner SB, Aguet M, Weissmann C. Normal development and behaviour of mice lacking the neuronal cell-surface PrP protein. Nature 1992;356:577–82.
[84] Bueler H, Aguzzi A, Sailer A, Greiner RA, Autenried P, Aguet M, Weissmann C. Mice devoid of PrP are resistant to scrapie. Cell 1993;73:1339–47.
[85] Sailer A, Bueler H, Fischer M, Aguzzi A, Weissmann C. No propagation of prions in mice devoid of PrP. Cell 1994;77:967–8.
[86] Prusiner SB. Prions. Proc Natl Acad Sci USA 1998;95:13363–83.
[87] Bradley R, Wilesmith JW. Epidemiology and control of bovine spongiform encephalopathy (BSE). Br Med Bull 1993;49:932–59.
[88] Brown P. On the origins of BSE. Lancet 1998;352:252–3.
[89] Taylor DM, Woodgate SL. Bovine spongiform encephalopathy: the causal role of ruminant-derived protein in cattle diets. Rev Sci Tech 1997;16:187–98.
[90] Millstone E, Van Zwanenberg P. Politics of expert advice: lessons from the early history of the BSE saga. Sci Public Policy 2001;28:99–112.
[91] Hill AF, Desbruslais M, Joiner S, Sidle KC, Gowland I, Collinge J, Doey LJ, Lantos P. The same prion strain causes vCJD and BSE. Nature 1997;389:448–50. 526.
[92] Raymond GJ, Hope J, Kocisko DA, Priola SA, Raymond LD, Bossers A, Ironside J, Will RG, Chen SG, Petersen RB, Gambetti P, Rubenstein R, Smits MA, Lansbury Jr PT, Caughey B. Molecular assessment of the potential transmissibilities of BSE and scrapie to humans. Nature 1997;388:285–8.
[93] Scott MR, Will R, Ironside J, Nguyen HO, Tremblay P, DeArmond SJ, Prusiner SB. Compelling transgenetic evidence for transmission of bovine spongiform encephalopathy prions to humans. Proc Natl Acad Sci USA 1999;96:15137–42.
[94] Ironside JW. Pathology of variant Creutzfeldt-Jakob disease. Arch Virol Suppl 2000:143–51.
[95] Diack AB, Head MW, McCutcheon S, Boyle A, Knight R, Ironside JW, Manson JC, Will RG. Variant CJD. 18 years of research and surveillance. Prion 2014;8:286–95.
[96] Diack AB, Will RG, Manson JC. Public health risks from subclinical variant CJD. PLoS Pathog 2017;13:e1006642.
[97] Collinge J, Whitfield J, McKintosh E, Beck J, Mead S, Thomas DJ, Alpers MP. Kuru in the 21st century–an acquired human prion disease with very long incubation periods. Lancet 2006;367:2068–74.
[98] Rudge P, Jaunmuktane Z, Adlard P, Bjurstrom N, Caine D, Lowe J, Norsworthy P, Hummerich H, Druyeh R, Wadsworth JD, Brandner S, Hyare H, Mead S, Collinge J. Iatrogenic CJD due to pituitary-derived growth hormone with genetically determined incubation times of up to 40 years. Brain J Neurol 2015;138:3386–99.
[99] Riek R, Hornemann S, Wider G, Billeter M, Glockshuber R, Wuthrich K. NMR structure of the mouse prion protein domain PrP(121-231). Nature 1996;382:180–2.
[100] Zahn R, Liu A, Luhrs T, Riek R, von Schroetter C, Lopez Garcia F, Billeter M, Calzolai L, Wider G, Wuthrich K. NMR solution structure of the human prion protein. Proc Natl Acad Sci USA 2000;97:145–50.

[101] Jackson GS, Clarke AR. Mammalian prion proteins. Curr Opin Struct Biol 2000;10:69–74.

[102] Prusiner SB, McKinley MP, Bowman KA, Bolton DC, Bendheim PE, Groth DF, Glenner GG. Scrapie prions aggregate to form amyloid-like birefringent rods. Cell 1983;35:349–58.

[103] Prusiner SB, Groth DF, Bolton DC, Kent SB, Hood LE. Purification and structural studies of a major scrapie prion protein. Cell 1984;38:127–34.

[104] Jackson GS, Hosszu LL, Power A, Hill AF, Kenney J, Saibil H, Craven CJ, Waltho JP, Clarke AR, Collinge J. Reversible conversion of monomeric human prion protein between native and fibrilogenic conformations. Science 1999;283:1935–7.

[105] Leffers KW, Wille H, Stohr J, Junger E, Prusiner SB, Riesner D. Assembly of natural and recombinant prion protein into fibrils. Biol Chem 2005;386:569–80.

[106] Tattum MH, Cohen-Krausz S, Thumanu K, Wharton CW, Khalili-Shirazi A, Jackson GS, Orlova EV, Collinge J, Clarke AR, Saibil HR. Elongated oligomers assemble into mammalian PrP amyloid fibrils. J Mol Biol 2006;357:975–85.

[107] Legname G, Baskakov IV, Nguyen HO, Riesner D, Cohen FE, DeArmond SJ, Prusiner SB. Synthetic mammalian prions. Science 2004;305:673–6.

[108] Wang F, Wang X, Yuan CG, Ma J. Generating a prion with bacterially expressed recombinant prion protein. Science 2010;327:1132–5.

[109] Schmidt C, Fizet J, Properzi F, Batchelor M, Sandberg MK, Edgeworth JA, Afran L, Ho S, Badhan A, Klier S, Linehan JM, Brandner S, Hosszu LL, Tattum MH, Jat P, Clarke AR, Klohn PC, Wadsworth JD, Jackson GS, Collinge J. A systematic investigation of production of synthetic prions from recombinant prion protein. Open Biol 2015;5:150165.

[110] Bendheim PE, Barry RA, DeArmond SJ, Stites DP, Prusiner SB. Antibodies to a scrapie prion protein. Nature 1984;310:418–21.

[111] Hill AF, Collinge J. Subclinical prion infection. Trends Microbiol 2003;11:578–84.

[112] Jackson GS, Collinge J. The molecular pathology of CJD: old and new variants. Mol Pathol 2001;54:393–9.

[113] Rudd PM, Merry AH, Wormald MR, Dwek RA. Glycosylation and prion protein. Curr Opin Struct Biol 2002;12:578–86.

[114] Head MW, Ironside JW. Review: Creutzfeldt-Jakob disease: prion protein type, disease phenotype and agent strain. Neuropathol Appl Neurobiol 2012;38:296–310.

[115] Wadsworth JD, Hill AF, Beck JA, Collinge J. Molecular and clinical classification of human prion disease. Br Med Bull 2003;66:241–54.

[116] Parchi P, de Boni L, Saverioni D, Cohen ML, Ferrer I, Gambetti P, Gelpi E, Giaccone G, Hauw JJ, Hoftberger R, Ironside JW, Jansen C, Kovacs GG, Rozemuller A, Seilhean D, Tagliavini F, Giese A, Kretzschmar HA. Consensus classification of human prion disease histotypes allows reliable identification of molecular subtypes: an inter-rater study among surveillance centres in Europe and USA. Acta Neuropathol 2012;124:517–29.

[117] Collinge J, Clarke AR. A general model of prion strains and their pathogenicity. Science 2007;318:930–6.

[118] Pattison IH, Millson GC. Scrapie produced experimentally in goats with special reference to the clinical syndrome. J Comp Pathol 1961;71:101–9.

[119] Aguzzi A, Polymenidou M. Mammalian prion biology: one century of evolving concepts. Cell 2004;116:313–27.

[120] Harper JD, Lansbury Jr PT. Models of amyloid seeding in Alzheimer's disease and scrapie: mechanistic truths and physiological consequences of the time-dependent solubility of amyloid proteins. Annu Rev Biochem 1997;66:385–407.

[121] Jarrett JT, Lansbury Jr PT. Seeding "one-dimensional crystallization" of amyloid: a pathogenic mechanism in Alzheimer's disease and scrapie? Cell 1993;73:1055–8.

[122] Gajdusek DC. Transmissible and non-transmissible amyloidoses: autocatalytic post-translational conversion of host precursor proteins to beta-pleated sheet configurations. J Neuroimmunol 1988;20:95–110.

[123] Kretzschmar HA, Giese A, Brown DR, Herms J, Keller B, Schmidt B, Groschup M. Cell death in prion disease. J Neural Transm Suppl 1997;50:191–210.
[124] Brandner S, Isenmann S, Raeber A, Fischer M, Sailer A, Kobayashi Y, Marino S, Weissmann C, Aguzzi A. Normal host prion protein necessary for scrapie-induced neurotoxicity. Nature 1996;379:339–43.
[125] Solforosi L, Criado JR, McGavern DB, Wirz S, Sanchez-Alavez M, Sugama S, DeGiorgio LA, Volpe BT, Wiseman E, Abalos G, Masliah E, Gilden D, Oldstone MB, Conti B, Williamson RA. Cross-linking cellular prion protein triggers neuronal apoptosis in vivo. Science 2004;303:1514–6.
[126] Stefani M, Dobson CM. Protein aggregation and aggregate toxicity: new insights into protein folding, misfolding diseases and biological evolution. J Mol Med (Berl) 2003;81:678–99.
[127] Watts JC, Prusiner SB. Mouse models for studying the formation and propagation of prions. J Biol Chem 2014;289:19841–9.
[128] Aguzzi A. Prions and the immune system: a journey through gut, spleen, and nerves. Adv Immunol 2003;81:123–71.
[129] Klein MA, Frigg R, Flechsig E, Raeber AJ, Kalinke U, Bluethmann H, Bootz F, Suter M, Zinkernagel RM, Aguzzi A. A crucial role for B cells in neuroinvasive scrapie. Nature 1997;390:687–90.
[130] Caughey B, Race RE, Chesebro B. Detection of prion protein mRNA in normal and scrapie-infected tissues and cell lines. J Gen Virol 1988;69(Pt 3):711–6.
[131] Chandler RL. Encephalopathy in mice produced by inoculation with scrapie brain material. Lancet 1961;1:1378–9.
[132] Klohn PC, Stoltze L, Flechsig E, Enari M, Weissmann C. A quantitative, highly sensitive cell-based infectivity assay for mouse scrapie prions. Proc Natl Acad Sci USA 2003;100:11666–71.
[133] Li J, Browning S, Mahal SP, Oelschlegel AM, Weissmann C. Darwinian evolution of prions in cell culture. Science 2010;327:869–72.
[134] Tuite MF, Lindquist SL. Maintenance and inheritance of yeast prions. Trends Genet 1996;12:467–71.
[135] True HL. The battle of the fold: chaperones take on prions. Trends Genet 2006;22:110–7.
[136] Wickner RB. [URE3] as an altered URE2 protein: evidence for a prion analog in *Saccharomyces cerevisiae*. Science 1994;264:566–9.
[137] Chien P, Weissman JS, DePace AH. Emerging principles of conformation-based prion inheritance. Annu Rev Biochem 2004;73:617–56.
[138] Gibbs Jr CJ, Gajdusek DC. Amyotrophic lateral sclerosis, Parkinson's disease, and the amyotrophic lateral sclerosis-Parkinsonism-dementia complex on Guam: a review and summary of attempts to demonstrate infection as the aetiology. J Clin Pathol Suppl (R Coll Pathol) 1972;6:132–40.
[139] Godec MS, Asher DM, Masters CL, Kozachuk WE, Friedland RP, Gibbs Jr CJ, Gajdusek DC, Rapoport SI, Schapiro MB. Evidence against the transmissibility of Alzheimer's disease. Neurology 1991;41:1320.
[140] Braak H, Del Tredici K, Rub U, de Vos RA, Jansen Steur EN, Braak E. Staging of brain pathology related to sporadic Parkinson's disease. Neurobiol Aging 2003;24:197–211.
[141] Bancher C, Braak H, Fischer P, Jellinger KA. Neuropathological staging of Alzheimer lesions and intellectual status in Alzheimer's and Parkinson's disease patients. Neurosci Lett 1993;162:179–82.
[142] Chiti F, Dobson CM. Protein misfolding, functional amyloid, and human disease. Annu Rev Biochem 2006;75:333–66.
[143] Kordower JH, Chu Y, Hauser RA, Freeman TB, Olanow CW. Lewy body-like pathology in long-term embryonic nigral transplants in Parkinson's disease. Nat Med 2008;14:504–6.
[144] Li JY, Englund E, Holton JL, Soulet D, Hagell P, Lees AJ, Lashley T, Quinn NP, Rehncrona S, Bjorklund A, Widner H, Revesz T, Lindvall O, Brundin P. Lewy bodies in grafted neurons in subjects with Parkinson's disease suggest host-to-graft disease propagation. Nat Med 2008;14:501–3.
[145] Olanow CW, Prusiner SB. Is Parkinson's disease a prion disorder? Proc Natl Acad Sci USA 2009;106:12571–2.

[146] Volpicelli-Daley LA, Luk KC, Patel TP, Tanik SA, Riddle DM, Stieber A, Meaney DF, Trojanowski JQ, Lee VM. Exogenous alpha-synuclein fibrils induce Lewy body pathology leading to synaptic dysfunction and neuron death. Neuron 2011;72:57–71.

[147] Luk KC, Kehm V, Carroll J, Zhang B, O'Brien P, Trojanowski JQ, Lee VM. Pathological alpha-synuclein transmission initiates Parkinson-like neurodegeneration in nontransgenic mice. Science 2012;338:949–53.

[148] Desplats P, Lee HJ, Bae EJ, Patrick C, Rockenstein E, Crews L, Spencer B, Masliah E, Lee SJ. Inclusion formation and neuronal cell death through neuron-to-neuron transmission of alpha-synuclein. Proc Natl Acad Sci USA 2009;106:13010–5.

[149] Brundin P, Melki R. Prying into the prion hypothesis for Parkinson's disease. J Neurosci Off J Soc Neurosci 2017;37:9808–18.

[150] Surmeier DJ, Obeso JA, Halliday GM. Parkinson's disease is not simply a prion disorder. J Neurosci Off J Soc Neurosci 2017;37:9799–807.

[151] Walsh DM, Selkoe DJ. A critical appraisal of the pathogenic protein spread hypothesis of neurodegeneration. Nat Rev Neurosci 2016;17:251–60.

[152] Clavaguera F, Bolmont T, Crowther RA, Abramowski D, Frank S, Probst A, Fraser G, Stalder AK, Beibel M, Staufenbiel M, Jucker M, Goedert M, Tolnay M. Transmission and spreading of tauopathy in transgenic mouse brain. Nat Cell Biol 2009;11:909–13.

[153] Frost B, Jacks RL, Diamond MI. Propagation of tau misfolding from the outside to the inside of a cell. J Biol Chem 2009;284:12845–52.

[154] Kfoury N, Holmes BB, Jiang H, Holtzman DM, Diamond MI. Trans-cellular propagation of Tau aggregation by fibrillar species. J Biol Chem 2012;287:19440–51.

[155] Meyer-Luehmann M, Coomaraswamy J, Bolmont T, Kaeser S, Schaefer C, Kilger E, Neuenschwander A, Abramowski D, Frey P, Jaton AL, Vigouret JM, Paganetti P, Walsh DM, Mathews PM, Ghiso J, Staufenbiel M, Walker LC, Jucker M. Exogenous induction of cerebral beta-amyloidogenesis is governed by agent and host. Science 2006;313:1781–4.

[156] Eisele YS, Obermuller U, Heilbronner G, Baumann F, Kaeser SA, Wolburg H, Walker LC, Staufenbiel M, Heikenwalder M, Jucker M. Peripherally applied abeta-containing inoculates induce cerebral beta-amyloidosis. Science 2010;330:980–2.

[157] Jaunmuktane Z, Mead S, Ellis M, Wadsworth JD, Nicoll AJ, Kenny J, Launchbury F, Linehan J, Richard-Loendt A, Walker AS, Rudge P, Collinge J, Brandner S. Evidence for human transmission of amyloid-beta pathology and cerebral amyloid angiopathy. Nature 2015;525:247–50.

[158] Ritchie DL, Adlard P, Peden AH, Lowrie S, Le Grice M, Burns K, Jackson RJ, Yull H, Keogh MJ, Wei W, Chinnery PF, Head MW, Ironside JW. Amyloid-beta accumulation in the CNS in human growth hormone recipients in the UK. Acta Neuropathol 2017;134:221–40.

[159] Duyckaerts C, Sazdovitch V, Ando K, Seilhean D, Privat N, Yilmaz Z, Peckeu L, Amar E, Comoy E, Maceski A, Lehmann S, Brion JP, Brandel JP, Haik S. Neuropathology of iatrogenic Creutzfeldt-Jakob disease and immunoassay of French cadaver-sourced growth hormone batches suggest possible transmission of tauopathy and long incubation periods for the transmission of abeta pathology. Acta Neuropathol 2017;135(2):201–12.

[160] Abbott A. The red-hot debate about transmissible Alzheimer's. Nature 2016;531:294–7.

[161] Lauren J, Gimbel DA, Nygaard HB, Gilbert JW, Strittmatter SM. Cellular prion protein mediates impairment of synaptic plasticity by amyloid-beta oligomers. Nature 2009;457:1128–32.

[162] Balducci C, Beeg M, Stravalaci M, Bastone A, Sclip A, Biasini E, Tapella L, Colombo L, Manzoni C, Borsello T, Chiesa R, Gobbi M, Salmona M, Forloni G. Synthetic amyloid-beta oligomers impair long-term memory independently of cellular prion protein. Proc Natl Acad Sci USA 2010;107:2295–300.

[163] Ferreira DG, Temido-Ferreira M, Miranda HV, Batalha VL, Coelho JE, Szego EM, Marques-Morgado I, Vaz SH, Rhee JS, Schmitz M, Zerr I, Lopes LV, Outeiro TF. Alpha-synuclein interacts with PrP(C) to induce cognitive impairment through mGluR5 and NMDAR2B. Nat Neurosci 2017;20:1569–79.

[164] Aguzzi A, Lakkaraju AKK, Frontzek K. Toward therapy of human prion diseases. Annu Rev Pharmacol Toxicol 2017;58:331–51.

[165] Sanders WL, Dunn TL. Creutzfeldt-Jakob disease treated with amantidine. A report of two cases. J Neurol Neurosurg Psychiatry 1973;36:581–4.

[166] David AS, Grant R, Ballantyne JP. Unsuccessful treatment of Creutzfeldt-Jakob disease with acyclovir. Lancet 1984;1:512–3.

[167] Trevitt CR, Collinge J. A systematic review of prion therapeutics in experimental models. Brain J Neurol 2006;129:2241–65.

[168] Heppner FL, Musahl C, Arrighi I, Klein MA, Rulicke T, Oesch B, Zinkernagel RM, Kalinke U, Aguzzi A. Prevention of scrapie pathogenesis by transgenic expression of anti-prion protein antibodies. Science 2001;294:178–82.

[169] White AR, Enever P, Tayebi M, Mushens R, Linehan J, Brandner S, Anstee D, Collinge J, Hawke S. Monoclonal antibodies inhibit prion replication and delay the development of prion disease. Nature 2003;422:80–3.

[170] Sigurdsson EM, Brown DR, Daniels M, Kascsak RJ, Kascsak R, Carp R, Meeker HC, Frangione B, Wisniewski T. Immunization delays the onset of prion disease in mice. Am J Pathol 2002;161:13–7.

[171] Moreno JA, Radford H, Peretti D, Steinert JR, Verity N, Martin MG, Halliday M, Morgan J, Dinsdale D, Ortori CA, Barrett DA, Tsaytler P, Bertolotti A, Willis AE, Bushell M, Mallucci GR. Sustained translational repression by eIF2alpha-P mediates prion neurodegeneration. Nature 2012;485:507–11.

[172] Moreno JA, Halliday M, Molloy C, Radford H, Verity N, Axten JM, Ortori CA, Willis AE, Fischer PM, Barrett DA, Mallucci GR. Oral treatment targeting the unfolded protein response prevents neurodegeneration and clinical disease in prion-infected mice. Sci Transl Med 2013;5:206ra138.

[173] Radford H, Moreno JA, Verity N, Halliday M, Mallucci GR. PERK inhibition prevents tau-mediated neurodegeneration in a mouse model of frontotemporal dementia. Acta Neuropathol 2015;130:633–42.

[174] Doh-Ura K, Iwaki T, Caughey B. Lysosomotropic agents and cysteine protease inhibitors inhibit scrapie-associated prion protein accumulation. J Virol 2000;74:4894–7.

[175] Korth C, May BC, Cohen FE, Prusiner SB. Acridine and phenothiazine derivatives as pharmacotherapeutics for prion disease. Proc Natl Acad Sci USA 2001;98:9836–41.

[176] Nakajima M, Yamada T, Kusuhara T, Furukawa H, Takahashi M, Yamauchi A, Kataoka Y. Results of quinacrine administration to patients with Creutzfeldt-Jakob disease. Dement Geriatr Cogn Disord 2004;17:158–63.

[177] Collinge J, Gorham M, Hudson F, Kennedy A, Keogh G, Pal S, Rossor M, Rudge P, Siddique D, Spyer M, Thomas D, Walker S, Webb T, Wroe S, Darbyshire J. Safety and efficacy of quinacrine in human prion disease (PRION-1 study): a patient-preference trial. Lancet Neurol 2009;8:334–44.

[178] Mead S, Ranopa M, Gopalakrishnan GS, Thompson AG, Rudge P, Wroe S, Kennedy A, Hudson F, MacKay A, Darbyshire JH, Collinge J, Walker AS. PRION-1 scales analysis supports use of functional outcome measures in prion disease. Neurology 2011;77:1674–83.

CHAPTER

THE MOTOR NEURON DISEASES AND AMYOTROPHIC LATERAL SCLEROSIS

5

CHAPTER OUTLINE

The Molecular and Clinical Pathology of Neurodegenerative Disease. https://doi.org/10.1016/B978-0-12-811069-0.00005-7

5.1 INTRODUCTION

Amyotrophic lateral sclerosis, known as motor neuron disease in the United Kingdom and often called Lou Gehrig disease in the United States [1] (Box 5.1), is a degenerative disorder that involves the progressive loss of motor neurons in the central nervous system. Acting to connect the brain with muscles, motor neurons are critical for the control of muscular function; the death of these cells leads to a catastrophic reduction of muscle innervation, resulting in muscle wasting and rapid loss of mobility. The term amyotrophic lateral sclerosis, first applied by Jean-Martin Charcot [2], derives from Greek and refers to a loss of muscle feeding (amyotrophic), the lateral cells of the spinal column and scarring

BOX 5.1 WHAT IS IN A NAME (1)?

What is in a name (1): Lou Gehrig and his disease. Henry Louis Gehrig was a baseball player for the New York Yankees and was one of the most successful batsmen of his generation—hitting 493 home runs. He was diagnosed with amyotrophic lateral sclerosis in 1939, dying of the disease in 1941 at the age of 37. His death raised public awareness of motor neuron disease, and to this day, amyotrophic lateral sclerosis is also known as Lou Gehrig disease, especially in the United States.

(sclerosis) (Box 5.2). The use of the term motor neuron disease is more recent, dating from the 1930s and the writings of the eminent British neurologist Russell Brain, the author of *Diseases of the Nervous System* [3]. The eventual consequence of the degeneration of motor neurons in the central nervous system is loss of autonomic function, including lung function, and death. Similar to many other neurodegenerative diseases, significant advances have been made in our understanding of amyotrophic lateral sclerosis in the past several decades, partly driven by the revolution in molecular genetics. In particular, an increasing appreciation of the genetic component in the etiology of amyotrophic lateral sclerosis has revealed that there are important overlaps between this disorder and neurodegeneration in the brain, leading to dementia (see Chapter 2). This has led to a revolution in the way that the motor neuron diseases are viewed as clinical entities [4] and provided a window on the processes that lead to regional specificity and selective vulnerability in neurodegeneration. Despite these advances, there is no disease-modifying treatment for this devastating disorder, although experimental therapies for some forms of motor neuron degeneration are beginning to show promise.

BOX 5.2 WHAT IS IN A NAME (2)?

What is in a name (2): Charcot and amyotrophic lateral sclerosis. Jean-Martin Charcot, shown here in the painting "Une Lecon Clinique a la Salpetriere" by Pierre Brouillet of one of his renowned clinical lectures at the Hopital Salpetriere in Paris, stands as one of the key figures in the history of neurology. His work to define and categorize diseases of the brain still influences diagnosis in clinical practice, and he was responsible for the derivation of the name amyotrophic lateral sclerosis for a subset of motor neuron diseases.

5.2 CLINICAL PRESENTATION AND CLASSIFICATION

The term motor neuron disease refers to a group of disorders that cause progressive weakness of limb and bulbar muscles because of degeneration of the motor neurons. Amyotrophic lateral sclerosis is the most common form of motor neuron disease, and the terms amyotrophic lateral sclerosis and motor neuron disease are sometimes used interchangeably.

The hallmark of amyotrophic lateral sclerosis is involvement of both upper and lower motor neurons. Lower motor neurons are located in the anterior horn of the spinal cord, hence the term anterior horn cell disease, which is used occasionally to describe the disease. The lower motor neurons in the brainstem, the motor nuclei of the cranial nerves, may also be affected. Lower motor neuron involvement is characterized by weakness, muscle wasting, and fasciculation (or muscle twitch). Upper motor neurons are located in the motor cortex and give rise to the corticospinal and corticobulbar tracts. Loss of upper motor neurons results in spasticity and brisk reflexes. Generally, a mix of both upper and lower motor neuron signs is seen in motor neuron disease although some patients have a mainly upper motor neuron form, whereas others have a predominantly lower motor neuron presentation (Fig. 5.1).

The symptoms of motor neuron disease typically begin with focal weakness in one limb. Distal muscles are frequently affected early in the disease course, and patients may complain of hand clumsiness or foot drop. Head drop is seen if the neck extensor muscles are involved. Muscle cramps and stiffness may occur with upper motor neuron involvement and this can be painful. Symptoms begin in the bulbar muscles in 20% of cases causing dysarthria and dysphagia. These patients tend to have a worse prognosis [5].

Patients who have upper motor neuron involvement of cranial nerves IX, X, and XII may develop a pseudobulbar affect where there is dysregulation of the motor output of emotion with excessive crying and laughing.

Regardless of onset, the weakness almost inevitably spreads to involve most muscles including the diaphragm, with death occurring on average 20–48 months post symptom onset, most often due to respiratory failure [6]. The diagnosis of motor neuron disease is predominantly clinical, based on a history of progressive weakness that spreads to different body segments with time. Sensory symptoms and signs should be absent, and the clinician will look for a mix of upper and lower motor neuron signs on examination. Electrophysiology analyses can support the diagnosis. Nerve conduction studies will show that the sensory nerve action potentials are normal, but electromyography of affected muscles will show active denervation and reinnervation. Compound motor nerve action potential amplitudes may be reduced in severely atrophied muscles. For a diagnosis of motor neuron disease to be made, it is important that other disorders that can cause mixed upper and lower motor neuron signs, such as cervical spondylosis, are excluded and patients will generally have a magnetic resonance image of their cervical spine. The revised El Escorial criteria were developed to ascertain clinical certainty about the diagnosis of motor neuron disease and are mainly used for research purposes (Box 5.3) [7,8].

5.2.1 EPIDEMIOLOGY

The incidence of motor neuron disease is approximately 2–3 per 100,000 with a slightly increased risk in males [9]. The majority of cases are sporadic in nature, but around 5% of cases are familial. There have been significant advances in recent years with the identification of multiple different genes that are associated with amyotrophic lateral sclerosis (see Section 5.4), and an increasing appreciation of the contribution that genetics makes to the etiology of motor neuron degeneration [10].

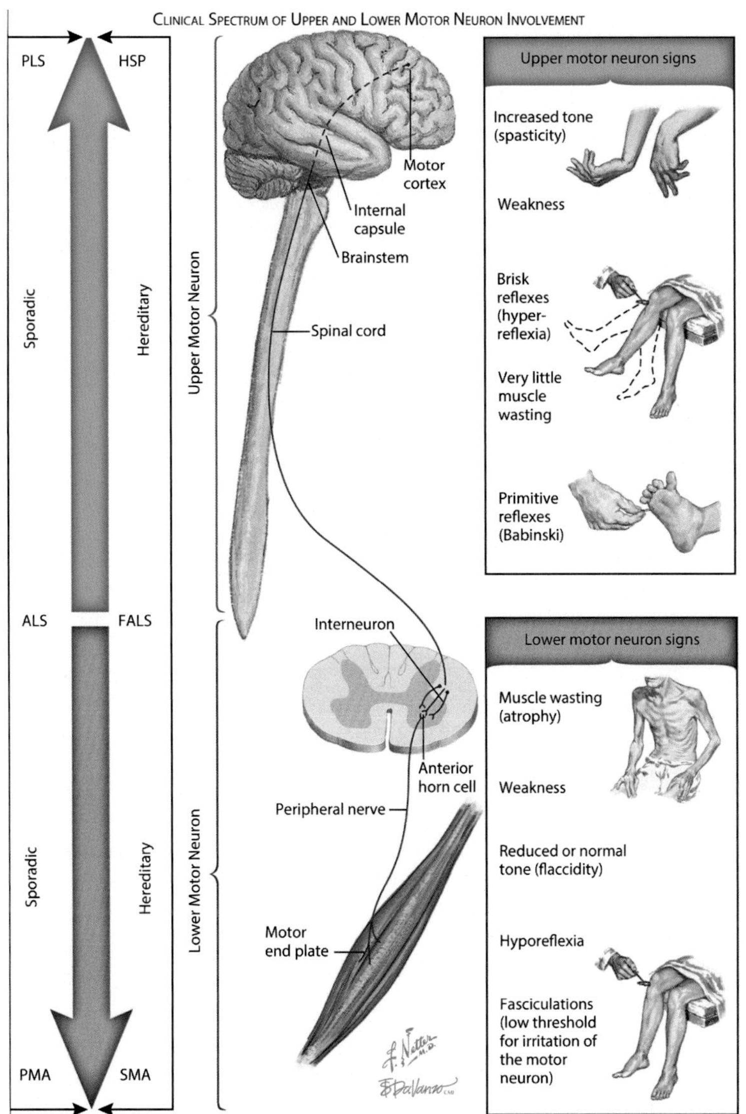

FIGURE 5.1

Upper and lower motor neuron diseases. The clinical phenotype of motor neuron diseases can be directly linked to the spatial distribution of motor neuron degeneration.

Image courtesy of Netter Images.

BOX 5.3 THE EL ESCORIAL CRITERIA

The revised El Escorial criteria for the diagnosis of amyotrophic lateral sclerosis

The presence of:

(A: 1) evidence of lower motor neuron (LMN) degeneration by clinical, electrophysiological or neuropathologic examination

(A: 2) evidence of upper motor neuron (UMN) degeneration by clinical examination

(A: 3) progressive spread of symptoms or signs within a region or to other regions, as determined by history or examination

The absence of:

(B:1) electrophysiological or pathological evidence of other disease processes that might explain the signs of LMN and/or UMN degeneration, and

(B:2) neuroimaging evidence of other disease processes that might explain the observed clinical and electrophysiological signs.

The revised El Escorial criteria for diagnosis of amyotrophic lateral sclerosis. The El Escorial criteria provide a framework for the differential diagnosis of amyotrophic lateral sclerosis, providing positive inclusion criteria (in green) and negative exclusion criteria (in red).

There are a number of examples of localized increases in incidence for motor neuron disease, with two standouts being the pacific island of Guam and the Kii peninsula in Japan [11,12]. The pathology of these disorders diverges from more common forms of amyotrophic lateral sclerosis, and there may be distinct etiologies underlying their development (see Sections 5.3 and 5.4). These conditions are also found in association with a form of parkinsonism–dementia complex, emphasizing the distinct nature of these localized disorders. There is also evidence linking traumatic brain injury to the incidence of amyotrophic lateral sclerosis, linking to the cause of chronic traumatic encephalopathy (see Chapter 2) [13].

5.2.2 CLINICAL DIVERSITY/RELATED DISORDERS

Amyotrophic lateral sclerosis is characterized by involvement of both upper and lower motor neurons. However, there is a spectrum with some patients having predominantly upper motor neuron presentations and others having a predominantly lower motor neuron presentation.

5.2.3 PROGRESSIVE BULBAR PALSY

Progressive bulbar palsy accounts for 15%–30% of cases of amyotrophic lateral sclerosis and is an exclusively bulbar presentation at diagnosis. Because of the early involvement of muscles in swallowing and respiration, it has a poor prognosis with a median survival of 2 years.

5.2.4 PRIMARY LATERAL SCLEROSIS

Primary lateral sclerosis accounts for 5% of cases of motor neuron disease and is an exclusively upper motor neuron form of the disease. There is selective involvement of corticospinal and corticopontine motor neurons causing spastic muscle stiffness without the typical lower motor neuron signs of fasciculations and atrophy. It tends to present in older patients, and there may be prominent bulbar involvement. It is generally more slowly progressive than classical amyotrophic lateral sclerosis [14].

5.2.5 PRIMARY MUSCULAR ATROPHY

Primary muscular atrophy, in contrast, is an exclusively lower motor neuron presentation. However, subclinical upper motor neuron involvement may be identified on electrophysiology, and patients may develop upper motor neuron signs as the disease progresses [15]. Survival tends to be longer than that observed in classical amyotrophic lateral sclerosis [16].

5.2.6 SPINAL MUSCULAR ATROPHY

Spinal muscular atrophy is a distinct clinical entity characterized by the degeneration of lower motor neurons [17]. It is a purely genetic disorder, caused by autosomal recessive mutations in the *SMN1* gene on chromosome 5 [18]. These mutations have a variable clinical presentation, caused partly by the presence of a paralog of *SMN1* also found on chromosome 5, *SMN2*. *SMN2* codes for a nearly identical protein but is subject to variable alternative splicing resulting in a range of transcripts, some of which are able to compensate for loss of *SMN1* but many of which do not. Spinal muscular atrophy can be divided into four types, with an onset ranging from just after birth through to adulthood, and dependent on the compensatory expression from *SMN2* [19]. The symptoms include muscle weakness and hypotonia, caused by the degeneration of motor neurons in the spinal cord. For type I, symptoms develop soon after birth and are severe, with very little motor control. Type I spinal muscular atrophy is fatal within 24 months of birth. Types II–IV are of decreasing severity, with individuals presenting with type IV developing disease in adolescence or early adulthood and with only mild motor symptoms [17].

5.2.7 OVERLAP WITH FRONTOTEMPORAL DEMENTIA

Amyotrophic lateral sclerosis was previously considered to be a pure motor disorder; however, in recent years, it has become apparent that cognitive impairment is seen in patients with amyotrophic lateral sclerosis. Indeed, up to 40% of patients with amyotrophic lateral sclerosis will have some cognitive impairment and 14% will meet the criteria for dementia [20]. The dementia that is typically observed in amyotrophic lateral sclerosis is frontal variant frontotemporal dementia, with executive function particularly affected. Language may also be impaired. In additon to the clinical overlaps between frontotemporal dementia and motor neuron disease, there are pathological and radiological overlaps between the two disorders [21]. The balance between motor neuron and cortical degeneration is central to the clinical presentation (Fig. 5.2) and is currently poorly understood.

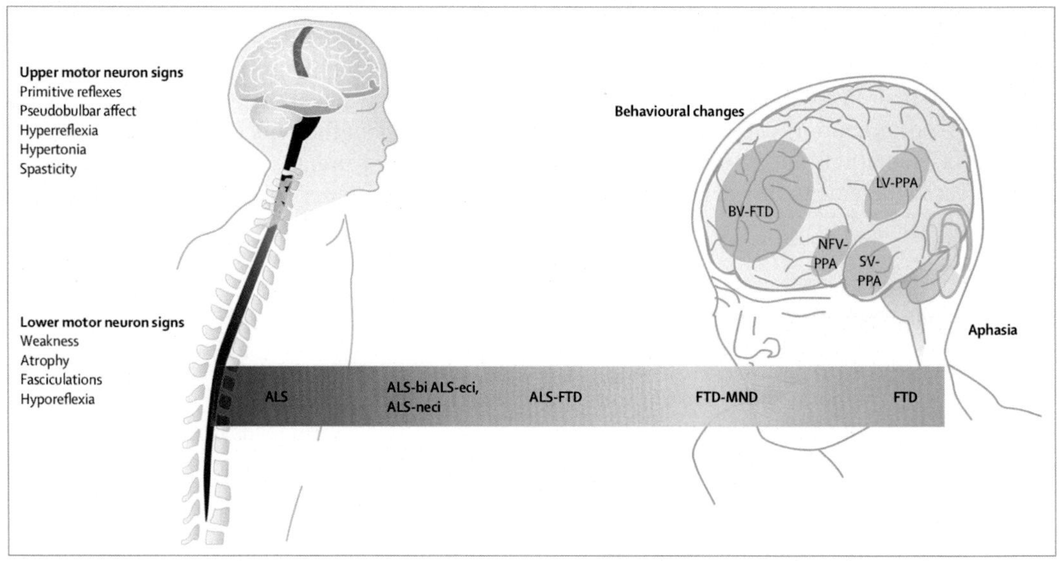

FIGURE 5.2

The amyotrophic lateral sclerosis/frontotemporal dementia clinical spectrum. Recent advances in our understanding of the genetics of motor neuron dysfunction have highlighted that the spectrum of degeneration extends beyond motor neurons and includes cortical pathology leading to dementia. *ALS*, amyotrophic lateral sclerosis; *ALS-bi*, amyotrophic lateral sclerosis with behavioral changes; *ALS-eci*, amyotrophic lateral sclerosis with evidence of executive dysfunction; *ALS-neci*, amyotrophic lateral sclerosis with no executive dysfunction but impairment in other cognitive domains; *bvFTD*, behavioral variant frontotemporal dementia; *FTD*, frontotemporal dementia; *lvFTD*, logopenic variant frontotemporal dementia; *MND*, motor neuron disease; *nfvFTD*, nonfluent variant frontotemporal dementia; *svFTD*, semantic variant frontotemporal dementia.

Image taken from van Es MA, Hardiman O, Chio A, Al-Chalabi A, Pasterkamp RJ, Veldink JH, van den Berg LH. Amyotrophic lateral sclerosis. Lancet 2017;390:2084–98.

5.2.8 TREATMENT

Motor neuron disease is a progressive disease, and unfortunately, there is currently no cure. The mainstay of management revolves around supportive care. It is important that there is early involvement of a multidisciplinary team including physiotherapists, occupational therapists, and speech and language therapists. Motor neuron disease is a devastating diagnosis, and patients must be given support regarding the difficult decisions that must be made regarding end-of-life care. These decisions can include whether to have a percutaneous endoscopic gastrostomy tube inserted to supplement nutrition when swallowing becomes difficult and whether to have respiratory support with noninvasive ventilation. Speech synthesizers and other communication aids are important as speech becomes affected. Palliative care teams provide invaluable support in making decisions about interventions that patients wish to have and can help alleviate the physical and psychological distress in the terminal stages of the disease [22].

Two drugs have been approved by the Food and Drug Administration as disease-modifying treatments for ALS. Riluzole blocks glutamate release and prevents excessive motor neuron firing. It has been shown to confer only a modest survival advantage of 2–3 months [23].

Edaravone is a free radical scavenger that was originally developed as a treatment for acute ischemic stroke. A beneficial effect in an ALS rating scale has been recently shown in a small subset of patients compared with placebo [24]. Edaravone is administered intravenously, which may limit its use because of practical implications.

5.3 PATHOLOGY

The pathological condition observed in amyotrophic lateral sclerosis and other motor neuron diseases is characterized primarily, as indicated by their names, by the progressive loss of motor neurons. The precise clinical presentation of a patient depends on where this loss occurs—centered on upper motor neurons, lower motor neurons, or both (see Fig. 5.1). In addition to motor neuron dysfunction and loss, the extent to which cognitive dysfunction and frontotemporal dementia is observed can be related to damage and cell loss in the frontal lobes of the brain (see Fig. 5.2). The cellular pathologies associated with amyotrophic lateral sclerosis were for many years unclear, although the identification of monogenic forms of disease caused by mutations in genes such as *SOD1* and *TARDBP* has led to a much better understanding of this [25,26]. It is now recognized that the majority of patients developing amyotrophic lateral sclerosis display intracellular inclusions containing TDP-43 protein (Fig. 5.3) [27]. Indeed, the identification of *TARDBP* as a key player in amyotrophic lateral sclerosis derived from the isolation of TDP-43 protein from the brains of individuals with ubiquitinated inclusions [28]. These inclusions also form a critical part of the pathological and etiological overlap with frontotemporal dementia, where TDP-43 inclusions are an important component of the pathological spectrum [29].

Beyond TDP-43, a number of other proteins have been linked to cellular pathology in patients with amyotrophic lateral sclerosis. Prominent among these are C9ORF72, FUS, and superoxide dismutase 1. All these are the products of genes that are mutated in familial forms of disease, with mutations in *C9ORF72* in particular representing a common genetic cause of disease (see Section 5.4). In terms of the specific pathology in these cases, patients with mutations in *C9ORF72* present with TDP-43 pathology but have additional neuronal cytoplasmic inclusions made up of a dipeptide product (a repeating glycine/alanine polypeptide) of the *C9ORF72* gene [30]. For patients harboring mutations in the *FUS* gene, the primary cellular pathological condition observed is intracellular aggregates of the FUS protein [31]. Finally, for patients with mutations in *SOD1*, the protein aggregate pathological condition observed consists of misfolded superoxide dismutase 1 protein [32]. In addition, there are rare forms of inherited disease with an overlapping clinical presentation where the pathology does not match. The neuropathological spectrum and categorization of these disorders is shown in Fig. 5.4.

Beyond specific protein aggregates, a number of other changes are characteristic of the pathological condition observed in the central nervous system of individuals with amyotrophic lateral sclerosis. Notable among these are the presence of Bunina bodies [33] and reactive astrocytes [34]. Bunina bodies are intracellular inclusions of unclear origin and makeup but are common in the lower motor neurons of individuals with amyotrophic lateral sclerosis, as well as being reported in the brains of individuals with frontotemporal dementia [35]. Activated astrocytes are a common feature of neurodegeneration across a wide range of disorders (e.g., see Chapter 2 and the neuropathology of dementia) and are thought to represent a reaction of the brain to the degenerative process, while also potentially

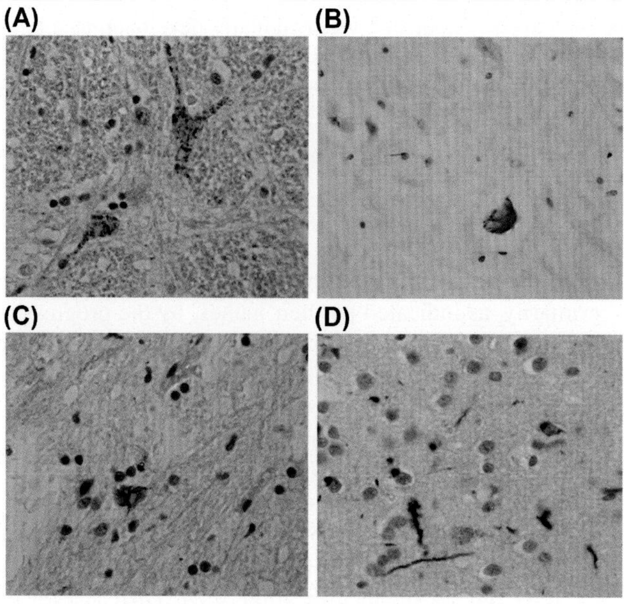

FIGURE 5.3

TDP-43 pathology in motor neuron diseases and beyond. (A) TDP-43 aggregates and nuclear clearing in the spinal cord of patient with idiopathic amyotrophic lateral sclerosis. (B) TDP-43 skein-like inclusion in the spinal cord of a *TARDP* mutation carrier. (C) TDP-43 pathology in the spinal cord of a patient with a mutation in *C9ORF72*. (D) TDP-43 inclusions in a patient with frontotemporal dementia.

Image taken from Jorge Gomez-Deza, Christopher E. Shaw, Chapter 5 - Amyotrophic Lateral Sclerosis and Other TDP-43 Proteinopathies, Editor(s): Michael S. Wolfe, The Molecular and Cellular Basis of Neurodegenerative Diseases, Academic Press; 2018. p. 99–115.

	Amyotrophic lateral sclerosis and frontotemporal dementia		
Aggregate	TDP-43	FUS	SOD1
Clinical entities	ALS-TDP FTD-TDP	ALS-FUS FTD-FUS NIFID BIBD	ALS-SOD1
Genes	*TARDBP* *C9ORF72* *VCP* *GRN* *ANG*	*FUS*	*SOD1*

FIGURE 5.4

The spectrum of pathologies observed in amyotrophic lateral sclerosis and frontotemporal dementia. *ALS*, amyotrophic lateral sclerosis; *BIBD*, basophilic inclusion body disease; *FTD*, frontotemporal dementia; *NIFID*, neuronal intermediate filament inclusion disease. Categorization derived from the schematic in Ref. [30].

exacerbating cytotoxic pathways [36]. The role of astrocytes, as well as other glial cells, in the process that leads to motor neuron degeneration in amyotrophic lateral sclerosis has been a subject of much interest for a number of years (see Section 5.4).

As noted earlier (Section 5.2), the epidemiology and clinical presentations of the amyotrophic lateral sclerosis/parkinsonism–dementia complexes reported on the Micronesian island of Guam and the Kii peninsula in Japan differ markedly from other motor neuron diseases. This is also true of the neuropathology, with both of these spectrum disorders predominantly presenting with neurofibrillary tangle pathology similar to that observed in Alzheimer disease (Chapter 2) and progressive supranuclear palsy (Chapter 3) [37,38]. This is also the case for chronic traumatic encephalopathy presenting with motor neuron degeneration, although there is also evidence of TDP-43 accumulation in these cases [39,40].

5.4 MOLECULAR MECHANISMS OF DEGENERATION

Our understanding of the molecular mechanisms that lead to neurodegeneration in amyotrophic lateral sclerosis and related disorders has undergone a revolution over the past two decades, driven primarily by the identification of monogenic forms of disease. More recently, and similar to many other neurodegenerative disorders, genome-wide association studies have extended our comprehension of the genetic basis for risk of developing amyotrophic lateral sclerosis. In tandem, the genes identified by these studies have cast light on the molecular changes that precede and predispose to degeneration [41].

5.4.1 THE GENETICS OF MOTOR NEURON DISEASES

Until relatively recently, the genetics of motor neuron degeneration were uncomplicated, with only one major identified Mendelian locus for the disease. This has changed significantly over the past 10 years, with a large number of monogenic, inherited forms of disease being identified (Fig. 5.5). As will be discussed in the following sections, the functions of these genes appear to coalesce around a number of pathways, leading to neuronal cell death.

The first gene to be identified as harboring causative mutations for amyotrophic lateral sclerosis was the *SOD1* gene in 1993 [42]. *SOD1* codes for superoxide dismutase 1, a protein that converts superoxide free radicals into nonreactive oxygen species [43]. In the original study, missense coding mutations in the gene, located on chromosome 21, were identified by position cloning in a series of 13 families with autosomal dominant amyotrophic lateral sclerosis. Many other mutations in *SOD1* have since been identified in families with inherited motor neuron degeneration around the world, with mutations in this gene causing around 10% of all familial amyotrophic lateral sclerosis cases [44].

For many years, *SOD1* was one of the few clear genes directly linked to the onset of amyotrophic lateral sclerosis. This began to change following the sequencing of the first draft of the human genome as technological advances facilitated the rapid screening of cases with inherited disease. Mutations were identified in a number of genes including *VCP* [45] and *DCTN1* [46], although for each of these the number of families carrying these mutations was low. Notably, mutations in *DCTN1* were subsequently also identified in Perry syndrome, expanding the clinical spectrum of disease

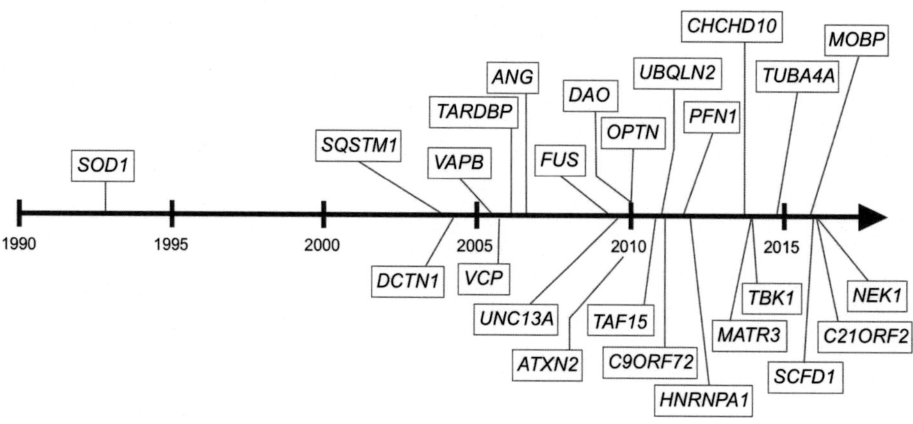

FIGURE 5.5

Timeline for gene identification in amyotrophic lateral sclerosis.

associated with this gene [47]. The identification of a key clearly causative gene for amyotrophic lateral sclerosis came, however, from a more traditional approach. Since the 1980s, it had been recognized that surviving motor neurons in individuals with amyotrophic lateral sclerosis were often characterized by the presence of ubiquitylated cytoplasmic inclusions [48,49]. In 2006, Neumann and coworkers identified the main protein constituent of these inclusions as TDP-43, encoded by the *TARDBP* gene on chromosome 1 [28], a result swiftly corroborated by Arai and colleagues [50]. Soon after the identification of this protein, mutations were described in individuals with familial and sporadic motor neuron disease [51,52], linking together the pathology and genetic etiology of these cases. Since 2008, it has become clear that TDP-43 inclusions are found across a wide range of neurodegenerative disorders, not just those with mutations in the *TARDBP* gene, and that a significant proportion of Mendelian cases of amyotrophic lateral sclerosis are caused by mutations in this gene [53].

Soon after the identification of mutations in *TARDBP*, mutations in the *FUS* gene on chromosome 16, coding for the FUS (fused in sarcoma) protein, were identified in a British family with autosomal dominant inheritance of amyotrophic lateral sclerosis as well as a number of other unrelated families [54,55]. In these families, intraneuronal inclusions consisting of misfolded FUS protein are observed [56], defining a clinico-pathological subgroup within familial amyotrophic lateral sclerosis. Intriguingly, there is substantial functional overlap between FUS and TDP-43, both of which are involved in the processing of RNA within the cell (see Section 5.4.5) [57].

A major advance in our understanding of the genetics of motor neuron degeneration (and, indeed, neurodegeneration more broadly) came with the identification of hexanucleotide repeat mutations in the *C9ORF72* gene, located on chromosome 9 and deriving its name from being the 72nd open reading frame of undetermined function on that chromosome in 2011 [58,59]. These mutations, resulting in an increased number of repeats of the nucleotide sequence GGGGCC, are located in a noncoding region of the gene [60]. In healthy individuals, these repeats range in number from 2 to 20—whereas in families with inherited neurodegeneration, the number ranges from several hundred to many thousands.

Similar to other nucleotide repeat disorders (e.g., Huntington disease, see Chapter 6), the onset of disease is linked with a substantial increase in these repeats. The identification of mutations in the *C9ORF72* gene provided strong evidence of a critical link between the underlying etiology of amyotrophic lateral sclerosis and frontotemporal dementia, as mutations in this gene were quickly recognized as being a common genetic cause of both disorders, with many patients presenting with both dementia and motor neuron degeneration. Importantly, the pathological condition observed in these cases has a clear TDP-43 component—linking to another key player in the genetics of amyotrophic lateral sclerosis—as well as intraneuronal inclusions of a dipeptide encoded by the repeat expansion [30]. *C9ORF72* mutations also highlight one of the challenges in terms of uncovering genetic forms of disease. The nature of the repeat expansion in this gene introduces a technical hurdle with regard to characterizing the precise number of repeats, a task that traditional Sanger sequencing approaches are ill suited to.

Over the past decade, a number of genes linked to the cellular process of macroautophagy have been identified as being mutated in amyotrophic lateral sclerosis, providing evidence of a key molecular pathway being disrupted in disease (see Section 5.4.3) [61]. These include *OPTN* in 2010 [62], *SQSTM1* [63,64], and *TBK1* [65,66], and there is now extensive cellular evidence suggesting that at least one of the physiological roles that is fulfilled by C9ORF72 relates to the regulation of macroautophagy [67].

As noted earlier (Section 5.2.6), spinal muscular atrophy is a distinct genetic clinic-pathological entity caused by loss of function in the *SMN1* gene located on chromosome 5 [18]. Mutations cause disease in an autosomal recessive fashion, with the variable clinical phenotype linked to expression of the *SMN2* gene (also located on chromosome 5) [68]. This gene produces a truncated transcript with functional homology to the product of the *SMN1* gene; however, some individuals possess alleles of the gene that express longer transcripts allowing for compensation of the loss of *SMN1* expression [69].

The increasing application of genome-wide analysis has had major benefits in terms of identifying genes strongly linked to amyotrophic lateral sclerosis (e.g., *TBK1*, which was partly identified by an exome sequencing strategy) but has had an even bigger impact on our ability to assess common genetic variation that increases lifetime risk of developing amyotrophic lateral sclerosis. Genome-wide association studies, comparing and contrasting genetic variant frequency in cases and controls, has uncovered a large number of loci associated with increased risk [70]. These have been combined with sophisticated bioinformatic analyses of genetic risk burden to nominate specific genes [71], or using cross-ethnic comparisons to refine and stratify genes associated with risk [72]. The genes nominated by genome-wide association studies have shed light on the links between Mendelian disease and sporadic amyotrophic lateral sclerosis, both in terms of shared pathways and shared genes. As for other neurodegenerative disorders, this has resulted in an increasingly detailed genetic architecture for disease (Fig. 5.6). A summary of the genes linked to amyotrophic lateral sclerosis is shown in Table 5.1.

The clinical heterogeneity of motor neuron disease, coupled to the complexities of the overlap with frontotemporal dementia, provides a new frontier in genetic analysis of these disorders [73]. In particular, why is it that individuals with mutations in the same gene can develop one form of neurodegeneration or another—as seen in *C9ORF72* mutation carriers. Significant efforts are now underway to understand multigenic inheritance of risk for developing motor neuron/cortical degeneration [74,75], as well as genetic modifiers for established Mendelian loci [76].

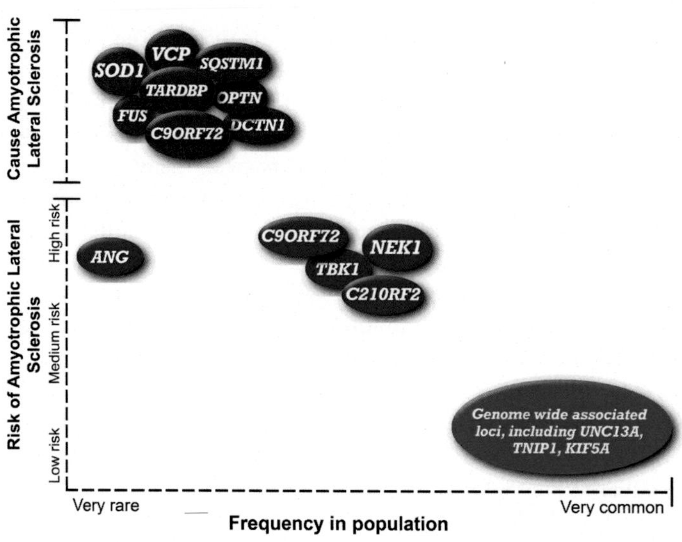

FIGURE 5.6

The genetic architecture of motor neuron diseases, showing gene variant frequency in the population on the x-axis and lifetime risk of disease on the y-axis.

5.4.2 PROTEIN AGGREGATION AND MOTOR NEURONS

As noted in Section 5.3, the accumulation of protein aggregates in the motor neurons of individuals who develop amyotrophic lateral sclerosis is now recognized as a key feature of the pathology of this disorder, as is also the case for frontotemporal dementia. It is also clear from the genetics of these disorders that there is a very close link between several of the genes that are mutated in familial amyotrophic lateral sclerosis and the proteins that are deposited in the central nervous systems of individuals developing disease. Protein aggregation and associated cytotoxicity is a common theme across a number of neurodegenerative diseases (as discussed in Chapters 2, 3, and 4 relating to dementia, Parkinson disease, and the prion disorders), and there is now a consensus that this is one of the key molecular mechanisms driving motor neuron death in amyotrophic lateral sclerosis. Significant experimental efforts have been undertaken to understand the precise role of protein aggregation and selective vulnerability in amyotrophic lateral sclerosis, efforts that began with the examination of the impact of mutations in superoxide dismutase 1, the first protein to be directly linked to inherited motor neuron disease. Following the identification of point mutations in this gene in 1993, detailed characterization of the biochemical, cellular, and organismal impact of these variants was carried out. This revealed that many mutated forms of superoxide dismutase 1 possessed wild-type enzymatic activity, making it unlikely that it was a loss of this function that was leading to disease. Combined with data from *SOD1* knockout mice revealing that these animals had no overt phenotype [77], this focused attention on a putative toxic gain of function—with increased misfolding and protein aggregation of particular interest because of the cellular pathological condition observed in *SOD1* mutation carriers. Looking across recombinant protein and cellular/animal models for *SOD1* mutants, there is a consistent increase in the aggregation of superoxide dismutase 1, resulting in the formation of large fibrillar structures that share some

Table 5.1 Summary of Genes Implicated in Amyotrophic Lateral Sclerosis and Frontotemporal Dementia

Gene	Chromosome	Phenotype	Function
TARDBP	1	ALS, FTD	RNA biology
DCTN1	2	ALS, Perry syndrome	Axonal transport
ALS2	2	ALS, PLS, HSP	Vesicle trafficking
TUBA4A	2	ALS, FTD	Cytoskeletal organization, axonal transport
CHMP2B	3	FTD, ALS	Proteostasis
NEK1	4	ALS	Cytoskeletal organization, DNA damage
MATR3	5	ALS, FTD, myopathy	RNA biology
SQSTM1	5	ALS, Paget disease, ataxia, dystonia	Autophagy, proteostasis
HNRNPA2B1	7	ALS, myopathy, CI, Paget disease	RNA biology
ELP3	8	ALS	RNA biology
C9ORF72	9	ALS, FTD	Autophagy, proteostasis, RNA biology
VCP	9	ALS, FTD, CMT, Paget disease	Proteostasis
SETX	9	ALS, ataxia	RNA biology
OPTN	10	ALS	Autophagy, proteostasis
HNRNPA1	12	ALS, myopathy, CI, Paget disease	RNA biology
TBK1	12	ALS, FTD	Autophagy, proteostasis
ATXN2	12	ALS, ataxia	RNA biology
KIF5A	12	ALS, HSP	Axonal transport
ANG	14	ALS	Angiogenesis
SPG11	15	ALS, HSP, CMT	Autophagy, DNA damage
CCNF	16	ALS, FTD	Proteostasis
FUS	16	ALS, essential tremor	RNA biology
PFN1	17	ALS	Cytoskeletal organization, axonal transport
VAPB	20	ALS, SMA	Proteostasis
SOD1	21	ALS	Proteostasis, oxidative stress
C21ORF2	21	ALS, FTD	Cytoskeletal organization
CHCHD10	22	ALS, FTD, ataxia, myopathy, SMA	Mitochondrial function
NEFH	22	ALS, CMT	Axonal transport
UBQLN2	X	ALS	Proteostasis

ALS, *amyotrophic lateral sclerosis;* CI, *cognitive impairment;* CMT, *Charcot–Marie–Tooth;* FTD, *frontotemporal dementia;* HSP, *hereditary spastic paraplegia;* PLS, *primary lateral sclerosis;* SMA, *spinal muscular atrophy.*

features with other proteins deposited in neurodegenerative disorders such as amyloid beta and alpha synuclein (in Alzheimer and Parkinson disease, respectively) [78–81]. These models have been used extensively in drug development efforts.

The identification of mutations in the *TARDBP* and *FUS* genes, in addition to the characterization of protein aggregates made up of the protein products of these genes, opened a new chapter in the role of protein aggregation in amyotrophic lateral sclerosis and its links to other neurodegenerative diseases. Both TDP-43 and FUS proteins are found in intracellular aggregates in amyotrophic lateral sclerosis, with TDP-43 pathology found beyond individuals with mutations in the *TARDBP* gene and extending into dementia. For both proteins, mutations result in an increase in protein aggregation and in cytotoxicity—providing a causal link between protein aggregates and neurodegeneration [54,82]. However, for TDP-43, these aggregates do not form classical amyloid deposits [83]. Further evidence of the role of protein aggregates in amyotrophic lateral sclerosis/frontotemporal dementia has also come from studies of the most frequently mutated gene in these disorders: *C9ORF72*. Although the hexanucleotide repeat that is expanded in *C9ORF72* mutation carriers sits in a noncoding region of the gene, studies in cellular and animal models have revealed that this expansion undergoes repeat-associated non-ATG translation, with a dipeptide glycine/alanine repeat aggregation-prone polypeptide produced [84,85].

As for other aggregated proteins in neurodegenerative diseases, the precise mechanism whereby the presence of these aggregates leads to cell death has yet to be determined. There is increasing evidence that the early steps in the process of aggregation leading to the formation of large protein aggregates are intrinsically cytotoxic and may have a disproportionate impact on vulnerable cells in the central nervous system [86]. Motor neurons in particular may be selectively vulnerable because of their energetic requirements and long axonal processes [87]. As discussed further in Section 5.4.4, there is increasing interest in the possibility that aggregates of proteins found in amyotrophic lateral sclerosis may be able to transmit from cell to cell and propagate in a fashion analogous to that seen in the prion diseases.

5.4.3 MACROAUTOPHAGY AS A PATHWAY IN AMYOTROPHIC LATERAL SCLEROSIS

The protein products of a number of genes causatively linked to or associated with lifetime risk of developing amyotrophic lateral sclerosis have been implicated in the cellular process of macroautophagy. Macroautophagy is one of the key processes used by eukaryotic cells to remove waste and recycle cellular contents [88], and there is extensive experimental evidence linking dysfunction of this system to neurodegeneration [89]. Interest in the possibility that there might be a causal link between macroautophagy and amyotrophic lateral sclerosis at a genetic level first emerged with the identification of mutations in several genes with links to this process. The *SQSTM1* gene, coding for p62, is a critical component of the labeling and sequestering of cargo for autophagosomes [90]. Mutations in this gene cause a range of inherited disorders, such as Paget's disease of bone [91], and are linked to familial forms of amyotrophic lateral sclerosis [64]. Mutations in *OPTN*, coding for the protein optineurin, are also found in familial amyotrophic lateral sclerosis [62]. Optineurin is thought to play a similar cellular role to p62, labeling aggregates for destruction via macroautophagy, with a loss of function in the *OPTN* gene, resulting in disruption of this process [92]. Optineurin has also been linked to a specialized form of macroautophagy, mitophagy (responsible for the removal of damaged mitochondria), a process that has been strongly implicated in Parkinson's disease [93,94]. The identification of mutations in *TBK1*, coding for tank binding kinase 1, in patients with amyotrophic lateral sclerosis resulting in haploinsufficiency provided yet another link to macroautophagy [66]. Tank binding kinase 1 has been demonstrated to regulate the activity of optineurin by direct phosphorylation [92], putting these genetic forms of amyotrophic lateral sclerosis in the same pathway. Finally, recent data suggest that C9ORF72 may have a physiological role in regulating macroautophagy and the endolysosomal system—acting as a GDP/GTP exchange factor to alter the initiation of autophagy [95–97]. Whether this cellular role for C9ORF72 is in addition to the protein aggregation observed with dipeptide products of the repeat expansion in a disease situation, or that loss of C9ORF72 protein function (and thence autophagy dysregulation) because of altered translation in disease is the key event in etiology, has not been determined. Together, these experimental data suggest that macroautophagy has an important role to play in the etiology of amyotrophic lateral sclerosis and contributes a discrete pathway to disease (Fig. 5.7) [61]. Given the crucial role played by autophagy in the degradation of protein aggregates in the cell, one possibility is that dysfunction of this system in motor neurons could result in an increased propensity for cellular aggregates to form—however, why this would specifically impact on motor neurons (or cortical neurons in frontotemporal dementia) is not clear.

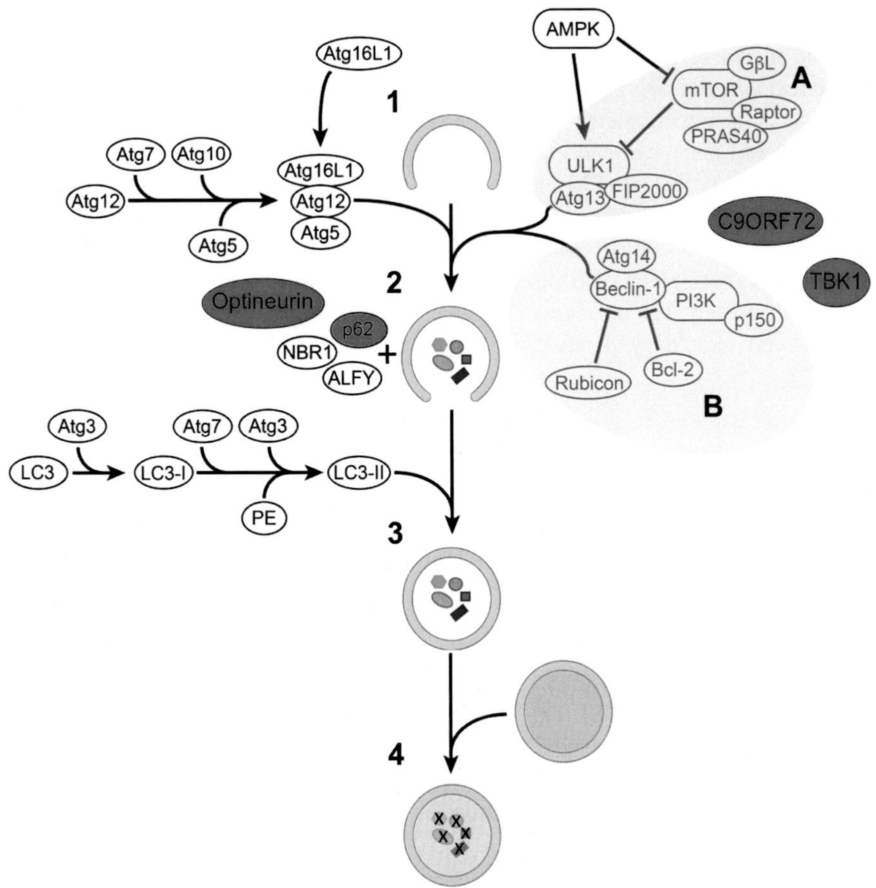

FIGURE 5.7

Macroautophagy as a pathway to motor neuron degeneration. The initiation of macroautophagy involves the initiation of phagophore formation (1), followed by maturation (2), closure (3), and fusion with the lysosomes (4). The canonical pathways involved in regulating autophagy are the mTOR/ULK1 pathway (A) and the Beclin/PI3 kinase pathway (B). Genes implicated in amyotrophic lateral sclerosis - highlighted in red - have been linked to the regulation of ULK1 activity (TBK1 and C9ORF72), and with the labeling and sequestration of cargo (p62, optineurin).

5.4.4 PRION-LIKE BEHAVIOR

As noted in previous chapters, an area of research of increasing interest across the field of neurodegeneration is that of prion-like behavior of protein aggregates in the central nervous system [98]. Prions are infectious protein aggregates that are capable of propagating between host organisms and transmitting aggregation phenotypes from cell to cell, resulting in cytotoxicity and neurodegeneration—a process that has been extensively characterized in the prion diseases such as scrapie and Creutzfeldt–Jakob disease [99]. There is now a substantial body of evidence suggesting that protein aggregates in Parkinson's and Alzheimer's diseases can act in a manner analogous to prions, albeit in the absence of data suggesting that person-to-person transmission of aggregates plays a role in the normal etiology of these disorders [100,101]. Given the important role of protein aggregation in amyotrophic lateral sclerosis, across TDP-43, FUS, superoxide dismutase 1, and C9ORF72, a number of groups have sought to test whether aggregates from these proteins can behave in a similar fashion to prions [102,103]. This has been examined for superoxide dismutase 1 [104,105], TDP-43 [106–108], and FUS [109,110], with some evidence that aggregates formed from these proteins can propagate between cells. Notably, both TDP-43 and FUS possess prion-like regions with homology to yeast and human prion proteins [111]. As for Alzheimer and Parkinson, the precise role that this prion-like behavior plays in disease is controversial [112]. The researchers involved in the original demonstration that kuru and Creutzfeldt-Jakob disease (two human prion diseases, see Chapter 4 for more details) could be transmitted to nonhuman primates did investigate the possibility that amyotrophic lateral sclerosis and related syndromes could also be transmitted but found no direct evidence of this [113,114].

5.4.5 DISRUPTION OF RNA FUNCTION

A cellular function that is linked to, and potentially disrupted in, amyotrophic lateral sclerosis via a number of genes implicated in the disorder is that of RNA processing and biology [115]. The potential link between motor neuron degeneration and RNA biology arose through the identification of mutations in *TARDBP* and *FUS*, both of which are RNA binding proteins and are localized to the nucleus under physiological conditions [116]. The normal function of TDP-43 is to regulate transcription and control exon skipping [117], whereas that of FUS is to shuttle RNA transcripts between the nucleus and the cytoplasm [118]. The impact of mutations on these proteins is to increase their propensity to aggregate, so one possible route to cell death is a direct cytotoxic impact of these aggregates (as outlined in Section 5.4.2). A consequence of the sequestration of TDP-43 or FUS in cytoplasmic aggregates is, however, that the normal function of these proteins is disrupted, leading to cell-wide alterations in RNA processing [119,120]. There is also evidence that dysfunction of TDP-43 and FUS can lead to the formation of stress granules [121,122], intracellular inclusions that form in response to stress to regulate gene expression and have been implicated in pathways to neurodegeneration [123]. An additional connection to altered RNA biology is provided by mutations in the *C9ORF72* gene, with the hexanucleotide repeat expansions in the noncoding regions of the gene resulting in large mRNA transcripts and dipeptide repeat polypeptides that can be recruited into RNA foci and stress granules, resulting in knock-on alterations in gene expression [124,125]. Taken together, these alterations have been proposed as a key pathway to motor neuron dysfunction and cell death in amyotrophic lateral sclerosis.

5.4.6 INTRACELLULAR TRAFFICKING

The long length of the axonal projections from many motor neurons, and the concomitant energetic requirements that these entail, has long been highlighted as a possible source of selective vulnerability in amyotrophic lateral sclerosis [87]. The molecular genetics of human motor neuron degeneration has highlighted a number of genes whose function is directly relevant to transportation along these long axons. Mutations in the *DCTN1* gene, encoding dynactin subunit 1, were identified in patients with amyotrophic lateral sclerosis in 2004 [46]. Dynactin subunit 1 is a critical component of the dynactin complex, a multiprotein assembly with a molecular mass in excess of a megadalton, with a critical role in cellular transport [126]. Intriguingly, the identification of mutations in human *DCTN1* followed on from the identification of mutations in the mouse *Dnchc1* gene associated with a motor neuron degenerative phenotype—*Dnchc1* codes for the dynein heavy chain protein, which interacts directly with dynactin to facilitate intracellular transport [127]. These results highlighted intracellular transport, and in particular retrograde axonal transport, as being critical for motor neuron survival—a cellular process that had previously been identified in mouse models for amyotrophic lateral sclerosis linked to mutations in *SOD1* [128]. More recently, the identification of *KIF5A*, coding for the kinesin heavy chain isoform A, as a risk locus for sporadic amyotrophic lateral sclerosis, has reemphasized the importance of cellular transport [71]. Kinesin is a cellular machine that acts to move cargo around the cell [129], with the function of this complex disrupted by the loss of *KIF5A* [130]. Mutations in this gene had previously been identified in a rare form of hereditary spastic paraplegia, a neurodegenerative movement disorder with some phenotypic overlap with amyotrophic lateral sclerosis [131]. The localization of genetic risk for amyotrophic lateral sclerosis to the *KIF5A* locus, combined with the identification of mutations in dynactin and evidence from mouse models, suggests that axonal transport is another potential pathway to motor neuron degeneration [132].

5.4.7 NON-CELL AUTONOMOUS DEGENERATION

An area of great interest across neurodegenerative diseases, and one where amyotrophic lateral sclerosis has made a significant contribution, is how the different cell types present in the central nervous system contribute to the disease process. Although neurodegenerative diseases are characterized, and (to an extent) defined by the loss of neurons in the central nervous system, it has been clear for many years that other cells resident in the central nervous system also have role to play in the etiology of these diseases [133]. In particular, the role of astrocytes and microglia in reacting to the degeneration of neurons, and potentially exacerbating the situation or contributing to a hostile environment, has been examined in some detail [134]. Genetically engineered animal models for genetic forms of neurodegeneration provide a powerful tool for studying this, allowing the directed expression of mutant genes in specific central nervous system cell types. Using this approach, a number of fascinating studies directing expression of mutated superoxide dismutase 1 to microglia [135] or astrocytes [136], but not to motor neurons, have demonstrated that mutant protein produced in these cells can result in motor neuron cell death despite these neurons expressing only wild-type protein. This has led to the development of a noncell autonomous model for *SOD1* ALS [137], a model with potentially far-reaching implications for neurodegenerative disease research. More recently, human stem cell–derived models for motor neuron function have replicated these results for mutant *SOD1* [138], reinforcing interest in exploring this paradigm in other genetic forms of amyotrophic lateral sclerosis, as well as other neurodegenerative disorders [139]. To date, however, it is not clear whether this model is directly applicable outside of *SOD1* linked disease.

5.4.8 SPINAL MUSCULAR ATROPHY

As a purely genetic disorder, spinal muscular atrophy can be linked directly to dysfunction of the *SMN1* gene—providing a clear pathway to causation that is somewhat more simple than in the genetically more heterogeneous amyotrophic lateral sclerosis. Spinal muscular atrophy is an autosomal recessive disease and is linked to demonstrable loss of function mutations in *SMN1*. The function of SMN1 protein is essential for the survival of lower motor neurons, although the precise reasons for this are as yet unclear. SMN1 has been implicated in the formation of Cajal coiled bodies, also known as Gem bodies [140,141]. It has been implicated in RNA processing, potentially linking across to other genes implicated in motor neuron degeneration [142,143].

5.4.9 GUAM AND THE KII PENINSULA

As noted in Section 5.2, there are a number of examples from around the world of geographically and temporally localized raised incidence of amyotrophic lateral sclerosis. The two best studied of these are the amyotrophic lateral sclerosis/Parkinson disease/dementia complex of Guam, affecting the Chamorro tribe on this island, and a similar example from the Kii peninsula in Japan [11,12]. In both cases, the rate of amyotrophic lateral sclerosis was far in excess of that seen globally (in the case of Guam, 200 cases per 100,000 per annum—around 100 times higher than the global average). The Guamanian disorder, called Lytico-bodig by the Chamorro, was first reported in 1904 and reached a peak in the 1950s, with the incidence declining since then [12]. Manifesting as a tauopathy, with abundant neurofibrillary tangles in the central nervous systems of those developing either the motor neuron (amyotrophic lateral sclerosis) or dopaminergic (Parkinson's disease) dysfunction, there are clear distinctions between Lytico-bodig and idiopathic amyotrophic lateral sclerosis. In terms of the molecular mechanisms of degeneration, for both the Guam and Kii disorders, these remain a subject of vigorous debate [144]. For Lytico-bodig, genetic, infectious, and toxicological etiologies have been proposed. To date, and despite extensive studies, there is no evidence for a genetic cause for the disease [145]. In contrast to kuru, an infectious prion disease (see Chapter 4), there is no evidence for a directly infectious protein aggregate–driven mechanism for disease [113], although the possibility that this disorder may result as a consequence of other infectious diseases (as is thought to be the case for postencephalitic parkinsonism) is difficult to exclude [146,147]. One intriguing, and controversial, proposal is that Lytico-bodig disease results from exposure to an environmental toxin. This hypothesis posits that the neuronal degeneration observed in individuals with this disorder is due to a direct toxic effect of a compound called β-methylamino-L-alanine, or BMAA [148,149]. This compound—produced by cyanobacteria—is found in small quantities in the seed of the cycad plant, seeds that were used by the Chamorro for making food. A number of in vivo studies have demonstrated that BMAA has a neurotoxic impact and mimics some of the cellular phenotypes seen in Lytico-bodig [150], and the gradual replacement of cycad seeds in the diet of the Chamorro has been put forward as a reason for the declining incidence of this disorder. This possible explanation for Lytico-bodig disease was popularized by the neurologist Oliver Sachs in his book *The Island of the Colorblind* [151]. The small quantities of BMAA that might have made their way into the diet of the Chamorro, however, are much lower than the quantities required to cause neurotoxicity in animal models [152] raising questions as to the causative link between BMAA and Lytico-bodig. A possible explanation for this was put forward in 2003 by Cox and coworkers, who suggested that BMAA might have been concentrated in the tissue of flying foxes

BOX 5.4 OF FLYING FOXES AND CYCADS

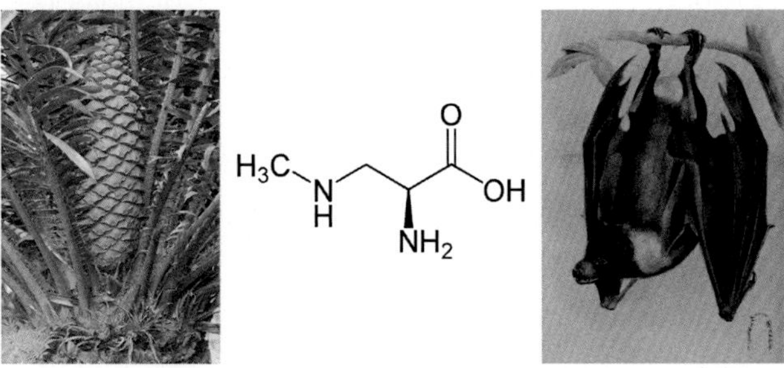

Of flying foxes and cycads. The amyotrophic lateral sclerosis/Parkinson disease/dementia complex of Guam is an enduring medical mystery. The causes of the disease, along with the reasons for the decrease in incidence since the 1950s, are still a matter of debate. A recurring hypothesis for its origin is that the neurodegeneration found in these patients is a result of exposure to β-methylamino-L-alanine, or BMAA (shown here, center). Exposure to this compound can be traced back to the seeds of the Cycad trees (shown on the left) found on Guam, entering the human diet either through the use of Cycad seeds for making flour or, possibly, through the concentration of BMAA in the flesh of flying foxes native to Guam (shown on the right). Whether either of these mechanisms could have resulted in exposure to neurotoxic levels of BMAA remains controversial. *Image of a cycad seed reproduced from Wikimedia commons through a creative commons licence.*

native to the island of Guam (*Pteropus tokudae*), a species that is now thought to be extinct [153]. Up until the 1960s, however, these flying foxes formed a part of the Chamorro diet—again potentially explaining the decrease in incidence observed over the past 50 years (Box 5.4) [154]. Conclusive proof of this being the driving force behind Lytico-bodig, as well as an explanation as to why some patients develop motor neuron dysfunction and others develop parkinsonism, remains elusive.

5.5 THERAPIES

The motor neuron diseases are devastating disorders, often with a rapid disease course and with frequently direct fatal consequences. As such, significant efforts have been directed toward developing therapies to target these diseases [155]. In common with most other neurodegenerative diseases, however, the history of drug discovery for amyotrophic lateral sclerosis, frontotemporal dementia, and other motor neuron diseases has been one characterized, for the most part, by frustration.

5.5.1 EXISTING DISEASE-MODIFYING THERAPIES

Two drugs have been licensed by the Food and Drug Administration for use as disease-modifying treatments in amyotrophic lateral sclerosis: riluzole and edaravone [156]. Riluzole acts to block glutamate receptor activity and reduce motor neuron firing in the central nervous system [23]. Treatment with this drug results in extended survival in patients with amyotrophic lateral sclerosis, with life expectancy increasing by 2–3 months compared with placebo.

Edaravone is a free radical scavenger and is thought to reduce oxidative stress associated with cytotoxicity and neurodegeneration found in the spinal cords of individuals with amyotrophic lateral sclerosis [24]. Originally developed for ischemic stroke, human trials for motor neuron dysfunction were initiated by Mitsubishi Tanabe Pharmaceuticals following promising results in a model for *SOD1* motor neuron disease [157,158]. In human trials, treatment with edaravone resulted in an improvement in disease rating scales for a subset of patients with amyotrophic lateral sclerosis, leading to the drug being given a license by the Japanese drug authorities in 2015 and orphan drug status by the Food and Drug Administration [24,156].

5.5.2 SYMPTOMATIC TREATMENTS

Symptomatic treatments for amyotrophic lateral sclerosis can provide significant relief to patients with the disorder, improving quality of life for individuals suffering from the disease [159]. Coordinated care of patients to reduce the impact of difficulties with movement, personal care, and nutrition can have significant benefits [160,161]. A particular challenge in amyotrophic lateral sclerosis is difficulty in breathing experienced upon degeneration of motor neurons controlling the muscles responsible for lung function, resulting in respiratory failure. Noninvasive ventilation can provide significant benefits in terms of survival but requires significant support for patients [162].

5.5.3 DRUG TRIAL FAILURES

Until relatively recently, drug development in the motor neuron disease field was heavily weighted toward *SOD1* linked disease because of the availability of animal and cellular models, such as the G93A *SOD1* mutant mouse, to test putative disease-modifying therapies [163]. A number of candidate therapies targeting processes disrupted in these animals have progressed to clinical trials in humans based on promising preclinical data demonstrating neuroprotection or disease modification. To date, however, and with the exception of edaravone, none of these drugs have resulted in clinical benefits for patients with amyotrophic lateral sclerosis.

5.5.4 SPINAL MUSCULAR ATROPHY AND GENE THERAPY

One area where significant progress has been made is in the development of gene therapy treatments for spinal muscular atrophy. The geneticically determined nature of this disorder makes it an ideal candidate for gene therapy, with the loss-of-function mechanism driven by mutations in *SMN1*, providing a potential route to treatment by replacement of the lost SMN1 protein [164]. This, in addition to the severe nature of the infant onset form (type I) and the presence of a potential compensatory gene (*SMN2*), led to the development of several gene therapy approaches. These included an adeno-associated virus expressing full-length SMN1 (developed by AveXis) and an antisense oligonucleotide approach to increase the expression of full-length protein from *SMN2* (nusinersen, developed by Ionis pharmaceuticals in partnership with Biogen) [165,166]. To directly target the central nervous system, the antisense oligonucleotides were administered by intrathecal injection directly into the cerebral spinal fluid, whereas the adeno-associated virus therapy was administered by intravenous injection. Both therapies underwent clinical trials for type I spinal muscular atrophy in 2016/17, reporting in November that year [166,167]. The results were impressive, with significant gains in motor control and reduced symptoms in both trials. The beneficial impact in the nusinersen trial was such that the trial was terminated prematurely to allow all participants access to the therapy, which has now been licensed by the Food and Drug Administration for use in spinal muscular atrophy. A subsequently reported trial for nusinersen in children with type II and type III spinal muscular atrophy (exhibiting later onset and longer survival) also demonstrated beneficial impacts [168]. Although both treatments have been demonstrated to provide clinical benefits and extended survival, there are a number of unknowns with regard to the long-term tolerability of these therapies, and the long-term prognosis for patients under treatment. For the adeno-associated virus approach, this has the benefit of consisting of one dose of virus, although how long expression of the exogenous *SMN1* transgene continues is unknown. For nusinersen, any clinical benefits will be dependent on repeated intrathecal injections of the antisense drug, and whether the increased expression of full-length *SMN2* can be maintained is not yet clear. Despite these caveats, and the very specific, genetically defined patient population, these trials have provided hope that analogous approaches to other forms of motor neuron disease (as well as neurodegeneration more broadly) may provide clinical benefits [164,169].

5.5.5 EXPERIMENTAL THERAPIES AND FUTURE DIRECTIONS

A number of experimental therapies are in development or undergoing clinical trials for amyotrophic lateral sclerosis and closely related disorders. Following the success of gene therapy approaches for spinal muscular atrophy, there is considerable interest in the use of similar approaches to target genes that are known to contribute to motor neuron degeneration, including *SOD1*, *TARDBP*, and *C9ORF72*. Antisense oligonucleotides targeting *SOD1* expression (BIIB067), developed by Ionis pharmaceuticals, have undergone initial phase I clinical trials and are currently under further evaluation [170]. For genetically defined populations, where disease is caused by mutations in these genes, a gene therapy approach—and especially one that looks to reduce the impact of a toxic gain of function—may hold promise [171]. Whether such an approach will provide benefits to idiopathic motor neuron disease, and how early treatment would have to be provided to result in any clinical benefits, is unclear. Similar to other neurodegenerative diseases, the complex etiology and variable progression of amyotrophic lateral sclerosis, along with the pathological process occurring within the central nervous system, mean that there are significant challenges to developing novel disease-modifying therapies. As for Alzheimer's disease, it is likely that careful patient population stratification [172], and the development of more reliable biomarkers [173,174], as well as modified therapeutic trial design, will be beneficial in terms of identifying efficacious treatments [175].

5.6 CONCLUSIONS

The motor neuron diseases and closely related dementias are devastating neurodegenerative disorders, with poor prognosis and extremely limited treatment options. The last decade has witnessed impressive advances in our understanding of the underlying molecular mechanisms that drive motor neuron degeneration in patients with amyotrophic lateral sclerosis, as well as revealing a genetic and phenotypic complex disease spectrum across the motor neuron diseases and frontotemporal dementia. Although there has been limited success in developing disease-modifying therapies for these disorders, the recent advances in gene therapy showcased by treatments for spinal muscular atrophy provide some hope that new therapies currently under clinical appraisal may yield breakthroughs in treatment.

REFERENCES

[1] Nathanson M. Lou Gehrig: a brief commentary. Neurology 1986;36:1349.

[2] Charcot JM. Deux cas d'atrophie musculaire progressive avec lesions de la substance grise et des faisceaux antero-lateraux de la moelle epiniere. Arch Physiol (Paris) 1869;2:744–60.

[3] Brain WRBB. Diseases of the nervous system. [S.l.]. Oxford University Press; 1933.

[4] van Es MA, Hardiman O, Chio A, Al-Chalabi A, Pasterkamp RJ, Veldink JH, van den Berg LH. Amyotrophic lateral sclerosis. Lancet 2017;390:2084–98.

[5] Turner MR, Scaber J, Goodfellow JA, Lord ME, Marsden R, Talbot K. The diagnostic pathway and prognosis in bulbar-onset amyotrophic lateral sclerosis. J Neurol Sci 2010;294:81–5.

[6] Chio A, Logroscino G, Hardiman O, Swingler R, Mitchell D, Beghi E, Traynor BG, Eurals C. Prognostic factors in ALS: a critical review. Amyotroph Lateral Scler 2009;10:310–23.

[7] Brooks BR, Miller RG, Swash M, Munsat TL, World Federation of Neurology Research Group on Motor Neuron Diseases. El Escorial revisited: revised criteria for the diagnosis of amyotrophic lateral sclerosis. Amyotroph Lateral Scler Other Motor Neuron Disord 2000;1:293–9.

[8] Brooks BR. El Escorial World Federation of Neurology criteria for the diagnosis of amyotrophic lateral sclerosis. Subcommittee on Motor Neuron Diseases/Amyotrophic Lateral Sclerosis of the World Federation of Neurology Research Group on Neuromuscular Diseases and the El Escorial "Clinical limits of amyotrophic lateral sclerosis" workshop contributors. J Neurol Sci 1994;124(Suppl.):96–107.

[9] Chio A, Logroscino G, Traynor BJ, Collins J, Simeone JC, Goldstein LA, White LA. Global epidemiology of amyotrophic lateral sclerosis: a systematic review of the published literature. Neuroepidemiology 2013;41:118–30.

[10] Guerreiro R, Brás J, Hardy J. SnapShot: genetics of ALS and FTD. Cell 2015;160:798.e1.

[11] Araki S, Iwahashi Y, Kuroiwa Y. Epidemiological study of amyotrophic lateral sclerosis and allied disorders in the Kii Peninsula (Japan). J Neurol Sci 1967;4:279–87.

[12] Steele JC, Guzman T. Observations about amyotrophic lateral sclerosis and the parkinsonism-dementia complex of Guam with regard to epidemiology and etiology. Can J Neurol Sci 1987;14:358–62.

[13] Chen H, Richard M, Sandler DP, Umbach DM, Kamel F. Head injury and amyotrophic lateral sclerosis. Am J Epidemiol 2007;166:810–6.

[14] Gordon PH, Cheng B, Katz IB, Pinto M, Hays AP, Mitsumoto H, Rowland LP. The natural history of primary lateral sclerosis. Neurology 2006;66:647–53.

[15] Liewluck T, Saperstein DS. Progressive muscular atrophy. Neurol Clin 2015;33:761–73.

[16] Kim WK, Liu X, Sandner J, Pasmantier M, Andrews J, Rowland LP, Mitsumoto H. Study of 962 patients indicates progressive muscular atrophy is a form of ALS. Neurology 2009;73:1686–92.

[17] Lunn MR, Wang CH. Spinal muscular atrophy. Lancet 2008;371:2120–33.

[18] Lefebvre S, Burglen L, Reboullet S, Clermont O, Burlet P, Viollet L, Benichou B, Cruaud C, Millasseau P, Zeviani M, et al. Identification and characterization of a spinal muscular atrophy-determining gene. Cell 1995;80:155–65.

[19] Pearn J. Classification of spinal muscular atrophies. Lancet 1980;1:919–22.

[20] Beeldman E, Raaphorst J, Klein Twennaar M, de Visser M, Schmand BA, de Haan RJ. The cognitive profile of ALS: a systematic review and meta-analysis update. J Neurol Neurosurg Psychiatry 2016;87:611–9.

[21] Phukan J, Elamin M, Bede P, Jordan N, Gallagher L, Byrne S, Lynch C, Pender N, Hardiman O. The syndrome of cognitive impairment in amyotrophic lateral sclerosis: a population-based study. J Neurol Neurosurg Psychiatry 2012;83:102–8.

[22] Ng L, Khan F, Young CA, Galea M. Symptomatic treatments for amyotrophic lateral sclerosis/motor neuron disease. Cochrane Database Syst Rev 2017;1:CD011776.

[23] Miller RG, Mitchell JD, Moore DH. Riluzole for amyotrophic lateral sclerosis (ALS)/motor neuron disease (MND). Cochrane Database Syst Rev 2012:CD001447.

[24] Writing Group, Edaravone (MCI-186) ALS Study Group. Safety and efficacy of edaravone in well defined patients with amyotrophic lateral sclerosis: a randomised, double-blind, placebo-controlled trial. Lancet Neurol 2017;16:505–12.

[25] Saberi S, Stauffer JE, Schulte DJ, Ravits J. Neuropathology of amyotrophic lateral sclerosis and its variants. Neurol Clin 2015;33:855–76.

[26] Al-Chalabi A, Jones A, Troakes C, King A, Al-Sarraj S, van den Berg LH. The genetics and neuropathology of amyotrophic lateral sclerosis. Acta Neuropathol 2012;124:339–52.

[27] Kwong LK, Neumann M, Sampathu DM, Lee VM, Trojanowski JQ. TDP-43 proteinopathy: the neuropathology underlying major forms of sporadic and familial frontotemporal lobar degeneration and motor neuron disease. Acta Neuropathol 2007;114:63–70. Epub 2007/05/12.

[28] Neumann M, Sampathu DM, Kwong LK, Truax AC, Micsenyi MC, Chou TT, Bruce J, Schuck T, Grossman M, Clark CM, McCluskey LF, Miller BL, Masliah E, Mackenzie IR, Feldman H, Feiden W, Kretzschmar HA, Trojanowski JQ, Lee VM. Ubiquitinated TDP-43 in frontotemporal lobar degeneration and amyotrophic lateral sclerosis. Science 2006;314:130–3. Epub 2006/10/07.

[29] Heyburn L, Moussa CE. TDP-43 in the spectrum of MND-FTLD pathologies. Mol Cell Neurosci 2017;83:46–54.

[30] Mackenzie IR, Frick P, Neumann M. The neuropathology associated with repeat expansions in the C9ORF72 gene. Acta Neuropathol 2014;127:347–57.

[31] Urwin H, Josephs KA, Rohrer JD, Mackenzie IR, Neumann M, Authier A, Seelaar H, Van Swieten JC, Brown JM, Johannsen P, Nielsen JE, Holm IE, Consortium FR, Dickson DW, Rademakers R, Graff-Radford NR, Parisi JE, Petersen RC, Hatanpaa KJ, White 3rd CL, Weiner MF, Geser F, Van Deerlin VM, Trojanowski JQ, Miller BL, Seeley WW, van der Zee J, Kumar-Singh S, Engelborghs S, De Deyn PP, Van Broeckhoven C, Bigio EH, Deng HX, Halliday GM, Kril JJ, Munoz DG, Mann DM, Pickering-Brown SM, Doodeman V, Adamson G, Ghazi-Noori S, Fisher EM, Holton JL, Revesz T, Rossor MN, Collinge J, Mead S, Isaacs AM. FUS pathology defines the majority of tau- and TDP-43-negative frontotemporal lobar degeneration. Acta Neuropathol 2010;120:33–41.

[32] Ince PG, Tomkins J, Slade JY, Thatcher NM, Shaw PJ. Amyotrophic lateral sclerosis associated with genetic abnormalities in the gene encoding Cu/Zn superoxide dismutase: molecular pathology of five new cases, and comparison with previous reports and 73 sporadic cases of ALS. J Neuropathol Exp Neurol 1998;57:895–904.

[33] Tomonaga M, Saito M, Yoshimura M, Shimada H, Tohgi H. Ultrastructure of the Bunina bodies in anterior horn cells of amyotrophic lateral sclerosis. Acta Neuropathol 1978;42:81–6.

[34] Kamo H, Haebara H, Akiguchi I, Kameyama M, Kimura H, McGeer PL. A distinctive distribution of reactive astroglia in the precentral cortex in amyotrophic lateral sclerosis. Acta Neuropathol 1987;74:33–8.

[35] Okamoto K, Mizuno Y, Fujita Y. Bunina bodies in amyotrophic lateral sclerosis. Neuropathology 2008;28:109–15.

[36] Yamanaka K, Komine O. The multi-dimensional roles of astrocytes in ALS. Neurosci Res 2018;126:31–8.

[37] Morris HR, Al-Sarraj S, Schwab C, Gwinn-Hardy K, Perez-Tur J, Wood NW, Hardy J, Lees AJ, McGeer PL, Daniel SE, Steele JC. A clinical and pathological study of motor neurone disease on Guam. Brain 2001;124:2215–22.

[38] Kuzuhara S, Kokubo Y, Sasaki R, Narita Y, Yabana T, Hasegawa M, Iwatsubo T. Familial amyotrophic lateral sclerosis and parkinsonism-dementia complex of the Kii Peninsula of Japan: clinical and neuropathological study and tau analysis. Ann Neurol 2001;49:501–11.

[39] McKee AC, Gavett BE, Stern RA, Nowinski CJ, Cantu RC, Kowall NW, Perl DP, Hedley-Whyte ET, Price B, Sullivan C, Morin P, Lee HS, Kubilus CA, Daneshvar DH, Wulff M, Budson AE. TDP-43 proteinopathy and motor neuron disease in chronic traumatic encephalopathy. J Neuropathol Exp Neurol 2010;69:918–29.

[40] McKee AC, Daneshvar DH, Alvarez VE, Stein TD. The neuropathology of sport. Acta Neuropathol 2014;127:29–51.
[41] Taylor JP, Brown Jr RH, Cleveland DW, Decoding ALS. From genes to mechanism. Nature 2016;539:197–206.
[42] Rosen DR, Siddique T, Patterson D, Figlewicz DA, Sapp P, Hentati A, Donaldson D, Goto J, O'Regan JP, Deng HX, et al. Mutations in Cu/Zn superoxide dismutase gene are associated with familial amyotrophic lateral sclerosis. Nature 1993;362:59–62.
[43] Valentine JS, Doucette PA, Zittin Potter S. Copper-zinc superoxide dismutase and amyotrophic lateral sclerosis. Annu Rev Biochem 2005;74:563–93.
[44] Byrne S, Walsh C, Lynch C, Bede P, Elamin M, Kenna K, McLaughlin R, Hardiman O. Rate of familial amyotrophic lateral sclerosis: a systematic review and meta-analysis. J Neurol Neurosurg Psychiatry 2011;82:623–7.
[45] Johnson JO, Mandrioli J, Benatar M, Abramzon Y, Van Deerlin VM, Trojanowski JQ, Gibbs JR, Brunetti M, Gronka S, Wuu J, Ding J, McCluskey L, Martinez-Lage M, Falcone D, Hernandez DG, Arepalli S, Chong S, Schymick JC, Rothstein J, Landi F, Wang YD, Calvo A, Mora G, Sabatelli M, Monsurro MR, Battistini S, Salvi F, Spataro R, Sola P, Borghero G, Consortium I, Galassi G, Scholz SW, Taylor JP, Restagno G, Chio A, Traynor BJ. Exome sequencing reveals VCP mutations as a cause of familial ALS. Neuron 2010;68:857–64.
[46] Munch C, Sedlmeier R, Meyer T, Homberg V, Sperfeld AD, Kurt A, Prudlo J, Peraus G, Hanemann CO, Stumm G, Ludolph AC. Point mutations of the p150 subunit of dynactin (DCTN1) gene in ALS. Neurology 2004;63:724–6.
[47] Konno T, Ross OA, Teive HAG, Slawek J, Dickson DW, Wszolek ZK. DCTN1-related neurodegeneration: Perry syndrome and beyond. Park Relat Disord 2017;41:14–24.
[48] Leigh PN, Anderton BH, Dodson A, Gallo JM, Swash M, Power DM. Ubiquitin deposits in anterior horn cells in motor neurone disease. Neurosci Lett 1988;93:197–203.
[49] Lowe J, Lennox G, Jefferson D, Morrell K, McQuire D, Gray T, Landon M, Doherty FJ, Mayer RJ. A filamentous inclusion body within anterior horn neurones in motor neurone disease defined by immunocytochemical localisation of ubiquitin. Neurosci Lett 1988;94:203–10.
[50] Arai T, Hasegawa M, Akiyama H, Ikeda K, Nonaka T, Mori H, Mann D, Tsuchiya K, Yoshida M, Hashizume Y, Oda T. TDP-43 is a component of ubiquitin-positive tau-negative inclusions in frontotemporal lobar degeneration and amyotrophic lateral sclerosis. Biochem Biophys Res Commun 2006;351:602–11.
[51] Kabashi E, Valdmanis PN, Dion P, Spiegelman D, McConkey BJ, Vande Velde C, Bouchard JP, Lacomblez L, Pochigaeva K, Salachas F, Pradat PF, Camu W, Meininger V, Dupre N, Rouleau GA. TARDBP mutations in individuals with sporadic and familial amyotrophic lateral sclerosis. Nat Genet 2008;40:572–4.
[52] Sreedharan J, Blair IP, Tripathi VB, Hu X, Vance C, Rogelj B, Ackerley S, Durnall JC, Williams KL, Buratti E, Baralle F, de Belleroche J, Mitchell JD, Leigh PN, Al-Chalabi A, Miller CC, Nicholson G, Shaw CE. TDP-43 mutations in familial and sporadic amyotrophic lateral sclerosis. Science 2008;319:1668–72.
[53] Mackenzie IR, Rademakers R, Neumann M. TDP-43 and FUS in amyotrophic lateral sclerosis and frontotemporal dementia. Lancet Neurol 2010;9:995–1007.
[54] Kwiatkowski Jr TJ, Bosco DA, Leclerc AL, Tamrazian E, Vanderburg CR, Russ C, Davis A, Gilchrist J, Kasarskis EJ, Munsat T, Valdmanis P, Rouleau GA, Hosler BA, Cortelli P, de Jong PJ, Yoshinaga Y, Haines JL, Pericak-Vance MA, Yan J, Ticozzi N, Siddique T, McKenna-Yasek D, Sapp PC, Horvitz HR, Landers JE, Brown Jr RH. Mutations in the FUS/TLS gene on chromosome 16 cause familial amyotrophic lateral sclerosis. Science 2009;323:1205–8.
[55] Vance C, Rogelj B, Hortobagyi T, De Vos KJ, Nishimura AL, Sreedharan J, Hu X, Smith B, Ruddy D, Wright P, Ganesalingam J, Williams KL, Tripathi V, Al-Saraj S, Al-Chalabi A, Leigh PN, Blair IP, Nicholson G, de Belleroche J, Gallo JM, Miller CC, Shaw CE. Mutations in FUS, an RNA processing protein, cause familial amyotrophic lateral sclerosis type 6. Science 2009;323:1208–11.

[56] Mackenzie IRA, Neumann M. Fused in sarcoma neuropathology in neurodegenerative disease. Cold Spring Harb Perspect Med 2017;7.

[57] Ederle H, Dormann D. TDP-43 and FUS en route from the nucleus to the cytoplasm. FEBS Lett 2017;591:1489–507.

[58] DeJesus-Hernandez M, Mackenzie IR, Boeve BF, Boxer AL, Baker M, Rutherford NJ, Nicholson AM, Finch NA, Flynn H, Adamson J, Kouri N, Wojtas A, Sengdy P, Hsiung GY, Karydas A, Seeley WW, Josephs KA, Coppola G, Geschwind DH, Wszolek ZK, Feldman H, Knopman DS, Petersen RC, Miller BL, Dickson DW, Boylan KB, Graff-Radford NR, Rademakers R. Expanded GGGGCC hexanucleotide repeat in noncoding region of C9ORF72 causes chromosome 9p-linked FTD and ALS. Neuron 2011;72:245–56.

[59] Renton AE, Majounie E, Waite A, Simon-Sanchez J, Rollinson S, Gibbs JR, Schymick JC, Laaksovirta H, van Swieten JC, Myllykangas L, Kalimo H, Paetau A, Abramzon Y, Remes AM, Kaganovich A, Scholz SW, Duckworth J, Ding J, Harmer DW, Hernandez DG, Johnson JO, Mok K, Ryten M, Trabzuni D, Guerreiro RJ, Orrell RW, Neal J, Murray A, Pearson J, Jansen IE, Sondervan D, Seelaar H, Blake D, Young K, Halliwell N, Callister JB, Toulson G, Richardson A, Gerhard A, Snowden J, Mann D, Neary D, Nalls MA, Peuralinna T, Jansson L, Isoviita VM, Kaivorinne AL, Holtta-Vuori M, Ikonen E, Sulkava R, Benatar M, Wuu J, Chio A, Restagno G, Borghero G, Sabatelli M, Consortium I, Heckerman D, Rogaeva E, Zinman L, Rothstein JD, Sendtner M, Drepper C, Eichler EE, Alkan C, Abdullaev Z, Pack SD, Dutra A, Pak E, Hardy J, Singleton A, Williams NM, Heutink P, Pickering-Brown S, Morris HR, Tienari PJ, Traynor BJ. A hexanucleotide repeat expansion in C9ORF72 is the cause of chromosome 9p21-linked ALS-FTD. Neuron 2011;72:257–68.

[60] Rohrer JD, Isaacs AM, Mizielinska S, Mead S, Lashley T, Wray S, Sidle K, Fratta P, Orrell RW, Hardy J, Holton J, Revesz T, Rossor MN, Warren JD. C9orf72 expansions in frontotemporal dementia and amyotrophic lateral sclerosis. Lancet Neurol 2015;14:291–301.

[61] Gao FB, Almeida S, Lopez-Gonzalez R. Dysregulated molecular pathways in amyotrophic lateral sclerosis-frontotemporal dementia spectrum disorder. EMBO J 2017;36:2931–50.

[62] Maruyama H, Morino H, Ito H, Izumi Y, Kato H, Watanabe Y, Kinoshita Y, Kamada M, Nodera H, Suzuki H, Komure O, Matsuura S, Kobatake K, Morimoto N, Abe K, Suzuki N, Aoki M, Kawata A, Hirai T, Kato T, Ogasawara K, Hirano A, Takumi T, Kusaka H, Hagiwara K, Kaji R, Kawakami H. Mutations of optineurin in amyotrophic lateral sclerosis. Nature 2010;465:223–6.

[63] Shimizu H, Toyoshima Y, Shiga A, Yokoseki A, Arakawa K, Sekine Y, Shimohata T, Ikeuchi T, Nishizawa M, Kakita A, Onodera O, Takahashi H. Sporadic ALS with compound heterozygous mutations in the SQSTM1 gene. Acta Neuropathol 2013;126:453–9.

[64] Fecto F, Yan J, Vemula SP, Liu E, Yang Y, Chen W, Zheng JG, Shi Y, Siddique N, Arrat H, Donkervoort S, Ajroud-Driss S, Sufit RL, Heller SL, Deng HX, Siddique T. SQSTM1 mutations in familial and sporadic amyotrophic lateral sclerosis. Arch Neurol 2011;68:1440–6.

[65] Freischmidt A, Wieland T, Richter B, Ruf W, Schaeffer V, Muller K, Marroquin N, Nordin F, Hubers A, Weydt P, Pinto S, Press R, Millecamps S, Molko N, Bernard E, Desnuelle C, Soriani MH, Dorst J, Graf E, Nordstrom U, Feiler MS, Putz S, Boeckers TM, Meyer T, Winkler AS, Winkelman J, de Carvalho M, Thal DR, Otto M, Brannstrom T, Volk AE, Kursula P, Danzer KM, Lichtner P, Dikic I, Meitinger T, Ludolph AC, Strom TM, Andersen PM, Weishaupt JH. Haploinsufficiency of TBK1 causes familial ALS and frontotemporal dementia. Nat Neurosci 2015;18:631–6.

[66] Cirulli ET, Lasseigne BN, Petrovski S, Sapp PC, Dion PA, Leblond CS, Couthouis J, Lu YF, Wang Q, Krueger BJ, Ren Z, Keebler J, Han Y, Levy SE, Boone BE, Wimbish JR, Waite LL, Jones AL, Carulli JP, Day-Williams AG, Staropoli JF, Xin WW, Chesi A, Raphael AR, McKenna-Yasek D, Cady J, Vianney de Jong JM, Kenna KP, Smith BN, Topp S, Miller J, Gkazi A, Consortium FS, Al-Chalabi A, van den Berg LH, Veldink J, Silani V, Ticozzi N, Shaw CE, Baloh RH, Appel S, Simpson E, Lagier-Tourenne C, Pulst SM, Gibson S, Trojanowski JQ, Elman L, McCluskey L, Grossman M, Shneider NA, Chung WK, Ravits

JM, Glass JD, Sims KB, Van Deerlin VM, Maniatis T, Hayes SD, Ordureau A, Swarup S, Landers J, Baas F, Allen AS, Bedlack RS, Harper JW, Gitler AD, Rouleau GA, Brown R, Harms MB, Cooper GM, Harris T, Myers RM, Goldstein DB. Exome sequencing in amyotrophic lateral sclerosis identifies risk genes and pathways. Science 2015;347:1436–41.

[67] Ugolino J, Ji YJ, Conchina K, Chu J, Nirujogi RS, Pandey A, Brady NR, Hamacher-Brady A, Wang J. Loss of C9orf72 enhances autophagic activity via deregulated mTOR and TFEB signaling. PLoS Genet 2016;12:e1006443.

[68] Ogino S, Wilson RB. Genetic testing and risk assessment for spinal muscular atrophy (SMA). Hum Genet 2002;111:477–500.

[69] Gavrilov DK, Shi X, Das K, Gilliam TC, Wang CH. Differential SMN2 expression associated with SMA severity. Nat Genet 1998;20:230–1.

[70] van Rheenen W, Shatunov A, Dekker AM, McLaughlin RL, Diekstra FP, Pulit SL, van der Spek RA, Vosa U, de Jong S, Robinson MR, Yang J, Fogh I, van Doormaal PT, Tazelaar GH, Koppers M, Blokhuis AM, Sproviero W, Jones AR, Kenna KP, van Eijk KR, Harschnitz O, Schellevis RD, Brands WJ, Medic J, Menelaou A, Vajda A, Ticozzi N, Lin K, Rogelj B, Vrabec K, Ravnik-Glavac M, Koritnik B, Zidar J, Leonardis L, Groselj LD, Millecamps S, Salachas F, Meininger V, de Carvalho M, Pinto S, Mora JS, Rojas-Garcia R, Polak M, Chandran S, Colville S, Swingler R, Morrison KE, Shaw PJ, Hardy J, Orrell RW, Pittman A, Sidle K, Fratta P, Malaspina A, Topp S, Petri S, Abdulla S, Drepper C, Sendtner M, Meyer T, Ophoff RA, Staats KA, Wiedau-Pazos M, Lomen-Hoerth C, Van Deerlin VM, Trojanowski JQ, Elman L, McCluskey L, Basak AN, Tunca C, Hamzeiy H, Parman Y, Meitinger T, Lichtner P, Radivojkov-Blagojevic M, Andres CR, Maurel C, Bensimon G, Landwehrmeyer B, Brice A, Payan CA, Saker-Delye S, Durr A, Wood NW, Tittmann L, Lieb W, Franke A, Rietschel M, Cichon S, Nothen MM, Amouyel P, Tzourio C, Dartigues JF, Uitterlinden AG, Rivadeneira F, Estrada K, Hofman A, Curtis C, Blauw HM, van der Kooi AJ, de Visser M, Goris A, Weber M, Shaw CE, Smith BN, Pansarasa O, Cereda C, Del Bo R, Comi GP, D'Alfonso S, Bertolin C, Soraru G, Mazzini L, Pensato V, Gellera C, Tiloca C, Ratti A, Calvo A, Moglia C, Brunetti M, Arcuti S, Capozzo R, Zecca C, Lunetta C, Penco S, Riva N, Padovani A, Filosto M, Muller B, Stuit RJ, Registry P, Group S, Registry S, Consortium FS, Consortium S, Group NS, Blair I, Zhang K, McCann EP, Fifita JA, Nicholson GA, Rowe DB, Pamphlett R, Kiernan MC, Grosskreutz J, Witte OW, Ringer T, Prell T, Stubendorff B, Kurth I, Hubner CA, Leigh PN, Casale F, Chio A, Beghi E, Pupillo E, Tortelli R, Logroscino G, Powell J, Ludolph AC, Weishaupt JH, Robberecht W, Van Damme P, Franke L, Pers TH, Brown RH, Glass JD, Landers JE, Hardiman O, Andersen PM, Corcia P, Vourc'h P, Silani V, Wray NR, Visscher PM, de Bakker PI, van Es MA, Pasterkamp RJ, Lewis CM, Breen G, Al-Chalabi A, van den Berg LH, Veldink JH. Genome-wide association analyses identify new risk variants and the genetic architecture of amyotrophic lateral sclerosis. Nat Genet 2016;48:1043–8.

[71] Nicolas A, Kenna KP, Renton AE, Ticozzi N, Faghri F, Chia R, Dominov JA, Kenna BJ, Nalls MA, Keagle P, Rivera AM, van Rheenen W, Murphy NA, van Vugt J, Geiger JT, Van der Spek RA, Pliner HA, Shankaracharya, Smith BN, Marangi G, Topp SD, Abramzon Y, Gkazi AS, Eicher JD, Kenna A, Consortium I, Mora G, Calvo A, Mazzini L, Riva N, Mandrioli J, Caponnetto C, Battistini S, Volanti P, La Bella V, Conforti FL, Borghero G, Messina S, Simone IL, Trojsi F, Salvi F, Logullo FO, D'Alfonso S, Corrado L, Capasso M, Ferrucci L, Genomic Translation for ALSCCMoreno CAM, Kamalakaran S, Goldstein DB, Consortium ALSS, Gitler AD, Harris T, Myers RM, Consortium NA, Phatnani H, Musunuri RL, Evani US, Abhyankar A, Zody MC, Answer ALSF, Kaye J, Finkbeiner S, Wyman SK, LeNail A, Lima L, Fraenkel E, Svendsen CN, Thompson LM, Van Eyk JE, Berry JD, Miller TM, Kolb SJ, Cudkowicz M, Baxi E, Clinical Research in ALS and Related Disorders for Therapeutic Development Consortium, Benatar M, Taylor JP, Rampersaud E, Wu G, Wuu J, Consortium S, Lauria G, Verde F, Fogh I, Tiloca C, Comi GP, Soraru G, Cereda C, French ALSC, Corcia P, Laaksovirta H, Myllykangas L, Jansson L, Valori M, Ealing J, Hamdalla H, Rollinson S, Pickering-Brown S, Orrell

RW, Sidle KC, Malaspina A, Hardy J, Singleton AB, Johnson JO, Arepalli S, Sapp PC, McKenna-Yasek D, Polak M, Asress S, Al-Sarraj S, King A, Troakes C, Vance C, de Belleroche J, Baas F, Ten Asbroek A, Munoz-Blanco JL, Hernandez DG, Ding J, Gibbs JR, Scholz SW, Floeter MK, Campbell RH, Landi F, Bowser R, Pulst SM, Ravits JM, MacGowan DJL, Kirby J, Pioro EP, Pamphlett R, Broach J, Gerhard G, Dunckley TL, Brady CB, Kowall NW, Troncoso JC, Le Ber I, Mouzat K, Lumbroso S, Heiman-Patterson TD, Kamel F, Van Den Bosch L, Baloh RH, Strom TM, Meitinger T, Shatunov A, Van Eijk KR, de Carvalho M, Kooyman M, Middelkoop B, Moisse M, McLaughlin RL, Van Es MA, Weber M, Boylan KB, Van Blitterswijk M, Rademakers R, Morrison KE, Basak AN, Mora JS, Drory VE, Shaw PJ, Turner MR, Talbot K, Hardiman O, Williams KL, Fifita JA, Nicholson GA, Blair IP, Rouleau GA, Esteban-Perez J, Garcia-Redondo A, Al-Chalabi A, Project Min EALSSC, Rogaeva E, Zinman L, Ostrow LW, Maragakis NJ, Rothstein JD, Simmons Z, Cooper-Knock J, Brice A, Goutman SA, Feldman EL, Gibson SB, Taroni F, Ratti A, Gellera C, Van Damme P, Robberecht W, Fratta P, Sabatelli M, Lunetta C, Ludolph AC, Andersen PM, Weishaupt JH, Camu W, Trojanowski JQ, Van Deerlin VM, Brown Jr RH, van den Berg LH, Veldink JH, Harms MB, Glass JD, Stone DJ, Tienari P, Silani V, Chio A, Shaw CE, Traynor BJ, Landers JE. Genome-wide analyses identify KIF5A as a novel ALS gene. Neuron 2018;97:1268–83.e6.

[72] Benyamin B, He J, Zhao Q, Gratten J, Garton F, Leo PJ, Liu Z, Mangelsdorf M, Al-Chalabi A, Anderson L, Butler TJ, Chen L, Chen XD, Cremin K, Deng HW, Devine M, Edson J, Fifita JA, Furlong S, Han YY, Harris J, Henders AK, Jeffree RL, Jin ZB, Li Z, Li T, Li M, Lin Y, Liu X, Marshall M, McCann EP, Mowry BJ, Ngo ST, Pamphlett R, Ran S, Reutens DC, Rowe DB, Sachdev P, Shah S, Song S, Tan LJ, Tang L, van den Berg LH, van Rheenen W, Veldink JH, Wallace RH, Wheeler L, Williams KL, Wu J, Wu X, Yang J, Yue W, Zhang ZH, Zhang D, Noakes PG, Blair IP, Henderson RD, McCombe PA, Visscher PM, Xu H, Bartlett PF, Brown MA, Wray NR, Fan D. Cross-ethnic meta-analysis identifies association of the GPX3-TNIP1 locus with amyotrophic lateral sclerosis. Nat Commun 2017;8:611.

[73] Nguyen HP, Van Broeckhoven C, van der Zee J. ALS genes in the genomic era and their implications for FTD. Trends Genet 2018;34:404–23.

[74] van Blitterswijk M, van Es MA, Hennekam EA, Dooijes D, van Rheenen W, Medic J, Bourque PR, Schelhaas HJ, van der Kooi AJ, de Visser M, de Bakker PI, Veldink JH, van den Berg LH. Evidence for an oligogenic basis of amyotrophic lateral sclerosis. Hum Mol Genet 2012;21:3776–84.

[75] Pang SY, Hsu JS, Teo KC, Li Y, Kung MHW, Cheah KSE, Chan D, Cheung KMC, Li M, Sham PC, Ho SL. Burden of rare variants in ALS genes influences survival in familial and sporadic ALS. Neurobiol Aging 2017;58(238):e9–15.

[76] Kramer NJ, Haney MS, Morgens DW, Jovicic A, Couthouis J, Li A, Ousey J, Ma R, Bieri G, Tsui CK, Shi Y, Hertz NT, Tessier-Lavigne M, Ichida JK, Bassik MC, Gitler AD. CRISPR-Cas9 screens in human cells and primary neurons identify modifiers of C9ORF72 dipeptide-repeat-protein toxicity. Nat Genet 2018;50:603–12.

[77] Reaume AG, Elliott JL, Hoffman EK, Kowall NW, Ferrante RJ, Siwek DF, Wilcox HM, Flood DG, Beal MF, Brown Jr RH, Scott RW, Snider WD. Motor neurons in Cu/Zn superoxide dismutase-deficient mice develop normally but exhibit enhanced cell death after axonal injury. Nat Genet 1996;13:43–7.

[78] Shaw BF, Valentine JS. How do ALS-associated mutations in superoxide dismutase 1 promote aggregation of the protein? Trends Biochem Sci 2007;32:78–85.

[79] Elam JS, Taylor AB, Strange R, Antonyuk S, Doucette PA, Rodriguez JA, Hasnain SS, Hayward LJ, Valentine JS, Yeates TO, Hart PJ. Amyloid-like filaments and water-filled nanotubes formed by SOD1 mutant proteins linked to familial ALS. Nat Struct Biol 2003;10:461–7.

[80] Chiti F, Dobson CM. Protein misfolding, functional amyloid, and human disease. Annu Rev Biochem 2006;75:333–66.

[81] Bruijn LI, Houseweart MK, Kato S, Anderson KL, Anderson SD, Ohama E, Reaume AG, Scott RW, Cleveland DW. Aggregation and motor neuron toxicity of an ALS-linked SOD1 mutant independent from wild-type SOD1. Science 1998;281:1851–4.

[82] Johnson BS, Snead D, Lee JJ, McCaffery JM, Shorter J, Gitler AD. TDP-43 is intrinsically aggregation-prone, and amyotrophic lateral sclerosis-linked mutations accelerate aggregation and increase toxicity. J Biol Chem 2009;284:20329–39.

[83] Neumann M, Kwong LK, Sampathu DM, Trojanowski JQ, Lee VM. TDP-43 proteinopathy in frontotemporal lobar degeneration and amyotrophic lateral sclerosis: protein misfolding diseases without amyloidosis. Arch Neurol 2007;64:1388–94. Epub 2007/10/10.

[84] Mori K, Weng SM, Arzberger T, May S, Rentzsch K, Kremmer E, Schmid B, Kretzschmar HA, Cruts M, Van Broeckhoven C, Haass C, Edbauer D. The C9orf72 GGGGCC repeat is translated into aggregating dipeptide-repeat proteins in FTLD/ALS. Science 2013;339:1335–8.

[85] Ash PE, Bieniek KF, Gendron TF, Caulfield T, Lin WL, Dejesus-Hernandez M, van Blitterswijk MM, Jansen-West K, Paul 3rd JW, Rademakers R, Boylan KB, Dickson DW, Petrucelli L. Unconventional translation of C9ORF72 GGGGCC expansion generates insoluble polypeptides specific to c9FTD/ALS. Neuron 2013;77:639–46.

[86] Chiti F, Dobson CM. Protein misfolding, amyloid formation, and human disease: a summary of progress over the last decade. Annu Rev Biochem 2017;86:27–68.

[87] Shaw PJ, Eggett CJ. Molecular factors underlying selective vulnerability of motor neurons to neurodegeneration in amyotrophic lateral sclerosis. J Neurol 2000;247(Suppl. 1):I17–27.

[88] Codogno P, Mehrpour M, Proikas-Cezanne T. Canonical and non-canonical autophagy: variations on a common theme of self-eating? Nat Rev Mol Cell Biol 2012;13:7–12.

[89] Menzies FM, Moreau K, Rubinsztein DC. Protein misfolding disorders and macroautophagy. Curr Opin Cell Biol 2011;23:190–7. Epub 2010/11/20.

[90] Bjorkoy G, Lamark T, Brech A, Outzen H, Perander M, Overvatn A, Stenmark H, Johansen T. p62/SQSTM1 forms protein aggregates degraded by autophagy and has a protective effect on huntingtin-induced cell death. J Cell Biol 2005;171:603–14.

[91] Hocking LJ, Lucas GJ, Daroszewska A, Mangion J, Olavesen M, Cundy T, Nicholson GC, Ward L, Bennett ST, Wuyts W, Van Hul W, Ralston SH. Domain-specific mutations in sequestosome 1 (SQSTM1) cause familial and sporadic Paget's disease. Hum Mol Genet 2002;11:2735–9.

[92] Korac J, Schaeffer V, Kovacevic I, Clement AM, Jungblut B, Behl C, Terzic J, Dikic I. Ubiquitin-independent function of optineurin in autophagic clearance of protein aggregates. J Cell Sci 2013;126:580–92.

[93] Wong YC, Holzbaur EL. Optineurin is an autophagy receptor for damaged mitochondria in parkin-mediated mitophagy that is disrupted by an ALS-linked mutation. Proc Natl Acad Sci USA 2014; 111:E4439–48.

[94] Deas E, Wood NW, Plun-Favreau H. Mitophagy and Parkinson's disease: the PINK1-parkin link. Biochim Biophys Acta 2011;1813:623–33. Epub 2010/08/26.

[95] Webster CP, Smith EF, Bauer CS, Moller A, Hautbergue GM, Ferraiuolo L, Myszczynska MA, Higginbottom A, Walsh MJ, Whitworth AJ, Kaspar BK, Meyer K, Shaw PJ, Grierson AJ, De Vos KJ. The C9orf72 protein interacts with Rab1a and the ULK1 complex to regulate initiation of autophagy. EMBO J 2016;35:1656–76.

[96] Farg MA, Sundaramoorthy V, Sultana JM, Yang S, Atkinson RA, Levina V, Halloran MA, Gleeson PA, Blair IP, Soo KY, King AE, Atkin JD. C9ORF72, implicated in amytrophic lateral sclerosis and frontotemporal dementia, regulates endosomal trafficking. Hum Mol Genet 2014;23:3579–95.

[97] Corrionero A, Horvitz HR. A C9orf72 ALS/FTD ortholog acts in endolysosomal degradation and lysosomal homeostasis. Curr Biol 2018;28:1522–35. e5.

[98] Prusiner SB. Cell biology. A unifying role for prions in neurodegenerative diseases. Science 2012; 336:1511–3.

[99] Collinge J, Clarke AR. A general model of prion strains and their pathogenicity. Science 2007;318:930–6. Epub 2007/11/10.
[100] Olanow CW, Prusiner SB. Is Parkinson's disease a prion disorder? Proc Natl Acad Sci USA 2009;106:12571–2.
[101] Abbott A. The red-hot debate about transmissible Alzheimer's. Nature 2016;531:294–7.
[102] Hock EM, Polymenidou M. Prion-like propagation as a pathogenic principle in frontotemporal dementia. J Neurochem 2016;138(Suppl. 1):163–83.
[103] Polymenidou M, Cleveland DW. The seeds of neurodegeneration: prion-like spreading in ALS. Cell 2011;147:498–508.
[104] Grad LI, Yerbury JJ, Turner BJ, Guest WC, Pokrishevsky E, O'Neill MA, Yanai A, Silverman JM, Zeineddine R, Corcoran L, Kumita JR, Luheshi LM, Yousefi M, Coleman BM, Hill AF, Plotkin SS, Mackenzie IR, Cashman NR. Intercellular propagated misfolding of wild-type Cu/Zn superoxide dismutase occurs via exosome-dependent and -independent mechanisms. Proc Natl Acad Sci USA 2014;111:3620–5.
[105] Munch C, O'Brien J, Bertolotti A. Prion-like propagation of mutant superoxide dismutase-1 misfolding in neuronal cells. Proc Natl Acad Sci USA 2011;108:3548–53.
[106] Nonaka T, Masuda-Suzukake M, Arai T, Hasegawa Y, Akatsu H, Obi T, Yoshida M, Murayama S, Mann DM, Akiyama H, Hasegawa M. Prion-like properties of pathological TDP-43 aggregates from diseased brains. Cell Rep 2013;4:124–34.
[107] Shimonaka S, Nonaka T, Suzuki G, Hisanaga S, Hasegawa M. Templated aggregation of TAR DNA-binding protein of 43 kDa (TDP-43) by seeding with TDP-43 peptide fibrils. J Biol Chem 2016;291:8896–907.
[108] Smethurst P, Newcombe J, Troakes C, Simone R, Chen YR, Patani R, Sidle K. In vitro prion-like behaviour of TDP-43 in ALS. Neurobiol Dis 2016;96:236–47.
[109] Feuillette S, Delarue M, Riou G, Gaffuri AL, Wu J, Lenkei Z, Boyer O, Frebourg T, Campion D, Lecourtois M. Neuron-to-neuron transfer of FUS in drosophila primary neuronal culture is enhanced by ALS-associated mutations. J Mol Neurosci 2017;62:114–22.
[110] Patel A, Lee HO, Jawerth L, Maharana S, Jahnel M, Hein MY, Stoynov S, Mahamid J, Saha S, Franzmann TM, Pozniakovski A, Poser I, Maghelli N, Royer LA, Weigert M, Myers EW, Grill S, Drechsel D, Hyman AA, Alberti S. A liquid-to-solid phase transition of the ALS protein FUS accelerated by disease mutation. Cell 2015;162:1066–77.
[111] Harrison AF, Shorter J. RNA-binding proteins with prion-like domains in health and disease. Biochem J 2017;474:1417–38.
[112] Walsh DM, Selkoe DJ. A critical appraisal of the pathogenic protein spread hypothesis of neurodegeneration. Nat Rev Neurosci 2016;17:251–60.
[113] Gibbs Jr CJ, Gajdusek DC. Amyotrophic lateral sclerosis, Parkinson's disease, and the amyotrophic lateral sclerosis-Parkinsonism-dementia complex on Guam: a review and summary of attempts to demonstrate infection as the aetiology. J Clin Pathol Suppl (R Coll Pathol) 1972;6:132–40.
[114] Salazar AM, Masters CL, Gajdusek DC, Gibbs Jr CJ. Syndromes of amyotrophic lateral sclerosis and dementia: relation to transmissible Creutzfeldt-Jakob disease. Ann Neurol 1983;14:17–26.
[115] Lagier-Tourenne C, Polymenidou M, Cleveland DW. TDP-43 and FUS/TLS: emerging roles in RNA processing and neurodegeneration. Hum Mol Genet 2010;19:R46–64.
[116] Colombrita C, Onesto E, Megiorni F, Pizzuti A, Baralle FE, Buratti E, Silani V, Ratti A. TDP-43 and FUS RNA-binding proteins bind distinct sets of cytoplasmic messenger RNAs and differently regulate their post-transcriptional fate in motoneuron-like cells. J Biol Chem 2012;287:15635–47.
[117] Wang HY, Wang IF, Bose J, Shen CK. Structural diversity and functional implications of the eukaryotic TDP gene family. Genomics 2004;83:130–9.
[118] Zinszner H, Sok J, Immanuel D, Yin Y, Ron D. TLS (FUS) binds RNA in vivo and engages in nucleo-cytoplasmic shuttling. J Cell Sci 1997;110(Pt 15):1741–50.

[119] Arnold ES, Ling SC, Huelga SC, Lagier-Tourenne C, Polymenidou M, Ditsworth D, Kordasiewicz HB, McAlonis-Downes M, Platoshyn O, Parone PA, Da Cruz S, Clutario KM, Swing D, Tessarollo L, Marsala M, Shaw CE, Yeo GW, Cleveland DW. ALS-linked TDP-43 mutations produce aberrant RNA splicing and adult-onset motor neuron disease without aggregation or loss of nuclear TDP-43. Proc Natl Acad Sci USA 2013;110:E736–45.
[120] Fujioka Y, Ishigaki S, Masuda A, Iguchi Y, Udagawa T, Watanabe H, Katsuno M, Ohno K, Sobue G. FUS-regulated region- and cell-type-specific transcriptome is associated with cell selectivity in ALS/FTLD. Sci Rep 2013;3:2388.
[121] McDonald KK, Aulas A, Destroismaisons L, Pickles S, Beleac E, Camu W, Rouleau GA, Vande Velde C. TAR DNA-binding protein 43 (TDP-43) regulates stress granule dynamics via differential regulation of G3BP and TIA-1. Hum Mol Genet 2011;20:1400–10.
[122] Gal J, Zhang J, Kwinter DM, Zhai J, Jia H, Jia J, Zhu H. Nuclear localization sequence of FUS and induction of stress granules by ALS mutants. Neurobiol Aging 2011;32. 2323.e27–40.
[123] Mahboubi H, Stochaj U. Cytoplasmic stress granules: dynamic modulators of cell signaling and disease. Biochim Biophys Acta 2017;1863:884–95.
[124] Boeynaems S, Bogaert E, Kovacs D, Konijnenberg A, Timmerman E, Volkov A, Guharoy M, De Decker M, Jaspers T, Ryan VH, Janke AM, Baatsen P, Vercruysse T, Kolaitis RM, Daelemans D, Taylor JP, Kedersha N, Anderson P, Impens F, Sobott F, Schymkowitz J, Rousseau F, Fawzi NL, Robberecht W, Van Damme P, Tompa P, Van Den Bosch L. Phase separation of C9orf72 dipeptide repeats perturbs stress granule dynamics. Mol Cell 2017;65:1044–55.e5.
[125] Peters OM, Cabrera GT, Tran H, Gendron TF, McKeon JE, Metterville J, Weiss A, Wightman N, Salameh J, Kim J, Sun H, Boylan KB, Dickson D, Kennedy Z, Lin Z, Zhang YJ, Daughrity L, Jung C, Gao FB, Sapp PC, Horvitz HR, Bosco DA, Brown SP, de Jong P, Petrucelli L, Mueller C, Brown Jr RH. Human C9ORF72 hexanucleotide expansion reproduces RNA foci and dipeptide repeat proteins but not neurodegeneration in BAC transgenic mice. Neuron 2015;88:902–9.
[126] Urnavicius L, Zhang K, Diamant AG, Motz C, Schlager MA, Yu M, Patel NA, Robinson CV, Carter AP. The structure of the dynactin complex and its interaction with dynein. Science 2015;347:1441–6.
[127] Hafezparast M, Klocke R, Ruhrberg C, Marquardt A, Ahmad-Annuar A, Bowen S, Lalli G, Witherden AS, Hummerich H, Nicholson S, Morgan PJ, Oozageer R, Priestley JV, Averill S, King VR, Ball S, Peters J, Toda T, Yamamoto A, Hiraoka Y, Augustin M, Korthaus D, Wattler S, Wabnitz P, Dickneite C, Lampel S, Boehme F, Peraus G, Popp A, Rudelius M, Schlegel J, Fuchs H, Hrabe de Angelis M, Schiavo G, Shima DT, Russ AP, Stumm G, Martin JE, Fisher EM. Mutations in dynein link motor neuron degeneration to defects in retrograde transport. Science 2003;300:808–12.
[128] Williamson TL, Cleveland DW. Slowing of axonal transport is a very early event in the toxicity of ALS-linked SOD1 mutants to motor neurons. Nat Neurosci 1999;2:50–6.
[129] Asbury CL. Kinesin: world's tiniest biped. Curr Opin Cell Biol 2005;17:89–97.
[130] Xia CH, Roberts EA, Her LS, Liu X, Williams DS, Cleveland DW, Goldstein LS. Abnormal neurofilament transport caused by targeted disruption of neuronal kinesin heavy chain KIF5A. J Cell Biol 2003;161:55–66.
[131] Reid E, Kloos M, Ashley-Koch A, Hughes L, Bevan S, Svenson IK, Graham FL, Gaskell PC, Dearlove A, Pericak-Vance MA, Rubinsztein DC, Marchuk DA. A kinesin heavy chain (KIF5A) mutation in hereditary spastic paraplegia (SPG10). Am J Hum Genet 2002;71:1189–94.
[132] De Vos KJ, Hafezparast M. Neurobiology of axonal transport defects in motor neuron diseases: opportunities for translational research? Neurobiol Dis 2017;105:283–99.
[133] McGeer PL, McGeer EG. Glial cell reactions in neurodegenerative diseases: pathophysiology and therapeutic interventions. Alzheimer Dis Assoc Disord 1998;12(Suppl. 2):S1–6.
[134] Pekny M, Pekna M, Messing A, Steinhauser C, Lee JM, Parpura V, Hol EM, Sofroniew MV, Verkhratsky A. Astrocytes: a central element in neurological diseases. Acta Neuropathol 2016;131:323–45.

[135] Boillee S, Yamanaka K, Lobsiger CS, Copeland NG, Jenkins NA, Kassiotis G, Kollias G, Cleveland DW. Onset and progression in inherited ALS determined by motor neurons and microglia. Science 2006;312:1389–92.

[136] Yamanaka K, Chun SJ, Boillee S, Fujimori-Tonou N, Yamashita H, Gutmann DH, Takahashi R, Misawa H, Cleveland DW. Astrocytes as determinants of disease progression in inherited amyotrophic lateral sclerosis. Nat Neurosci 2008;11:251–3.

[137] Boillee S, Vande Velde C, Cleveland DW. ALS: a disease of motor neurons and their nonneuronal neighbors. Neuron 2006;52:39–59.

[138] Di Giorgio FP, Carrasco MA, Siao MC, Maniatis T, Eggan K. Non-cell autonomous effect of glia on motor neurons in an embryonic stem cell-based ALS model. Nat Neurosci 2007;10:608–14.

[139] Chen H, Kankel MW, Su SC, Han SWS, Ofengeim D. Exploring the genetics and non-cell autonomous mechanisms underlying ALS/FTLD. Cell Death Differ 2018;25:646–60.

[140] Liu Q, Dreyfuss G. A novel nuclear structure containing the survival of motor neurons protein. EMBO J 1996;15:3555–65.

[141] Carvalho T, Almeida F, Calapez A, Lafarga M, Berciano MT, Carmo-Fonseca M. The spinal muscular atrophy disease gene product, SMN: a link between snRNP biogenesis and the Cajal (coiled) body. J Cell Biol 1999;147:715–28.

[142] Pellizzoni L, Yong J, Dreyfuss G. Essential role for the SMN complex in the specificity of snRNP assembly. Science 2002;298:1775–9.

[143] Meister G, Buhler D, Pillai R, Lottspeich F, Fischer U. A multiprotein complex mediates the ATP-dependent assembly of spliceosomal U snRNPs. Nat Cell Biol 2001;3:945–9.

[144] Steele JC. Parkinsonism-dementia complex of Guam. Mov Disord 2005;20(Suppl. 12):S99–107.

[145] Sieh W, Choi Y, Chapman NH, Craig UK, Steinbart EJ, Rothstein JH, Oyanagi K, Garruto RM, Bird TD, Galasko DR, Schellenberg GD, Wijsman EM. Identification of novel susceptibility loci for Guam neurodegenerative disease: challenges of genome scans in genetic isolates. Hum Mol Genet 2009;18:3725–38.

[146] Hudson AJ, Rice GP. Similarities of guamanian ALS/PD to post-encephalitic parkinsonism/ALS: possible viral cause. Can J Neurol Sci 1990;17:427–33.

[147] Reid AH, McCall S, Henry JM, Taubenberger JK. Experimenting on the past: the enigma of von Economo's encephalitis lethargica. J Neuropathol Exp Neurol 2001;60:663–70.

[148] Spencer PS, Nunn PB, Hugon J, Ludolph AC, Ross SM, Roy DN, Robertson RC. Guam amyotrophic lateral sclerosis-parkinsonism-dementia linked to a plant excitant neurotoxin. Science 1987;237:517–22.

[149] Duncan MW. beta-Methylamino-L-alanine (BMAA) and amyotrophic lateral sclerosis-parkinsonism dementia of the western Pacific. Ann NY Acad Sci 1992;648:161–8.

[150] Weiss JH, Koh JY, Choi DW. Neurotoxicity of beta-N-methylamino-L-alanine (BMAA) and beta-N-oxalylamino-L-alanine (BOAA) on cultured cortical neurons. Brain Res 1989;497:64–71.

[151] Sacks O. The island of the colour-blind: and, Cycad Island. Pan Macmillan; 1997.

[152] Duncan MW, Steele JC, Kopin IJ, Markey SP. 2-Amino-3-(methylamino)-propanoic acid (BMAA) in cycad flour: an unlikely cause of amyotrophic lateral sclerosis and parkinsonism-dementia of Guam. Neurology 1990;40:767–72.

[153] Cox PA, Banack SA, Murch SJ. Biomagnification of cyanobacterial neurotoxins and neurodegenerative disease among the Chamorro people of Guam. Proc Natl Acad Sci USA 2003;100:13380–3.

[154] Ince PG, Codd GA. Return of the cycad hypothesis - does the amyotrophic lateral sclerosis/parkinsonism dementia complex (ALS/PDC) of Guam have new implications for global health? Neuropathol Appl Neurobiol 2005;31:345–53.

[155] Blasco H, Patin F, Andres CR, Corcia P, Gordon PH. Amyotrophic Lateral Sclerosis, 2016: existing therapies and the ongoing search for neuroprotection. Expert Opin Pharmacother 2016;17:1669–82.

[156] Dash RP, Babu RJ, Srinivas NR. Two decades-long journey from riluzole to edaravone: revisiting the clinical pharmacokinetics of the only two amyotrophic lateral sclerosis therapeutics. Clin Pharmacokinet 2002;2:CD001447.

[157] Yoshida H, Yanai H, Namiki Y, Fukatsu-Sasaki K, Furutani N, Tada N. Neuroprotective effects of edaravone: a novel free radical scavenger in cerebrovascular injury. CNS Drug Rev 2006;12:9–20.

[158] Ito H, Wate R, Zhang J, Ohnishi S, Kaneko S, Ito H, Nakano S, Kusaka H. Treatment with edaravone, initiated at symptom onset, slows motor decline and decreases SOD1 deposition in ALS mice. Exp Neurol 2008;213:448–55.

[159] Miller RG, Jackson CE, Kasarskis EJ, England JD, Forshew D, Johnston W, Kalra S, Katz JS, Mitsumoto H, Rosenfeld J, Shoesmith C, Strong MJ, Woolley SC, Quality Standards Subcommittee of the American Academy of Neurology. Practice parameter update: the care of the patient with amyotrophic lateral sclerosis: multidisciplinary care, symptom management, and cognitive/behavioral impairment (an evidence-based review): report of the Quality Standards Subcommittee of the American Academy of Neurology. Neurology 2009;73:1227–33.

[160] Van den Berg JP, Kalmijn S, Lindeman E, Veldink JH, de Visser M, Van der Graaff MM, Wokke JH, Van den Berg LH. Multidisciplinary ALS care improves quality of life in patients with ALS. Neurology 2005;65:1264–7.

[161] Kellogg J, Bottman L, Arra EJ, Selkirk SM, Kozlowski F. Nutrition management methods effective in increasing weight, survival time and functional status in ALS patients: a systematic review. Amyotroph Lateral Scler Frontotemporal Degener 2018;19:7–11.

[162] Bourke SC, Tomlinson M, Williams TL, Bullock RE, Shaw PJ, Gibson GJ. Effects of non-invasive ventilation on survival and quality of life in patients with amyotrophic lateral sclerosis: a randomised controlled trial. Lancet Neurol 2006;5:140–7.

[163] McGoldrick P, Joyce PI, Fisher EM, Greensmith L. Rodent models of amyotrophic lateral sclerosis. Biochim Biophys Acta 2013;1832:1421–36.

[164] Groen EJN, Talbot K, Gillingwater TH. Advances in therapy for spinal muscular atrophy: promises and challenges. Nat Rev Neurol 2018;14:214–24.

[165] Corey DR. Nusinersen, an antisense oligonucleotide drug for spinal muscular atrophy. Nat Neurosci 2017;20:497–9.

[166] Mendell JR, Al-Zaidy S, Shell R, Arnold WD, Rodino-Klapac LR, Prior TW, Lowes L, Alfano L, Berry K, Church K, Kissel JT, Nagendran S, L'Italien J, Sproule DM, Wells C, Cardenas JA, Heitzer MD, Kaspar A, Corcoran S, Braun L, Likhite S, Miranda C, Meyer K, Foust KD, Burghes AHM, Kaspar BK. Single-dose gene-replacement therapy for spinal muscular atrophy. N Engl J Med 2017;377:1713–22.

[167] Finkel RS, Mercuri E, Darras BT, Connolly AM, Kuntz NL, Kirschner J, Chiriboga CA, Saito K, Servais L, Tizzano E, Topaloglu H, Tulinius M, Montes J, Glanzman AM, Bishop K, Zhong ZJ, Gheuens S, Bennett CF, Schneider E, Farwell W, De Vivo DC, Group ES. Nusinersen versus Sham control in infantile-onset spinal muscular atrophy. N Engl J Med 2017;377:1723–32.

[168] Mercuri E, Darras BT, Chiriboga CA, Day JW, Campbell C, Connolly AM, Iannaccone ST, Kirschner J, Kuntz NL, Saito K, Shieh PB, Tulinius M, Mazzone ES, Montes J, Bishop KM, Yang Q, Foster R, Gheuens S, Bennett CF, Farwell W, Schneider E, De Vivo DC, Finkel RS, Group CS. Nusinersen versus Sham control in later-onset spinal muscular atrophy. N Engl J Med 2018;378:625–35.

[169] Dickey AS, La Spada AR. Therapy development in Huntington disease: from current strategies to emerging opportunities. Am J Med Genet A 2018;176:842–61.

[170] Miller TM, Pestronk A, David W, Rothstein J, Simpson E, Appel SH, Andres PL, Mahoney K, Allred P, Alexander K, Ostrow LW, Schoenfeld D, Macklin EA, Norris DA, Manousakis G, Crisp M, Smith R, Bennett CF, Bishop KM, Cudkowicz ME. An antisense oligonucleotide against SOD1 delivered intrathecally for patients with SOD1 familial amyotrophic lateral sclerosis: a phase 1, randomised, first-in-man study. Lancet Neurol 2013;12:435–42.

[171] Armon C, Hardiman O. The beginning of precision medicine in ALS? Treatment to fit the genes. Neurology 2017;89:1850–1.

[172] Beghi E, Chio A, Couratier P, Esteban J, Hardiman O, Logroscino G, Millul A, Mitchell D, Preux PM, Pupillo E, Stevic Z, Swingler R, Traynor BJ, Van den Berg LH, Veldink JH, Zoccolella S, Eurals C. The epidemiology and treatment of ALS: focus on the heterogeneity of the disease and critical appraisal of therapeutic trials. Amyotroph Lateral Scler 2011;12:1–10.
[173] Barschke P, Oeckl P, Steinacker P, Ludolph A, Otto M. Proteomic studies in the discovery of cerebrospinal fluid biomarkers for amyotrophic lateral sclerosis. Expert Rev Proteomics 2017;14:769–77.
[174] Grolez G, Moreau C, Danel-Brunaud V, Delmaire C, Lopes R, Pradat PF, El Mendili MM, Defebvre L, Devos D. The value of magnetic resonance imaging as a biomarker for amyotrophic lateral sclerosis: a systematic review. BMC Neurol 2016;16:155.
[175] Bruijn L, Cudkowicz M, Group ALSCTW. Opportunities for improving therapy development in ALS. Amyotroph Lateral Scler Frontotemporal Degener 2014;15:169–73.

CHAPTER

HUNTINGTON'S CHOREA

6

CHAPTER OUTLINE

6.1 INTRODUCTION

Huntington's disease, also known as Huntington's chorea, stands distinct from the other neurodegenerative diseases dealt with in this textbook as a purely hereditary disorder. In contrast to Alzheimer or Parkinson diseases, where up to 90% of cases do not have an obvious cause, all cases of Huntington disease are the direct result of an underlying alteration in the genome of the patient—a mutation in the *HTT* gene located on chromosome 4. As a genetically defined disorder that correlates precisely with a clinical defined disease entity, investigations into the events that result in neurodegeneration resulting from mutations in *HTT* have provided important insights into how neuronal cells die—insights that have had significant implications for other neurodegenerative diseases.

The Molecular and Clinical Pathology of Neurodegenerative Disease. https://doi.org/10.1016/B978-0-12-811069-0.00006-9

Medical descriptions of what we now know as Huntington's disease date back several hundred years, with a number of early 19th-century reports of neurological disorders that match the clinical profile of Huntington disease [1–3]. It was the publication in 1872 of George Huntington's report "On Chorea," however, that assembled the clinical details of the disorder as a single disease entity (Box 6.1) [4]. The term chorea, derived from the Greek χορεία (*khoreía*, dancing), describes the involuntary movements that are the most prominent symptom of Huntington disease [5]. Huntington's original description is notable for recognizing that the disorder extended beyond a movement disorder, including changes in personality and in cognition, as well as observing that this was a hereditary disorder. It took over a 100 years to move from the observation that Huntington's disease was an inherited disorder to the identification of the causative mutation in 1993 (see Section 6.4.1 for a more detailed description of this). The cloning and study of the *HTT* gene, and genetic insights into the mechanisms driving disease, have been critical in increasing our understanding of the events that lead to

BOX 6.1 DOCTOR HUNTINGTON AND HIS DISEASE

THE

MEDICAL AND SURGICAL REPORTER.

No. 789.] PHILADELPHIA, APRIL 13, 1872. [Vol. XXVI.—No. 15.

ORIGINAL DEPARTMENT.

Communications.

ON CHOREA.

By George Huntington, M. D.,
Of Pomeroy, Ohio.

Essay read before the Meigs and Mason Academy of Medicine at Middleport, Ohio, February 15, 1872

Chorea is essentially a disease of the nervous system. The name "chorea" is given to the disease on account of the *dancing* propensities of those who are affected by it, and it is a very appropriate designation. The disease, as it is commonly seen, is by no means a dangerous or serious affection, however distressing it may be to the one suffering from it, or to his friends. Its most marked and characteristic feature is a clonic spasm affecting the voluntary muscles. There is no loss of sense or of volition attending these contractions, as there is in epilepsy; the will is there, but its power to perform is deficient, the desired movements are after a manner performed, but there seems to exist some hidden power, something that is playing tricks, as it were, upon the will, and in a measure thwarting and perverting its designs; and after the will has ceased to exert its power in any given direction, taking things into its own hands, and keeping the poor victim in a continual jigger as long as he remains awake, generally, though not always, granting a respite during sleep. The disease commonly begins by slight twitchings in the muscles of the face, which gradually increase in violence and variety. The eyelids are kept winking, the brows are corrugated, and then elevated, the nose is screwed first to the one side and then to the other, and the mouth is drawn in various directions, giving the patient the most ludicrous appearance imaginable. The upper extremities may be the first affected, or both simultaneously. All the voluntary muscles are liable to be affected, those of the face rarely being exempted.

If the patient attempt to protrude the tongue it is accomplished with a great deal of difficulty and uncertainty. The hands are kept rolling—first the palms upward, and then the backs. The shoulders are shrugged, and the feet and legs kept in perpetual motion; the toes are turned in, and then everted; one foot is thrown across the other, and then suddenly withdrawn, and, in short, every conceivable attitude and expression is assumed, and so varied and irregular are the motions gone through with, that a complete description of them would be impossible. Sometimes the muscles of the lower extremities are not affected, and I believe they never are *alone* involved. In cases of death from chorea, all the muscles of the body seem to have been affected, and the time required for recovery and degree of success in treatment seem to depend greatly upon the amount of muscular involvement. Romberg refers to two cases in which the muscles of *respiration were affected*.

The disease is generally confined to childhood, being most frequent between the ages of eight and fourteen years, and occurring oftener in girls than in boys. Dufosse and Rufz refer to 429 cases; 130 occurring in boys and 299 in girls. Watson mentions a collection of 1,029 cases, of whom 733 were *females*, giving a proportion of nearly 5 to 2. Dr. Watson also remarks upon the disease being most frequent among children of *dark* complexion, while the two authorities just alluded to, Dufosse and Rufz, give as their opinion that it is most frequent in children of *light* hair. In every case visiting the clinics

Doctor Huntington and his disease. George Huntington (right panel) was a physician based in New England and wrote his treatise "On Chorea" in 1872 (left panel) documenting patients he had observed with a hereditary movement disorder. Although there are a number of descriptions of hereditary chorea that predate his article, the comprehensive nature of George Huntington's essay stands out as capturing the key features of the disorder that is now named in his honor.*Image of George Huntington courtesy of the Wellcome collection.*

neurodegeneration in patients with Huntington's disease [6]. These insights, in turn, have led to the development of experimental treatments for disease, although to date there is no disease-modifying therapy that significantly extends life following onset of symptoms [7].

6.2 CLINICAL PRESENTATION

Huntington's disease is a progressive neurodegenerative disease characterized by a movement disorder (chorea), dementia, and behavioral and neuropsychiatric problems. Hereditary chorea was described by George Huntington in a paper entitled "On Chorea" in 1872 where he described rare cases of inherited chorea that were associated with behavioral problems that developed in adult life [4]. Unlike many of the other neurodegenerative disorders described in this book, Huntington disease has clearly defined genetics and is inherited in an autosomal dominant fashion. The mutation that results in the disease is an expansion of a CAG trinucleotide repeat in the Huntingtin gene (*HTT*) on chromosome 4, which results in the production of mutant protein Huntingtin [8]. The normal number of CAG repeats is 28 or less. Those who have more than 39 repeats are certain to develop Huntington's disease, whereas there is reduced penetrance in those who have between 36 and 39 repeats. If there are between 28 and 35 repeats, the person will not develop symptoms of Huntington disease but is at risk of passing on a larger repeat to their offspring that may be in the disease-causing range, due to a phenomenon known as anticipation. Like other triplet repeat disorders, Huntington disease exhibits anticipation with an expansion of the repeat number with successive generations. Anticipation occurs particularly when the mutation is passed down the paternal line.

In Western populations, Huntington disease has a prevalence of 5.7 per 100,00. It is less common in Asian and African populations [9]. The difference in prevalence is due to differences in the *HTT* gene; populations who have longer than average CAG repeats tend to have a higher prevalence of the disease [10].

The onset of symptoms in Huntington's disease is generally in adulthood with a mean age of onset of 40 years, although juvenile and elderly onset forms are described. The length of the CAG repeat is inversely correlated with age of onset [11]. The course of the disease is relentlessly progressive with death generally occurring 15–20 years after symptom onset. The juvenile onset form is defined as age of onset before the age of 20 years. These patients usually have more than 60 CAG repeats and often have more severe disease with a rigid-akinetic picture dominating rather than chorea.

6.2.1 MOVEMENT DISORDER

Chorea is the key feature of the movement disorder in Huntington's disease. The word chorea is derived from the Greek word for "dance" and is characterized by involuntary brief unpredictable arrhythmic movements that seem to flow from one muscle to the next. In the early stages of the disease, chorea may be mild and be misinterpreted as restlessness or fidgeting. As the disease progresses, the chorea becomes more widespread and more florid, interfering with speech swallowing, posture, and gait. This hyperkinetic phase of the movement disorder plateau with time is followed by a hypokinetic syndrome with bradykinesia, dystonia, and rigidity [12,13]. Other motor features of Huntington's disease include dyspraxia, poor balance, and poor postural control, which leads to falls [14]. Eye movements are abnormal and this may be the first clinical sign, with gaze impersistence and delayed initiation and slowing of saccades.

6.2.2 NEUROPSYCHIATRIC DISTURBANCE

Psychiatric symptoms are prominent in Huntington's disease. Depression, irritability, and apathy are seen frequently in the early stages and are often detectable during the premanifest stage of the disease, years before the onset of the movement disorder [15]. Frank psychosis and obsessive compulsive disorder are also recognized but less frequently. The suicide rate is higher than that in the general population [16]. Apathy is frequently seen and is particularly common in patients with advanced disease [17].

6.2.3 COGNITIVE

Cognitive decline is a major feature of Huntington's disease and, as is the case for psychiatric symptoms, subtle cognitive changes may be detectable many years before the movement disorder develops [18]. The cognitive decline in HD has a subcortical flavor with executive function particularly affected. Processing speed and a decline in visuospatial skills are also seen.

6.2.4 DIAGNOSIS

The diagnosis of Huntington's disease is based on a confirmed family history or positive genetic tests in association with the typical movement disorder. MRI imaging will show atrophy of the caudate. Positron emission tomography (PET) imaging can detect metabolic changes in the brains of people with Huntington's disease before symptom onset [19]. Genetic testing can be either diagnostic, in a patient who has motor symptoms of the disease, or predictive, which is done prior to disease onset in a patient who is at risk of inheriting the mutation. International recommendations have been developed to guide presymptomatic testing. These include counseling, a detailed examination to ensure that the patient is not symptomatic, a psychological review, and posttest follow-up [20,21]. Genetic testing is often requested to aid with reproductive decisions. Options for patients who are at risk of having inherited the *HTT* mutation who wish to start a family include prenatal diagnosis with termination in the case of the fetus being found to have the Huntington's disease mutation, or preimplantation genetic diagnosis where in vitro fertilization is performed and only embryos that do not have the mutation are implanted [22].

6.2.5 MANAGEMENT

Given the combination of symptoms seen in Huntington's disease, as well as the significant distress that an inherited disease causes to patients and their family, who may be at risk of developing the disease themselves, it is important that Huntington's disease is managed by a multidisciplinary team [23]. There is currently no treatment that has been shown to alter the clinical progression of the disease although this is an exciting and active field of research. In late 2017, it was reported that an antisense oligonucleotide, given intrathecally produced a dose-dependent reduction in the level of mutant Huntingtin protein in a phase 1/2a randomized controlled trial [24]. Further studies are underway to test whether this reduction in Huntingtin protein can slow progression of the disease. In day-to-day clinical practice currently, the aim is to maximize the quality of life for patients with Huntington disease and their families. Physiotherapy, occupational therapy, and speech and language therapy are of paramount importance, and patients need access to psychiatrists and psychologists. Genetic counselors play a vital role in helping patients understand the nature of inheritance, so they can make informed decisions about presymptomatic testing and reproductive choices.

A variety of pharmacological agents are used to improve the motor and psychiatric symptoms of Huntington's disease.

6.2.6 TREATMENT OF CHOREA

The only drug licensed to treat chorea is tetrabenazine. This drug selectively depletes central monoamines and was found to be effective in reducing chorea in a randomized controlled trial [25]. There is also evidence for a modified version of tetrabenazine, deutetrabenazine, which may be associated with fewer side effects [26]. Sulpiride and other neuroleptics may also alleviate chorea but are often associated with side effects including drowsiness and weight gain.

6.2.7 PSYCHIATRIC TREATMENT

Both pharmacological and nonpharmacological measures are used to treat the psychiatric manifestations of Huntington's disease. Nonpharmacological approaches include cognitive behavioral therapy and psychodynamic therapy, but their use may be limited if there is significant cognitive impairment. Pharmacological therapy can include antidepressants such as selective serotonin reuptake inhibitors (SSRIs) and others, such as venlafaxine and mirtazapine, which have both serotonergic and noradrenergic effects. Apathy is difficult to treat [27]. Different medications tried include bromocriptine, amantadine, levodopa, and atypical antipsychotics, but there is little evidence of what is most effective [28]. Environmental modifications, psychosocial support, and education for carers may help [29].

6.2.8 TREATMENT OF COGNITIVE DECLINE

There are mixed reports about the efficacy of acetylcholinesterase inhibitors in treating the cognitive decline in patients with Huntington disease with some reports that they are beneficial and other studies suggesting that they have no effect [30,31]. As the cognitive impairment inevitably progresses, social support for the patients and their carers becomes more important. By the end stage of the disease, patients generally require full-time care.

6.3 PATHOLOGY

Pathological examination of the brain in patients with Huntington chorea reveals a number of macroscopic and microscopic changes associated with disease. The first neuropathological reports date back to the final decade of the 19th century and the first decades of the 20th century, with a number of pathologists (including Alois Alzheimer) describing atrophy of the striatum in the brains of patients with Huntington's disease [32–34]. It is now clear that there is significant loss of brain volume in the latter stages of disease, with up to 20% of brain mass lost on average [35]. This is primarily a consequence of degeneration of neurons within the striatum, located within the midbrain, loss of which is so marked that this is clearly visible at a macroscopic level (Fig. 6.1A) [36]. The majority of cells lost in Huntington's disease are medium spiny neurons, cells that utilize gamma-aminobutyric acid as a neurotransmitter and are involved in the coordination and control of movement [37,38]. A number of the characteristics of these cells are thought to make medium spiny neurons differentially vulnerable to the

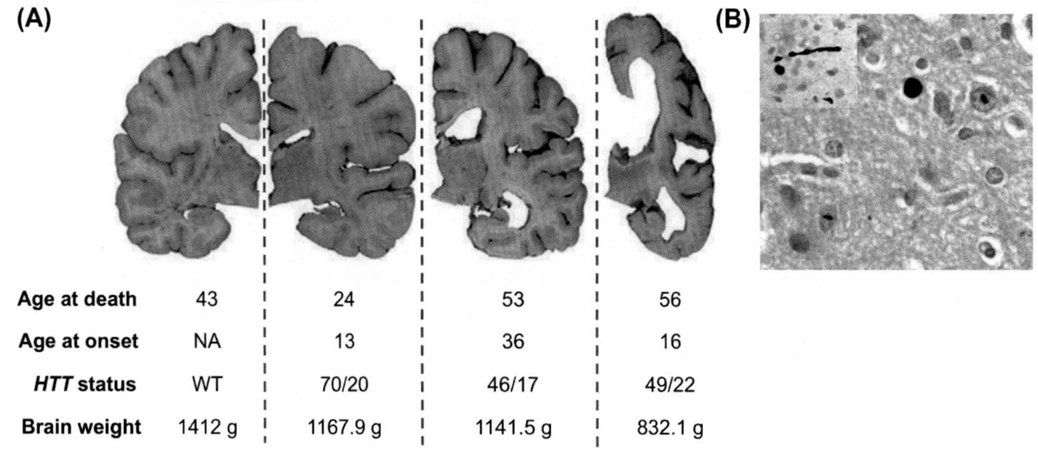

FIGURE 6.1

The macroscopic and microscopic neuropathology of Huntington's disease. (A) Atrophy of the striatum in the brains of individuals with mutations in the *HTT* gene, noting the polyglutamine repeat size, age at onset of symptoms, and duration of disease, compared to a neurologically normal brain. (B) Accumulation of ubiquitinated intranuclear inclusion bodies (main panel) and dystrophic neurites (inset panel top left) in the brain of a woman with 42 CAG repeats in the *HTT* gene. Original magnification 640×.

Both images are from Ref. Vonsattel JPG, Keller C, Ramirez EPC. Huntington's disease–neuropathology. In: Handbook of clinical neurology, vol. 100. Elsevier; 2011. p. 83–100.

disease process in Huntington's disease [39]. This loss of gamma-aminobutyric acid innervation can be directly linked to the chorea witnessed in individuals with Huntington disease [40]. In addition to neuronal loss, there are alterations in glial cell densities and activity, with evidence of increased numbers of oligodendrocytes [41], activated microglia [42], and astrocytes [43]. How these glial alterations contribute to the disease process is as yet unclear [44,45].

At a microscopic level, surviving cells within the striatum of patients with Huntington's disease accumulate intracellular inclusions of aggregated protein (Fig. 6.1B) [46,47]. These inclusions localize primarily to the nucleus, although aggregates can also be observed in the neuropil and can be detected using immunohistochemical analysis using antibodies specific for the Huntingtin protein or for ubiquitin [48–50].

6.4 MOLECULAR MECHANISMS OF DEGENERATION

The genetic nature of Huntington's disease means that a direct link can be drawn between the underlying nature of the mutations causing disease and the cellular changes that lead to neurodegeneration. The identification of the causative mutation in Huntington disease in 1993, therefore, opened up the possibility of understanding the disorder from genetics through to end-stage disease, and much progress has been made in regard to this over the past 25 years [51]. As discussed in Section 6.5, however, disease-modifying treatments have not yet moved from preclinical and experimental settings into clinical practice.

6.4.1 GENETICS

The hereditary nature of Huntington's disease was noted by George Huntington, as well as a number of other clinicians during the 19th century [4,52]. At the time of the writing of "On Chorea," however, the concept of genetics and the mechanisms driving inheritance were yet to be clearly enunciated. Indeed, the terms genetics and gene were not used until the first decade of the 20th century [53,54]. Very early in the development of the field of genetics, Huntington's disease was highlighted as a standout example of a hereditary disorder, passed on in a Mendelian fashion [55]. Progress in identifying the causative variant for Huntington's disease accelerated with the blossoming of molecular biology during the latter half of the 20th century, leading to advances in positional cloning and genetics that facilitated the localization of the causative gene to chromosome 4 [56,57]. The identification of mutations in the *HTT* gene, located on chromosome 4q16.3, followed in 1993 as a result of an international effort [8]. The *HTT* gene consists of 67 exons and codes for a large (>340 kDa) multidomain protein and includes a trinucleotide CAG (cytosine, adenine, guanosine) repeat in exon 1 coding for a stretch of glutamines in the N terminus of the protein. This trinucleotide repeat is polymorphic and was found to be expanded in individuals with Huntington's disease (Fig. 6.2). A detailed analysis of the *HTT* locus and variation in the CAG repeats swiftly followed the identification of this gene as being central to the pathogenesis of Huntington's disease. This revealed that the precise length of the repeat segment in exon of *HTT* was important for disease penetrance, age at onset of symptoms, and clinical presentation [58]. The normal (i.e., nonpathogenic) range for the CAG repeats in *HTT* is between 6 and 35. Huntington's disease is

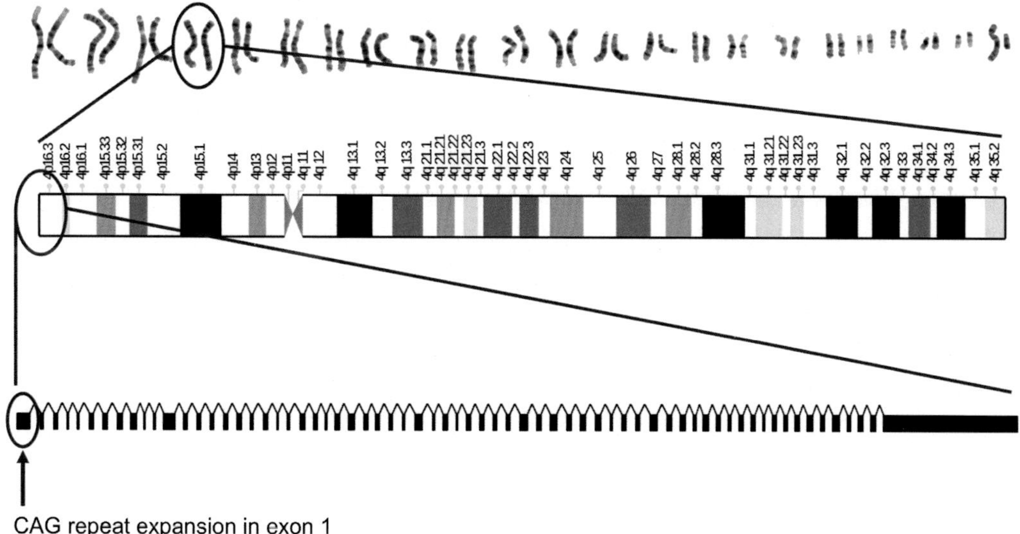

FIGURE 6.2

The location and structure of the *HTT* gene. Upper panel shows the human karyotype from chromosome 1 to the sex chromosomes left to right. The *HTT* gene is located on chromosome 4q16.3 and consists of 67 exons. The CAG repeat region is located in exon 1 at the 5′ end of the gene. *Images derived and modified from the Wikimedia commons collection using a creative commons licence.*

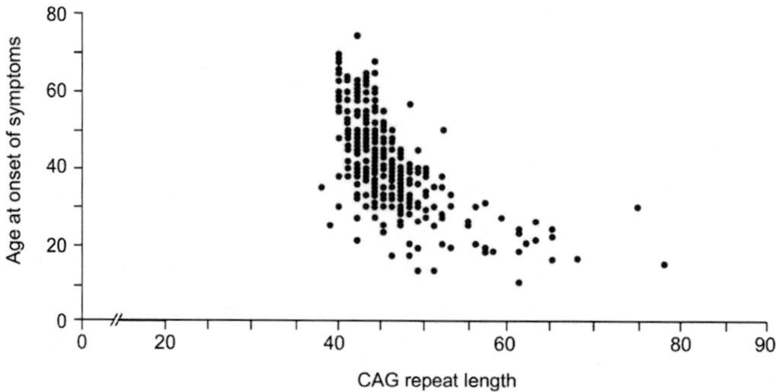

FIGURE 6.3

The relationship between CAG repeat number and age at onset of symptoms in Huntington's disease. Age at onset is shown on the y-axis, with number of pathogenic repeats displayed on the x-axis. Higher CAG repeat numbers are correlated with lower age of onset.

Adapted from Ref. Walker FO. Huntington's disease. Lancet 2007;369:218–28. Epub 2007/01/24.

fully penetrant in individuals with more than 40 repeats, with the repeat range between 36 and 39 associated with increased lifetime risk of disease [59]. Above 40 CAG repeats, there is a correlation between the length of the repeat and the age at which symptoms are first observed—with longer repeats associated with an earlier age at onset (Fig. 6.3) [11,60,61]. Huntington's disease displays an autosomal dominant mode of inheritance, requiring only one mutated allele to be present for disease to occur. Intriguingly, individuals homozygous for expanded repeats have been reported, but the impact on clinical presentation appears to be limited [62].

Trinucleotide repeats within the human genome are now recognized to be associated with a range of genetic disorders, including a number of neurological diseases (Box 6.2). These disorders are associated with increased size of repetitive nucleotide sequences in both coding and noncoding regions of genes, resulting in toxic gain of function and loss of function mechanisms of disease [63,64]. Mutations in these repeated DNA sequences are linked to the intrinsic instability of repetitive nucleotide stretches and the propensity of these sequences to increase in size during DNA replication [65]. The underlying mechanism of the origin of the mutations that lead to Huntington disease has a number of implications for the molecular biology of this disorder and through this clinical presentation. First, the propensity of the CAG repeat in *HTT* to increase in size through repeated DNA replication events in the germ line means that over several generations the CAG repeat number in a family with Huntington disease can increase. As there is a correlation between the length of repeat and age at onset, this can result in a phenomenon known as anticipation where the age at onset of symptoms within a kindred decreases through successive generations [66]. An additional implication of this phenomenon is that individuals with no family history of Huntington disease can develop the disorder in an apparently sporadic fashion if the number of CAG repeats in one of their *HTT* alleles increases from the normal to pathogenic range compared with their parents [67,68]. As genetic technology has advanced, a further consequence of this instability has become clear as analysis of repeat length in different tissues and cells from individuals harboring pathogenic repeat expansions in *HTT* has become available and revealed somatic variation in repeat length [69].

BOX 6.2 TRINUCLEOTIDE REPEAT DISORDERS

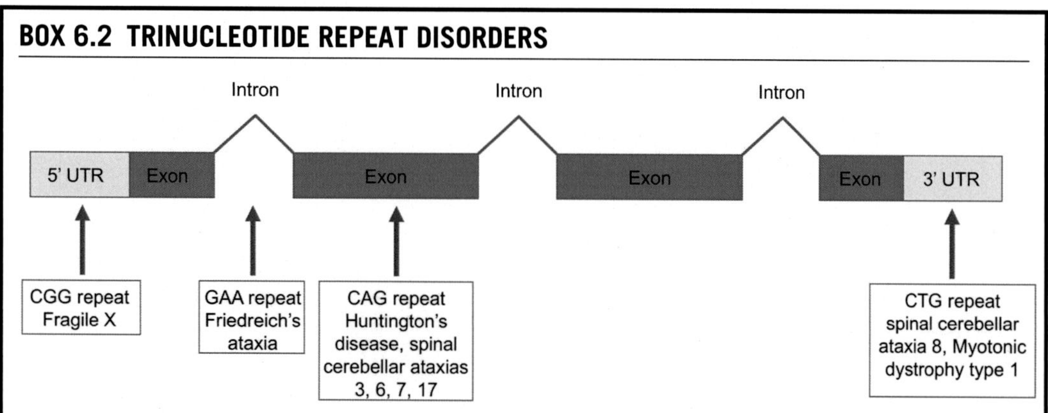

Trinucleotide repeat disorders. The trinucleotide repeat disorders are a group of inherited conditions caused by increased numbers of specific trinucleotide repeats within the genome. These are repetitive sequences within genes that increase from a normal range found across human populations into a pathogenic number of repeats. The precise sequence of the repeats, as well as their location within gene structures, varies across the disorders as shown in the panel. The underlying mechanisms causing repeat expansions are thought to be conserved; however, it is not clear as to whether there are shared mechanisms leading to disease. UTR, untranslated region. For a more detailed review of trinucleotide repeat disorders, see Ref. [63].

Although there is a correlation between repeat length and age at onset of symptoms, as shown in Fig. 6.3, there is substantial variation that is unrelated to the number of CAG repeats. The underlying cause of this variation is unclear but is likely to derive from a combination of genetic modifiers of disease in addition to the pathogenic allele of *HTT* [70,71]. Again, advances in genome-wide technologies have facilitated significant advances in understanding genetic modifiers of age at onset for Huntington's disease, and a number of studies have addressed this using a variety of approaches.

Important insights into how Huntington's disease onset can be modified have been provided by studying carefully characterized populations with greater than expected rates of Huntington's disease, a standout example of which is provided a group of families around Lake Maracaibo in Venezuela (Box 6.3) [72]. The elevated incidence of Huntington's disease in this area, first identified by Americo Negrette, a local physician, led to longitudinal studies of the natural history of Huntington's disease, including detailed analysis of age at onset [73,74]. A 2004 study of more than 450 patients with Huntington's disease from the region revealed a substantial component (59%) of the variation in age at onset was due to either genetic modifiers independent of the *HTT* gene or shared environmental exposure [75]. Using a candidate gene methodology, variants in the *GRIK2* gene, coding for the GluR6 receptor, were identified as modifiers of disease onset in Huntington's disease [76,77]. This has not, however, been replicated in a series of 2911 patients with Huntington's disease [78].

The most comprehensive genomic study to date utilized a genome-wide association approach to uncover loci associated with modulation of age at onset [79]. This revealed two loci that reached genome-wide significance, on chromosomes 15 and 8. Variation at these loci acted to reduce the age at which motor symptoms were first identified by 6 and 1.6 years, respectively. The mechanisms linking these loci to the pathogenic processes leading to neurodegeneration in Huntington's disease are unclear, although the chromosome 8 locus is close to the *MLH1* gene—involved in DNA repair pathways—that

BOX 6.3 LAKE MARACAIBO AND HUNTINGTON'S DISEASE

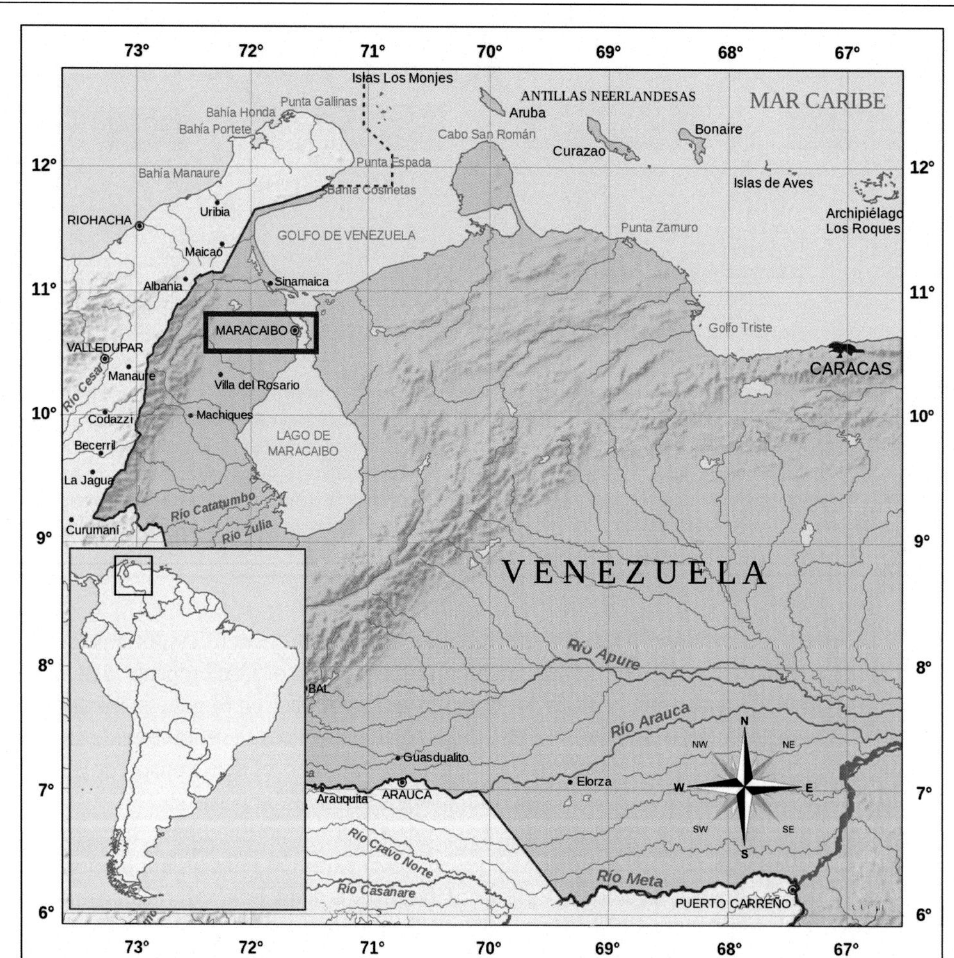

Lake Maracaibo and Huntington's disease. The area around Lake Maracaibo in the Northwest corner of Venezuela has played an important role in genetic and clinical research into Huntington's disease. The raised incidence of Huntington's disease in this area was first reported by Americo Negrette, a Venezuelan clinician, in the 1960s and it is now recognized that this area has the highest frequency of Huntington's mutation carriers in the world. It is thought that all of the carriers of pathogenic repeats in the *HTT* gene in this area are descended from an initial mutational event in a woman living on the edge of the lake in the early 19th century. By studying the genetics of this extended pedigree, and comparing this to clinical phenotype and age at onset, a huge amount has been learned about the genetic and phenotype heterogeneity of this disorder.*Images derived and modified from the Wikimedia commons collection using a creative commons licence.*

had previously been implicated in a mouse model for Huntington's disease [80]. Further genetic analysis has since replicated the potential link to this gene [81]. Additional studies have been carried out to analyze the impact of variation across the genome in the progression of Huntington's disease (the rate of motor and cognitive decline following the onset of symptoms) [82]. Using a cohort of more than 200 individuals registered to the TRACK-HD program, the authors identified a locus on chromosome 5 that was associated with accelerated disease progression. The *MSH3* gene, coding for a DNA mismatch repair enzyme and previously linked to Huntington's etiology in mouse models for the disease, was nominated as being the likely functional cause of this association [83].

6.4.2 POLYGLUTAMINE REPEATS AND PROTEIN BIOLOGY

The monogenic nature of Huntington's disease, and the molecular insight provided by the identification of the genetic cause of this disorder, has allowed extensive analysis of the biochemical and cellular impact of the CAG repeat expansions in the *HTT* gene [84]. A number of hypotheses have been proposed to explain the mechanistic link between CAG/polyglutamine repeats and neuronal cell death observed in patients with these mutations, including a toxic gain of function linked to aggregation state and a potential loss of function mechanism [85,86]. The *HTT* gene codes for a protein of more than 3100 amino acids and a predicted molecular mass of approximately 350 kDa. The protein has a number of HEAT repeat domains, thought to be important for protein–protein interactions and protein scaffolding [87]. The polyglutamine sequence that is expanded in individuals with Huntington's disease is located in the N terminal portion of the protein (Fig. 6.4A). The precise function of Huntingtin is unclear, although it has been implicated in vesicle trafficking through an interaction with microtubules. The large size of the Huntingtin protein has proved to be a major challenge for structural biologists attempting to gain structural insights into its function; however, recent advances in cryoelectron microscopy have revealed a 4 Å structure for Huntingtin in complex with one of its binding partners, Huntingtin-associated protein 40 or HAP40 (Fig. 6.4B) [88]. Although this provides important information regarding the three-dimensional arrangement of this protein, a clear indication as to the biochemical and cellular function's of this protein remains elusive.

With regard to alterations in protein biology related to increasing numbers of polyglutamines in the N terminal of the protein, a key observation was made by Max Perutz (who received the Nobel Prize in chemistry in 1962 for his work on the structure of hemoglobin) in 1994, not long after the identification of the *HTT* gene mutations. Using a combination of molecular modeling and structural analyses, he and his colleagues proposed that the extended repeats associated with disease could favor the formation of polar zippers in an extended beta pleated sheet conformation similar to that observed for amyloid aggregates [89]. The generation of a transgenic mouse model expressing a fragment of Huntingtin containing an expanded polyglutamine repeat sequence suggested that the presence of these repeats was neurotoxic [90]. Further research revealed that, similar to a number of other proteins involved in neurodegeneration, mutant Huntingtin was capable of forming structured aggregates both in vitro and in vivo [91] and that the presence of these correlated with neurodegeneration in a mouse model for Huntington's disease [92]. Importantly, it was also found that intranuclear inclusions can be identified in the striatum of individuals with the disease [46]. The structure of intracellular inclusion bodies made up of mutant Huntingtin has recently been revealed in exquisite detail using in situ cryoelectron microscopy analysis of cells expressing an N terminal fragment of Huntingtin containing 97 glutamines (Fig. 6.4C) [93].

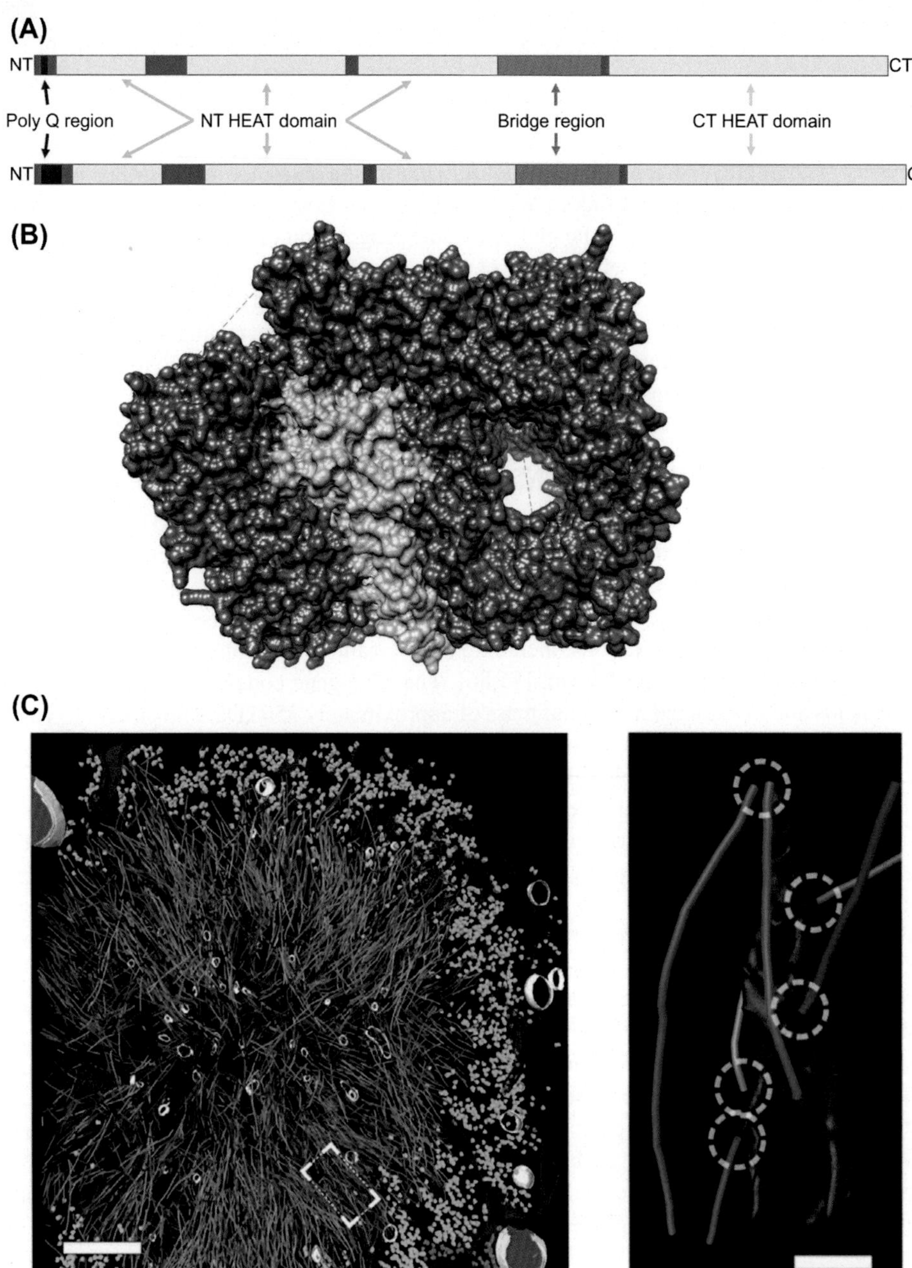

FIGURE 6.4

The structure of Huntingtin protein and aggregates. (A) Ideogram showing the domain organization of the Huntingtin protein, highlighting in red the polyglutamine repeats in the N-terminus of the protein. (B) The three-dimensional organization of Huntingtin (turquoise) in complex with Huntingtin-associated protein 40 (HAP40, shown in light brown). Image derived from structural coordinates in PDB accession 6ez8 [88] using Chimera. (C) The cryoelectron microscopy-derived structure of Huntingtin aggregates in situ, with aggregates shown in blue the endoplasmic reticulum shown in red, ribosomes in green, and mitochondria in yellow (scale bar 400 nM). Inset image shows points of contact between fibrils in blue and the endoplasmic reticulum in red (scale bar 50 nM).

Cryoelectron microscopy images in (C) are taken from Ref. Bauerlein FJB, Saha I, Mishra A, Kalemanov M, Martinez-Sanchez A, Klein R, Dudanova I, Hipp MS, Hartl FU, Baumeister W, Fernandez-Busnadiego R. In situ architecture and cellular interactions of PolyQ inclusions. Cell 2017;171:179–87. e10. Epub 2017/09/12.

Similar to other neurodegenerative disorders, including Alzheimer's and Parkinson's diseases, how the formation of protein aggregates relates to the cytotoxicity observed in patients carrying mutations in the *HTT* gene remains unclear and a matter of much debate [94]. In a seminal study published in 2004 by Arrasate and coworkers, the formation of visible intranuclear inclusions in primary neuronal cells expressing fragments of Huntingtin containing 72 polyglutamines (using a 17 repeat form of the protein as a control) followed longitudinally was found to be predictive of cell survival rather than cytotoxicity—and it was cells that did not form such aggregates that had an increased risk of death [95]. Further studies using automated microscopy revealed a quantifiable relationship between expression levels, polyglutamine expansion length, the formation of inclusion bodies, and cell death [96]. Intriguingly, intracellular aggregates in the brains of individuals with Huntington disease are more frequent in areas of the brain that are relatively spared compared to the striatum [50]. These data suggest that the formation of oligomeric, or prefibrillar, species of Huntingtin aggregate is the toxic culprit in the brains of individuals with Huntington's disease [47,97,98].

An area of interest across a wide spectrum of neurodegenerative disorders is that of prionlike behavior. Prions are infectious, transmissible protein aggregates that are able to perpetuate a misfolding process leading to cell death [99] and are directly linked to the pathogenesis of the prion disorders, including Creutzfeldt–Jakob disease and kuru (see Chapter 4) [100]. It has been suggested that some aspects of prionlike behavior are relevant to the misfolding and aggregation of proteins involved in other neurodegenerative disorders, including Huntington's disease [101]. This has been investigated using a range of models, both cellular and whole organism [102]. There is evidence that aggregates can transfer from cell to cell [103] and that human aggregates can induce aggregation in mouse models of disease [104]. How these studies relate to the disease process in the context of the human brain, and specifically in patients carrying a mutation in the *HTT* gene causing every cell to express an expanded polyglutamine containing protein, is unclear.

6.4.3 CELLULAR DYSFUNCTION

Although it is clear that protein aggregates are intimately involved in the disease process in Huntington's disease, the cellular consequences of the formation of oligomeric species of Huntingtin, or of larger fibrillar species, is a matter of some debate. Soon after the identification of the *HTT* gene, several studies reported that the protein product of *HTT* was expressed in a wide range of tissues and cells [105,106]. In the 25 years since, a number of cellular pathways have been identified as being disrupted by mutations in the *HTT* gene; however, the normal function of the Huntingtin protein (and whether the loss of this function plays a role in the etiology of the disease) remains to be determined [86]. The intranuclear nature of many of the inclusions observed in the brains of individuals with mutations in *HTT*, as well as in animal models for disease, led to a number of studies examining whether nuclear function—and, in particular, transcription—is disrupted in Huntington's disease [107]. Oligomers of Huntingtin bind to transcription factors in the nucleus, impacting on a range of genes and disrupting normal expression within cells [108].

A related aspect of nucleic acid biology is the linkage of DNA repeat mechanisms to alterations in the age at onset of movement symptoms in individuals with mutations in *HTT* [109]. There is evidence from genetic studies, and, in particular, genome-wide association studies examining factors that influence the age at onset of symptoms, that genes involved in DNA repair play an important role in modifying the disease process in Huntington's disease. One possible explanation for this is that differences in repair mechanisms driven by genetic variation can favor somatic expansion of CAG repeats, altering the aggregation properties of the resulting Huntingtin protein.

Outside of the nucleus, a number of cellular processes and pathways have been highlighted as being impacted by mutant Huntingtin. Similar to a number of other neurodegenerative disorders [110], mitochondrial biology is altered in the presence of *HTT* mutant alleles [111]. There is cellular evidence of alterations in trafficking of mitochondria because of the presence of mutant Huntingtin [112]. Linking in with the potential role of mutant Huntingtin in altering transcription, there is also evidence that expanded glutamine repeats in this protein can cause transcriptional repression of the key mitochondrial regulator gene *PPARGC1A*, coding for the transcription factor PGC1-α [113]. By altering mitochondrial location, function, and number, it has been proposed that mutant *HTT* could have a significant impact on oxidative stress in striatal neurons and result in cytotoxicity, although why this would differentially impact on the basal ganglia (as compared to other vulnerable neuronal populations in the central nervous system) is unclear [114].

Mutant Huntingtin has also been implicated in disruption of synaptic function and neurotransmission [115]. In particular, and in addition to disruption of striatal innervation due to neuronal cell death, there is evidence that mutant protein can act to interfere with *N*-methyl-D-aspartate receptor activity [116] and has been linked to increased GluR6 excitotoxicity [77]. It has been suggested that these alterations are an early event in the cellular pathogenesis of Huntington's disease.

As noted earlier (Section 6.4.2), there is clear evidence that mutations in the *HTT* gene result in changes in the propensity of Huntingtin protein to aggregate. The formation of protein aggregates has been demonstrated to cause cellular dysfunction and generalized dysregulation of proteostasis within the brain in a number of neurodegenerative diseases [117], and this has also been the subject of experimental investigation in Huntington's disease [118]. Mutant Huntingtin has been reported to have a global impact on protein folding within the cell, with consequences for cellular function and viability [119]. An additional consequence of the formation of protein aggregates of Huntingtin is that key proteins within the cell can be sequestered within these aggregates and caused a generalized reduction in cellular function and viability [120]. Two of the key cellular systems for protein degradation, the ubiquitin/proteasome system and autophagy, have also been reported to be disrupted in models for Huntington's disease. Mutant Huntingtin can cause global disruption of the ubiquitin/proteasome system [121], a disruption that is thought to be due to indirect inhibition of 26S proteasome subunit [122]. Investigations into Huntingtin dysfunction in transgenic mice as well as using human-derived neuronal cells carrying mutations revealed that mutant Huntingtin causes alterations in cargo recognition in macroautophagy, disrupting this process [123]. How closely these phenomena are linked, and whether these are a result of, or cause, alterations in protein folding across the cell, has not been fully clarified.

As is evident from the literature, there are a wide range of cellular processes that are altered because of the presence of mutant Huntingtin (summarized in Fig. 6.5). Whether these occur in parallel, or if one event/process disruption causes a cascade of consequences in the cell, remains a matter of debate. As is the case for other neurodegenerative diseases, a key question is why there is differential vulnerability across the brain—with striatal medium spiny neurons being particularly susceptible [124,125].

There is an increasing appreciation of the role that different brain cell types play in the pathobiology of Huntington's disease. Similar to many other neurodegenerative diseases, research into the roles of glial cells, in particular, has revealed that astrocytes and microglia are involved in the processes that lead to neuronal cell death in Huntington's disease [44,45]. Examples of this include

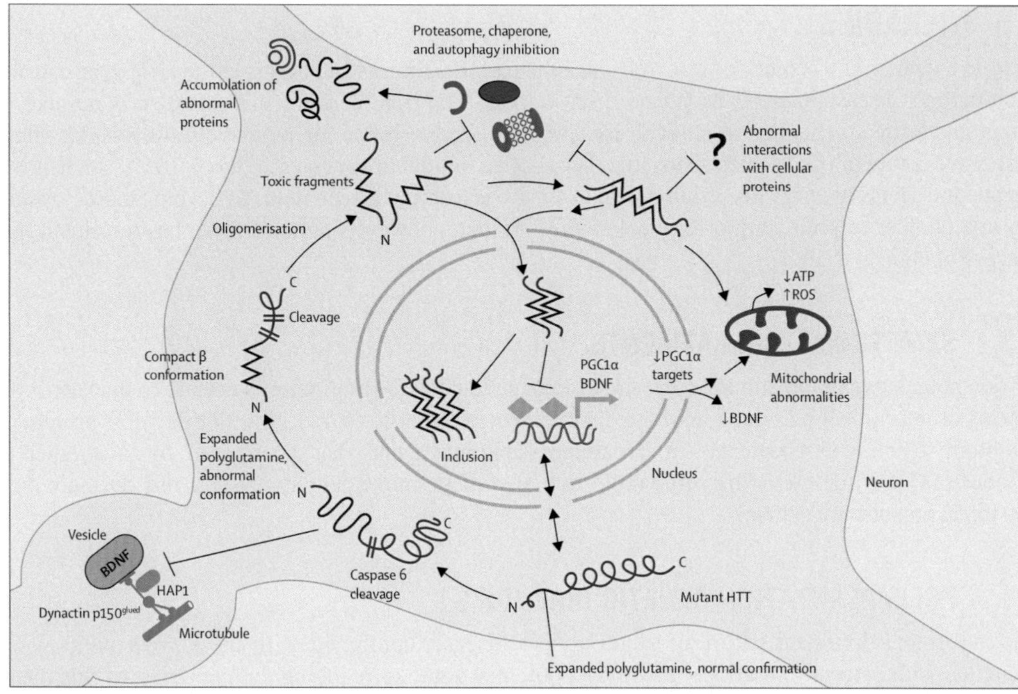

FIGURE 6.5

The cellular impact of mutations in *HTT*. Expanded polyglutamine repeats in Huntingtin have been implicated in dysfunction in a wide range of cellular processes, including transcriptional regulation, proteosomal activity, mitochondrial biology, macroautophagy, and trafficking.

Diagram is taken from Ref. Ross C.A., Tabrizi S.J.. Huntington's disease: from molecular pathogenesis to clinical treatment. Lancet Neurol. 2011;10:83–98. Epub 2010/12/18.

studies targeting expression of mutant Huntingtin to astrocyte cells in murine models for disease. This results in increased excitotoxicity [126] and neurological symptoms [127]. With regard to a microglial contribution to disease, there is evidence of aberrant microglial activation in the brains of individuals with Huntington's disease [42]. In cell and animal models for Huntington's disease, expression of Huntingtin containing expanded polyglutamine repeats results in the activation of microglia, suggesting that there is a direct impact of the presence of mutant protein on these cells [128]. Precisely how this activation interacts with the disease process in the brains of individuals with Huntington's disease is unclear.

An underinvestigated aspect of Huntington's disease is the impact of mutant *HTT* beyond the brain [129]. As Huntington disease is caused by a germ line mutation, and there is expression of the protein throughout a wide range of tissues, there is scope for cellular dysfunction linked to the mutated protein across the body. Although it is clear that there is enhanced susceptibility to cytotoxicity for neuronal cells (and, in particular, striatal neurons), gaining a deeper understanding of dysfunction in the periphery is likely to provide important insights relevant for holistic treatment of the disease [130].

6.5 THERAPIES

Despite a quarter of a century of research since the identification of mutations in the *HTT* gene causing Huntington's disease, there is as yet no disease-modifying treatment for this disorder. A number of symptomatic treatments are in clinical use, providing some relief for patients; however, the often-aggressive nature of the disease means that these have a limited impact on quality of life. As such, novel therapeutic agents are urgently required and extensive efforts have been made to develop these—resulting in a number of promising preclinical studies. To date, however, none of these has translated into successful clinical trials.

6.5.1 SYMPTOMATIC TREATMENTS

A wide range of symptomatic therapies are used in patients with Huntington's disease to manage some aspects of the clinical presentation of the disorder (reviewed in Refs. [23,28]). For the most prominent symptom, chorea, tetrabenazine—a dopamine depleting agent—has been used for a number of decades [131]. While providing some relief to hyperkinetic movement symptoms, this does not have any impact on disease course.

6.5.2 PREIMPLANTATION GENETIC DIAGNOSIS

The autosomal dominant nature of Huntington's disease, combined with the narrow window for reducing penetrance of mutations, provides a potential route to reducing the incidence of this disorder through preimplantation genetic diagnosis. This procedure combines in vitro fertilization technology with genetic sequencing of specific genes to identify fertilized embryos that do not carry deleterious genetic variants (Fig. 6.6) [132,133]. These embryos can then be implanted and allowed to proceed to full term, avoiding the inheritance of disease-related mutations. This approach has been used for a number of years as a method to prevent the passing on of pathogenic repeat expansions to subsequent generations [21,22]. The use of this technology is, however, illegal in some countries, and there are significant legal and ethical restraints and considerations that must be taken into account [134].

6.5.3 EXPERIMENTAL TREATMENTS

The clear need for disease-modifying treatments for Huntington's disease has led to a large number of experimental treatments undergoing clinical trials. To date, the results of these trials have been disappointing, with no clear impact on disease trajectory [135]. There are, however, a number of ongoing trials taking a range of approaches to target the impact of mutant Huntingtin [24,136].

Similar to several neurodegenerative disorders, there have been experimental efforts to replace the cells lost in the brains of individuals with Huntington's disease using neural transplants in open label trials. These demonstrate engraftment (that is, the neural cells successfully integrating in the host tissue) and some stabilization in clinical progression [137,138]. These benefits, however, have proved to be transient in nature and substantial issues remain to be overcome with the use of this approach in a more general clinical setting [139]. Again similar to other neurodegenerative disorders, and, in particular, Parkinson's disease, deep brain stimulation has been used as a means to combat chorea [140]. Although there is evidence that this may be beneficial for some patients, there are substantial

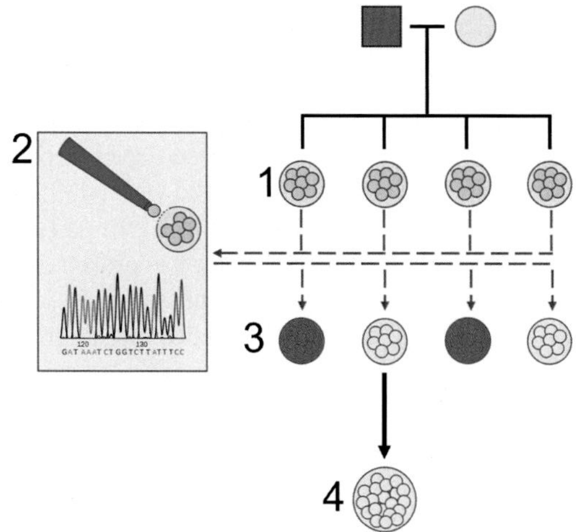

FIGURE 6.6

Preimplantation genetic diagnosis. Preimplantation genetic diagnosis uses a combination of in vitro fertilization and genetic screening to generate and screen zygotes for deleterious mutations. In this example, a couple where the male (indicated by a *red square*) carries a mutation in the *HTT* gene and the female (indicated by a *green circle*) is wild type that produces four embryos by in vitro fertilization (step 1). Individual cells are taken from each of these embryos and sequenced to determine *HTT* genotype (step 2). Those that do not carry the mutated allele are identified in green in step 3, with those harboring the mutation are shown in red. An embryo without the mutation is selected for implantation in the mother and will develop in utero (step 4), preventing the transmission of Huntington's disease to subsequent generations.

challenges to using deep brain stimulation more widely due to the invasive nature of the procedure and the requirement for highly trained specialist neurosurgical teams [141].

A number of small molecule approaches targeting pathways highlighted by research into the molecular mechanisms of Huntington's disease have yielded promising preclinical data in animal models for the disorder. These include compounds modulating proteostasis, for example macroautophagy [142], the activity of histone deacetylases [143,144], and, based on evidence for a dysregulated kynurenine pathway [145], kynurenine 3 monooxygenase inhibitors (Alzheimer's disease and Huntington's disease) [146]. Despite acting to ameliorate degeneration and slow or halt disease progression in animal models, these approaches have either not progressed to human clinical trials or failed once reaching this stage.

An area of experimental treatment research that is increasing in importance is the use of genetic approaches to reduce or remove mutant Huntingtin in the brains of mutation carriers [147]. This has built upon extensive preclinical studies using a number of genetic technologies to remove or knockdown expression in animal models for disease. These include the switching off of mutant gene expression in a conditional mouse model, resulting in the reversal of degeneration [148], and using RNAi to silence expression of mutant [149]. Importantly, experiments in mouse models for Huntington's disease have demonstrated that silencing of *HTT* expression can be targeted directly at mutant alleles [150]. This allows for continued expression of the wild-type allele, which in turn reduces any impact on the normal function of Huntingtin. Antisense oligonucleotides have also been demonstrated to lower Huntingtin levels in nonhuman primate brain [151]. An antisense oligonucleotide approach (designated IONIS-HTT$_{RX}$)

developed by Ionis Pharmaceuticals has now undergone preclinical analysis [152,153] and has gone through phase I safety trials involving individuals carrying mutations in *HTT* [154]. Although a detailed (and peer-reviewed) account of the results of this trial has yet to be published, it has been reported that this trial achieved safety and tolerability outcomes that support further development, paving the way for phase II trials [24]. Intriguingly, it was also reported that a dose-dependent decrease in mutant Huntingtin levels was observed in the cerebral spinal fluid of those treated with IONIS-HTT_{RX}—suggesting that the treatment is able to successfully engage the target gene. This trial is of particular note given developments in treatment for spinal muscular atrophy, where a gene therapy approach (Nusinersen, also developed by Ionis Pharmaceuticals) has recently successfully completed phase III trials and has been granted a license by the Food and Drug Administration (see Chapter 5) [155,156]. Although the therapeutic mechanism of Nusinersen is distinct from that of IONIS-HTT_{RX} (inducing expression of a gene transcript rather than suppressing expression of a mutated allele), the demonstration that a gene therapy approach can work in the context of the central nervous system has raised the prospect that analogous treatments may work in other neurodegenerative disorders. It is important to note, however, that major challenges remain with this approach. First, if the antisense oligonucleotide approach is successful in ameliorating symptoms or slowing disease, it is not clear whether repeated dosing would be required to maintain this impact. Second, the current route of administration is intrathecal (that is, directly into the cerebral spinal fluid, thereby bypassing the blood–brain barrier). This is an invasive procedure and may limit the potential use of this treatment in a standard clinical setting.

Using an alternative approach, preclinical efforts to use genome editing have proven successful in rodent models for Huntington's disease. These include using both zinc finger nuclease technology [157] and CRISPR-CAS9 [158] to artificially ablate mutant alleles. Although these technologies are beginning to make their way into clinical trials, they are still very much at an experimental medicine stage [159], so any efforts to translate these preclinical advances in the Huntington's field into clinical treatments will require a substantial commitment of both resources and time.

6.6 CONCLUSIONS

The genetically defined nature of Huntington's disease, combined with the identification and characterization of mutations in the *HTT* gene, has provided significant insights into the underlying mechanisms driving pathology in this disease. Although we have an increasingly sophisticated understanding of the events that lead to neuronal cell death in the brains of individuals with Huntington's disease, to date these advances have not been translated into clinical benefit for patients either with or at risk of disease. As noted in Section 6.5, however, there are a number of therapies in advanced preclinical or clinical trials—some of which are building on successful interventions in other neurological disorders.

FURTHER READING

Labbadia J, Morimoto RI. Huntington's disease: underlying molecular mechanisms and emerging concepts. Trends Biochem Sci 2013;38:378–85.

Ross CA, Tabrizi SJ. Huntington's disease: from molecular pathogenesis to clinical treatment. Lancet Neurol 2011;10:83–98.

Wild EJ, Tabrizi SJ. Therapies targeting DNA and RNA in Huntington's disease. Lancet Neurol 2017;16:837–47.

REFERENCES

[1] Dunglison R. The practice of medicine; or, a treatise on special pathology and therapeutics: Lea & Blanchard. 1842.

[2] Lyon IW. Chronic hereditary chorea. Am Med Times 1863;7:289–90.

[3] Lanska DJ. George Huntington (1850-1916) and hereditary chorea. J Hist Neurosci 2000;9:76–89. Epub 2001/03/10.

[4] Huntington G. On chorea. Med Surg Rep 1872;26:317–21.

[5] Vale TC, Cardoso F. Chorea: a journey through history. Tremor Other Hyperkinet Mov (NY) 2015;5. Epub 2015/06/10.

[6] Bates GP. History of genetic disease: the molecular genetics of Huntington disease - a history. Nat Rev Genet 2005;6:766–73. Epub 2005/09/02.

[7] Kieburtz K, Reilmann R, Olanow CW. Huntington's disease: current and future therapeutic prospects. Mov Disord 2018. Epub 2018/05/09.

[8] A novel gene containing a trinucleotide repeat that is expanded and unstable on Huntington's disease chromosomes. Huntingt Dis Collab Res Group Cell 1993;72:971–83.

[9] Pringsheim T, Wiltshire K, Day L, Dykeman J, Steeves T, Jette N. The incidence and prevalence of Huntington's disease: a systematic review and meta-analysis. Mov Dis 2012;27:1083–91. Epub 2012/06/14.

[10] Baig SS, Strong M, Quarrell OW. The global prevalence of Huntington's disease: a systematic review and discussion. Neurodegener Dis Manag 2016;6:331–43. Epub 2016/08/11.

[11] Duyao M, Ambrose C, Myers R, Novelletto A, Persichetti F, Frontali M, Folstein S, Ross C, Franz M, Abbott M, et al. Trinucleotide repeat length instability and age of onset in Huntington's disease. Nat Genet 1993;4:387–92. Epub 1993/08/01.

[12] van Vugt JP, van Hilten BJ, Roos RA. Hypokinesia in Huntington's disease. Mov Dis 1996;11:384–8. Epub 1996/07/01.

[13] Jacobs M, Hart EP, van Zwet EW, Bentivoglio AR, Burgunder JM, Craufurd D, Reilmann R, Saft C, Roos RA. Network RiotEHsD. Progression of motor subtypes in Huntington's disease: a 6-year follow-up study. J Neurol 2016;263:2080–5. Epub 2016/07/21.

[14] Ha AD, Fung VS. Huntington's disease. Curr Opin Neurol 2012;25:491–8. Epub 2012/07/10.

[15] Tabrizi SJ, Langbehn DR, Leavitt BR, Roos RA, Durr A, Craufurd D, Kennard C, Hicks SL, Fox NC, Scahill RI, Borowsky B, Tobin AJ, Rosas HD, Johnson H, Reilmann R, Landwehrmeyer B, Stout JC, investigators T-H. Biological and clinical manifestations of Huntington's disease in the longitudinal TRACK-HD study: cross-sectional analysis of baseline data. Lancet Neurol 2009;8:791–801. Epub 2009/08/04.

[16] Di Maio L, Squitieri F, Napolitano G, Campanella G, Trofatter JA, Conneally PM. Suicide risk in Huntington's disease. J Med Genet 1993;30:293–5. Epub 1993/04/01.

[17] van Duijn E, Craufurd D, Hubers AA, Giltay EJ, Bonelli R, Rickards H, Anderson KE, van Walsem MR, van der Mast RC, Orth M, Landwehrmeyer GB. European huntington's disease network behavioural phenotype working G. Neuropsychiatric symptoms in a European huntington's disease cohort (REGISTRY). J Neurol Neurosurg Psychiatry 2014;85:1411–8. Epub 2014/05/16.

[18] Stout JC, Paulsen JS, Queller S, Solomon AC, Whitlock KB, Campbell JC, Carlozzi N, Duff K, Beglinger LJ, Langbehn DR, Johnson SA, Biglan KM, Aylward EH. Neurocognitive signs in prodromal Huntington disease. Neuropsychology 2011;25:1–14. Epub 2010/10/06.

[19] Niccolini F, Politis M. Neuroimaging in Huntington's disease. World J Radiol 2014;6:301–12. Epub 2014/07/01.

[20] McColgan P, Tabrizi SJ. Huntington's disease: a clinical review. Eur J Neurol 2018;25:24–34. Epub 2017/08/18.

[21] Nance MA. Genetic counseling and testing for Huntington's disease: a historical review. Am J Med Genet B Neuropsychiatr Genet 2017;174:75–92. Epub 2016/05/14.

[22] Lashwood A, Flinter F. Clinical and counselling implications of preimplantation genetic diagnosis for Huntington's disease in the UK. Hum Fertil (Camb) 2001;4:235–8. Epub 2001/11/24.
[23] Frank S. Treatment of Huntington's disease. Neurotherapeutics 2014;11:153–60. Epub 2013/12/25.
[24] Rodrigues FB, Wild EJ. Huntington's disease clinical trials corner. J Huntingt Dis February 2018;2018(7):89–98. Epub 2018/02/27.
[25] Frank S. Tetrabenazine as anti-chorea therapy in Huntington disease: an open-label continuation study. Huntington Study Group/TETRA-HD Investigators. BMC Neurol 2009;9:62. Epub 2009/12/22.
[26] Rodrigues FB, Duarte GS, Costa J, Ferreira JJ, Wild EJ. Tetrabenazine versus deutetrabenazine for huntington's disease: twins or distant cousins?. Mov Disord Clin Pract 2017;4:582–5. Epub 2017/09/19.
[27] Fritz NE, Boileau NR, Stout JC, Ready R, Perlmutter JS, Paulsen JS, Quaid K, Barton S, McCormack MK, Perlman SL, Carlozzi NE. Huntington disease quality of life site I, coordinators. Relationships among apathy, health-related quality of life, and function in huntington's disease. J Neuropsychiatry Clin Neurosci 2018. appineuropsych17080173. Epub 2018/03/22.
[28] Mestre T, Ferreira J, Coelho MM, Rosa M, Sampaio C. Therapeutic interventions for symptomatic treatment in Huntington's disease. Cochrane Database Syst Rev 2009. CD006456. Epub 2009/07/10.
[29] Krishnamoorthy A, Craufurd D. Treatment of apathy in huntington's disease and other movement disorders. Curr Treat Options Neurol 2011;13:508–19. Epub 2011/07/30.
[30] Cubo E, Shannon KM, Tracy D, Jaglin JA, Bernard BA, Wuu J, Leurgans SE. Effect of donepezil on motor and cognitive function in Huntington disease. Neurology 2006;67:1268–71. Epub 2006/10/13.
[31] de Tommaso M, Difruscolo O, Sciruicchio V, Specchio N, Livrea P. Two years' follow-up of rivastigmine treatment in Huntington disease. Clin Neuropharmacol 2007;30:43–6. Epub 2007/02/03.
[32] Anton G. Über die Beteiligung der großen basalen Gehirnganglien bei Bewegungsstörungen und insbesondere bei Chorea. Jahrb f Psychiatr u Neurol 1896;14:41.
[33] Alzheimer A. Über die anatomische Grundlage der Huntingtonschen Chorea und der choreatischen Bewegungen uberhaupt. Neurol Cent 1911;30:891–2.
[34] Jergelsma G. Nue anatomische befunde bei paralysis agitans und bei chronischer progressive chorea. Neurol Cent 1908;27:995–6.
[35] Vonsattel JP, Myers RH, Stevens TJ, Ferrante RJ, Bird ED, Richardson Jr EP. Neuropathological classification of Huntington's disease. J Neuropathol Exp Neurol 1985;44:559–77. Epub 1985/11/01.
[36] Aylward EH, Li Q, Stine OC, Ranen N, Sherr M, Barta PE, Bylsma FW, Pearlson GD, Ross CA. Longitudinal change in basal ganglia volume in patients with Huntington's disease. Neurology 1997;48:394–9. Epub 1997/02/01.
[37] Ehrlich ME. Huntington's disease and the striatal medium spiny neuron: cell-autonomous and non-cell-autonomous mechanisms of disease. Neurotherapeutics 2012;9:270–84. Epub 2012/03/24.
[38] Graveland GA, Williams RS, DiFiglia M. Evidence for degenerative and regenerative changes in neostriatal spiny neurons in Huntington's disease. Science 1985;227:770–3. Epub 1985/02/15.
[39] Han I, You Y, Kordower JH, Brady ST, Morfini GA. Differential vulnerability of neurons in Huntington's disease: the role of cell type-specific features. J Neurochem 2010;113:1073–91. Epub 2010/03/20.
[40] Pearson SJ, Heathfield KW, Reynolds GP. Pallidal GABA and chorea in Huntington's disease. J Neural Transm Gen Sect 1990;81:241–6. Epub 1990/01/01.
[41] Myers RH, Vonsattel JP, Paskevich PA, Kiely DK, Stevens TJ, Cupples LA, Richardson Jr EP, Bird ED. Decreased neuronal and increased oligodendroglial densities in Huntington's disease caudate nucleus. J Neuropathol Exp Neurol 1991;50:729–42. Epub 1991/11/11.
[42] Sapp E, Kegel KB, Aronin N, Hashikawa T, Uchiyama Y, Tohyama K, Bhide PG, Vonsattel JP, DiFiglia M. Early and progressive accumulation of reactive microglia in the Huntington disease brain. J Neuropathol Exp Neurol 2001;60:161–72. Epub 2001/03/29.
[43] Selkoe DJ, Salazar FJ, Abraham C, Kosik KS. Huntington's disease: changes in striatal proteins reflect astrocytic gliosis. Brain Res 1982;245:117–25. Epub 1982/08/05.

[44] Crotti A, Glass CK. The choreography of neuroinflammation in Huntington's disease. Trends Immunol 2015;36:364–73. Epub 2015/05/24.
[45] Khakh BS, Beaumont V, Cachope R, Munoz-Sanjuan I, Goldman SA, Grantyn R. Unravelling and exploiting astrocyte dysfunction in huntington's disease. Trends Neurosci 2017;40:422–37. Epub 2017/06/06.
[46] DiFiglia M, Sapp E, Chase KO, Davies SW, Bates GP, Vonsattel JP, Aronin N. Aggregation of huntingtin in neuronal intranuclear inclusions and dystrophic neurites in brain. Science 1997;277:1990–3. Epub 1997/09/26.
[47] Arrasate M, Finkbeiner S. Protein aggregates in Huntington's disease. Exp Neurol 2012;238:1–11. Epub 2011/12/28.
[48] Waldvogel HJ, Kim EH, Tippett LJ, Vonsattel JP, Faull RL. The neuropathology of huntington's disease. Curr Top Behav Neurosci 2015;22:33–80. Epub 2014/10/11.
[49] Vonsattel JPG, Keller C, Ramirez EPC. Huntington's disease–neuropathology. In: Handbook of clinical neurology, 100. Elsevier; 2011. p. 83–100.
[50] Gutekunst CA, Li SH, Yi H, Mulroy JS, Kuemmerle S, Jones R, Rye D, Ferrante RJ, Hersch SM, Li XJ. Nuclear and neuropil aggregates in Huntington's disease: relationship to neuropathology. J Neurosci 1999;19:2522–34. Epub 1999/03/23.
[51] Ross CA, Tabrizi SJ. Huntington's disease: from molecular pathogenesis to clinical treatment. Lancet Neurol 2011;10:83–98. Epub 2010/12/18.
[52] Hoffmann J. Ueber chorea chronica progressiva (Huntington'sche Chorea, Chorea hereditaria). Arch für Pathol Anat Physiol für Klin Med 1888;111:513–48.
[53] Johannsen W. The genotype conception of heredity. Am Nat 1911;45:129–59.
[54] Shull GH. The "presence and absence" hypothesis. Am Nat 1909;43:410–9.
[55] Punnett R. Mendelian inheritance in man. Proc R Soc Med 1908;1:135–68.
[56] Pericak-Vance MA, Conneally PM, Merritt AD, Roos R, Norton Jr JA, Vance JM. Genetic linkage studies in Huntington disease. Cytogenet Cell Genet 1978;22:640–5. Epub 1978/01/01.
[57] Gusella JF, Wexler NS, Conneally PM, Naylor SL, Anderson MA, Tanzi RE, Watkins PC, Ottina K, Wallace MR, Sakaguchi AY, et al. A polymorphic DNA marker genetically linked to Huntington's disease. Nature 1983;306:234–8. Epub 1983/11/17.
[58] Walker FO. Huntington's disease. Lancet 2007;369:218–28. Epub 2007/01/24.
[59] Rubinsztein DC, Leggo J, Coles R, Almqvist E, Biancalana V, Cassiman JJ, Chotai K, Connarty M, Crauford D, Curtis A, Curtis D, Davidson MJ, Differ AM, Dode C, Dodge A, Frontali M, Ranen NG, Stine OC, Sherr M, Abbott MH, Franz ML, Graham CA, Harper PS, Hedreen JC, Hayden MR, et al. Phenotypic characterization of individuals with 30-40 CAG repeats in the Huntington disease (HD) gene reveals HD cases with 36 repeats and apparently normal elderly individuals with 36-39 repeats. Am J Hum Genet 1996;59:16–22. Epub 1996/07/01.
[60] Snell RG, MacMillan JC, Cheadle JP, Fenton I, Lazarou LP, Davies P, MacDonald ME, Gusella JF, Harper PS, Shaw DJ. Relationship between trinucleotide repeat expansion and phenotypic variation in Huntington's disease. Nat Genet 1993;4:393–7. Epub 1993/08/01.
[61] Andrew SE, Goldberg YP, Kremer B, Telenius H, Theilmann J, Adam S, Starr E, Squitieri F, Lin B, Kalchman MA, et al. The relationship between trinucleotide (CAG) repeat length and clinical features of Huntington's disease. Nat Genet 1993;4:398–403. Epub 1993/08/01.
[62] Wexler NS, Young AB, Tanzi RE, Travers H, Starosta-Rubinstein S, Penney JB, Snodgrass SR, Shoulson I, Gomez F, Ramos Arroyo MA, et al. Homozygotes for Huntington's disease. Nature 1987;326:194–7. Epub 1987/03/12.
[63] Orr HT, Zoghbi HY. Trinucleotide repeat disorders. Annu Rev Neurosci 2007;30:575–621. Epub 2007/04/10.
[64] La Spada AR, Taylor JP. Repeat expansion disease: progress and puzzles in disease pathogenesis. Nat Rev Genet 2010;11:247–58. Epub 2010/02/24.
[65] McMurray CT. Mechanisms of trinucleotide repeat instability during human development. Nat Rev Genet 2010;11:786–99. Epub 2010/10/19.

[66] Ranen NG, Stine OC, Abbott MH, Sherr M, Codori AM, Franz ML, Chao NI, Chung AS, Pleasant N, Callahan C, et al. Anticipation and instability of IT-15 (CAG)n repeats in parent-offspring pairs with Huntington disease. Am J Hum Genet 1995;57:593–602. Epub 1995/09/01.

[67] Myers RH, MacDonald ME, Koroshetz WJ, Duyao MP, Ambrose CM, Taylor SA, Barnes G, Srinidhi J, Lin CS, Whaley WL, et al. De novo expansion of a (CAG)n repeat in sporadic Huntington's disease. Nat Genet 1993;5:168–73. Epub 1993/10/01.

[68] De Rooij KE, De Koning Gans PA, Skraastad MI, Belfroid RD, Vegter-Van Der Vlis M, Roos RA, Bakker E, Van Ommen GJ, Den Dunnen JT, Losekoot M. Dynamic mutation in Dutch Huntington's disease patients: increased paternal repeat instability extending to within the normal size range. J Med Genet 1993;30:996–1002. Epub 1993/12/01.

[69] Telenius H, Kremer B, Goldberg YP, Theilmann J, Andrew SE, Zeisler J, Adam S, Greenberg C, Ives EJ, Clarke LA, et al. Somatic and gonadal mosaicism of the Huntington disease gene CAG repeat in brain and sperm. Nat Genet 1994;6:409–14. Epub 1994/04/01.

[70] Mo C, Hannan AJ, Renoir T. Environmental factors as modulators of neurodegeneration: insights from gene-environment interactions in Huntington's disease. Neurosci Biobehav Rev 2015;52:178–92. Epub 2015/03/15.

[71] Holmans PA, Massey TH, Jones L. Genetic modifiers of Mendelian disease: Huntington's disease and the trinucleotide repeat disorders. Hum Mol Genet 2017;26:R83–90. Epub 2017/10/05.

[72] Young AB, Shoulson I, Penney JB, Starosta-Rubinstein S, Gomez F, Travers H, Ramos-Arroyo MA, Snodgrass SR, Bonilla E, Moreno H, et al. Huntington's disease in Venezuela: neurologic features and functional decline. Neurology 1986;36:244–9. Epub 1986/02/01.

[73] Negrette A. Corea de Huntington: Estudio de Una Sola Familia a Traves de Varias Genereaciones. Maracaibo, Venezuela: Universidad de Zulia; 1955.

[74] Okun MS, Thommi N. Americo Negrette (1924 to 2003): diagnosing Huntington disease in Venezuela. Neurology 2004;63:340–3. Epub 2004/07/28.

[75] Wexler NS, Lorimer J, Porter J, Gomez F, Moskowitz C, Shackell E, Marder K, Penchaszadeh G, Roberts SA, Gayan J, Brocklebank D, Cherny SS, Cardon LR, Gray J, Dlouhy SR, Wiktorski S, Hodes ME, Conneally PM, Penney JB, Gusella J, Cha JH, Irizarry M, Rosas D, Hersch S, Hollingsworth Z, MacDonald M, Young AB, Andresen JM, Housman DE, De Young MM, Bonilla E, Stillings T, Negrette A, Snodgrass SR, Martinez-Jaurrieta MD, Ramos-Arroyo MA, Bickham J, Ramos JS, Marshall F, Shoulson I, Rey GJ, Feigin A, Arnheim N, Acevedo-Cruz A, Acosta L, Alvir J, Fischbeck K, Thompson LM, Young A, Dure L, O'Brien CJ, Paulsen J, Brickman A, Krch D, Peery S, Hogarth P, Higgins Jr DS, Landwehrmeyer B, Project US-VCR. Venezuelan kindreds reveal that genetic and environmental factors modulate Huntington's disease age of onset. Proc Natl Acad Sci USA 2004;101:3498–503. Epub 2004/03/03.

[76] MacDonald ME, Vonsattel JP, Shrinidhi J, Couropmitree NN, Cupples LA, Bird ED, Gusella JF, Myers RH. Evidence for the GluR6 gene associated with younger onset age of Huntington's disease. Neurology 1999;53:1330–2. Epub 1999/10/16.

[77] Rubinsztein DC, Leggo J, Chiano M, Dodge A, Norbury G, Rosser E, Craufurd D. Genotypes at the GluR6 kainate receptor locus are associated with variation in the age of onset of Huntington disease. Proc Natl Acad Sci USA 1997;94:3872–6. Epub 1997/04/15.

[78] Lee JH, Lee JM, Ramos EM, Gillis T, Mysore JS, Kishikawa S, Hadzi T, Hendricks AE, Hayden MR, Morrison PJ, Nance M, Ross CA, Margolis RL, Squitieri F, Gellera C, Gomez-Tortosa E, Ayuso C, Suchowersky O, Trent RJ, McCusker E, Novelletto A, Frontali M, Jones R, Ashizawa T, Frank S, Saint-Hilaire MH, Hersch SM, Rosas HD, Lucente D, Harrison MB, Zanko A, Abramson RK, Marder K, Sequeiros J, Landwehrmeyer GB. Registry Study of the European Huntington's Disease N, Shoulson I, Huntington Study Group Cp, Myers RH, MacDonald ME, Gusella JF. TAA repeat variation in the GRIK2 gene does not influence age at onset in Huntington's disease. Biochem Biophys Res Commun 2012;424:404–8. Epub 2012/07/10.

[79] Genetic Modifiers of Huntington's Disease C. Identification of genetic factors that modify clinical onset of Huntington's disease. Cell 2015;162:516–26. Epub 2015/08/02.

[80] Pinto RM, Dragileva E, Kirby A, Lloret A, Lopez E, St Claire J, Panigrahi GB, Hou C, Holloway K, Gillis T, Guide JR, Cohen PE, Li GM, Pearson CE, Daly MJ, Wheeler VC. Mismatch repair genes Mlh1 and Mlh3 modify CAG instability in Huntington's disease mice: genome-wide and candidate approaches. PLoS Genet 2013;9:e1003930. Epub 2013/11/10.

[81] Lee JM, Chao MJ, Harold D, Abu Elneel K, Gillis T, Holmans P, Jones L, Orth M, Myers RH, Kwak S, Wheeler VC, MacDonald ME, Gusella JF. A modifier of Huntington's disease onset at the MLH1 locus. Hum Mol Genet 2017;26:3859–67. Epub 2017/09/22.

[82] Hensman Moss DJ, Pardinas AF, Langbehn D, Lo K, Leavitt BR, Roos R, Durr A, Mead S, investigators T-H, investigators R, Holmans P, Jones L, Tabrizi SJ. Identification of genetic variants associated with Huntington's disease progression: a genome-wide association study. Lancet Neurol 2017;16:701–11. Epub 2017/06/24.

[83] Tome S, Manley K, Simard JP, Clark GW, Slean MM, Swami M, Shelbourne PF, Tillier ER, Monckton DG, Messer A, Pearson CE. MSH3 polymorphisms and protein levels affect CAG repeat instability in Huntington's disease mice. PLoS Genet 2013;9:e1003280. Epub 2013/03/08.

[84] Labbadia J, Morimoto RI. Huntington's disease: underlying molecular mechanisms and emerging concepts. Trends Biochem Sci 2013;38:378–85. Epub 2013/06/19.

[85] Gil JM, Rego AC. Mechanisms of neurodegeneration in Huntington's disease. Eur J Neurosci 2008;27:2803–20. Epub 2008/07/01.

[86] Cattaneo E, Zuccato C, Tartari M. Normal huntingtin function: an alternative approach to Huntington's disease. Nat Rev Neurosci 2005;6:919–30. Epub 2005/11/17.

[87] Andrade MA, Bork P. HEAT repeats in the Huntington's disease protein. Nat Genet 1995;11:115–6. Epub 1995/10/01.

[88] Guo Q, Bin H, Cheng J, Seefelder M, Engler T, Pfeifer G, Oeckl P, Otto M, Moser F, Maurer M, Pautsch A, Baumeister W, Fernandez-Busnadiego R, Kochanek S. The cryo-electron microscopy structure of huntingtin. Nature 2018;555:117–20. Epub 2018/02/22.

[89] Perutz MF, Johnson T, Suzuki M, Finch JT. Glutamine repeats as polar zippers: their possible role in inherited neurodegenerative diseases. Proc Natl Acad Sci USA 1994;91:5355–8. Epub 1994/06/07.

[90] Mangiarini L, Sathasivam K, Seller M, Cozens B, Harper A, Hetherington C, Lawton M, Trottier Y, Lehrach H, Davies SW, Bates GP. Exon 1 of the HD gene with an expanded CAG repeat is sufficient to cause a progressive neurological phenotype in transgenic mice. Cell 1996;87:493–506. Epub 1996/11/01.

[91] Scherzinger E, Lurz R, Turmaine M, Mangiarini L, Hollenbach B, Hasenbank R, Bates GP, Davies SW, Lehrach H, Wanker EE. Huntingtin-encoded polyglutamine expansions form amyloid-like protein aggregates in vitro and in vivo. Cell 1997;90:549–58. Epub 1997/08/08.

[92] Davies SW, Turmaine M, Cozens BA, DiFiglia M, Sharp AH, Ross CA, Scherzinger E, Wanker EE, Mangiarini L, Bates GP. Formation of neuronal intranuclear inclusions underlies the neurological dysfunction in mice transgenic for the HD mutation. Cell 1997;90:537–48. Epub 1997/08/08.

[93] Bauerlein FJB, Saha I, Mishra A, Kalemanov M, Martinez-Sanchez A, Klein R, Dudanova I, Hipp MS, Hartl FU, Baumeister W, Fernandez-Busnadiego R. In situ architecture and cellular interactions of PolyQ inclusions. Cell 2017;171:179–87. e10. Epub 2017/09/12.

[94] Chiti F, Dobson CM. Protein misfolding, amyloid formation, and human disease: a summary of progress over the last decade. Annu Rev Biochem 2017;86:27–68.

[95] Arrasate M, Mitra S, Schweitzer ES, Segal MR, Finkbeiner S. Inclusion body formation reduces levels of mutant huntingtin and the risk of neuronal death. Nature 2004;431:805–10.

[96] Miller J, Arrasate M, Shaby BA, Mitra S, Masliah E, Finkbeiner S. Quantitative relationships between huntingtin levels, polyglutamine length, inclusion body formation, and neuronal death provide novel insight into huntington's disease molecular pathogenesis. J Neurosci 2010;30:10541–50. Epub 2010/08/06.

[97] Takahashi T, Kikuchi S, Katada S, Nagai Y, Nishizawa M, Onodera O. Soluble polyglutamine oligomers formed prior to inclusion body formation are cytotoxic. Hum Mol Genet 2008;17:345–56. Epub 2007/10/20.

[98] Legleiter J, Mitchell E, Lotz GP, Sapp E, Ng C, DiFiglia M, Thompson LM, Muchowski PJ. Mutant huntingtin fragments form oligomers in a polyglutamine length-dependent manner in vitro and in vivo. J Biol Chem 2010;285:14777–90. Epub 2010/03/12.

[99] Collinge J, Clarke AR. A general model of prion strains and their pathogenicity. Science 2007;318:930–6. Epub 2007/11/10.

[100] Prusiner SB. Prions. Proc Natl Acad Sci USA 1998;95:13363–83.

[101] Jucker M, Walker LC. Self-propagation of pathogenic protein aggregates in neurodegenerative diseases. Nature 2013;501:45–51.

[102] Jansen AH, Batenburg KL, Pecho-Vrieseling E, Reits EA. Visualization of prion-like transfer in Huntington's disease models. Biochimica Biophys acta 2017;1863:793–800. Epub 2017/01/04.

[103] Ren PH, Lauckner JE, Kachirskaia I, Heuser JE, Melki R, Kopito RR. Cytoplasmic penetration and persistent infection of mammalian cells by polyglutamine aggregates. Nat Cell Biol 2009;11:219–25. Epub 2009/01/20.

[104] Jeon I, Cicchetti F, Cisbani G, Lee S, Li E, Bae J, Lee N, Li L, Im W, Kim M, Kim HS, Oh SH, Kim TA, Ko JJ, Aube B, Oueslati A, Kim YJ, Song J. Human-to-mouse prion-like propagation of mutant huntingtin protein. Acta Neuropathol 2016;132:577–92. Epub 2016/05/26.

[105] Li SH, Schilling G, Young 3rd WS, Li XJ, Margolis RL, Stine OC, Wagster MV, Abbott MH, Franz ML, Ranen NG, et al. Huntington's disease gene (IT15) is widely expressed in human and rat tissues. Neuron 1993;11:985–93. Epub 1993/11/01.

[106] Strong TV, Tagle DA, Valdes JM, Elmer LW, Boehm K, Swaroop M, Kaatz KW, Collins FS, Albin RL. Widespread expression of the human and rat Huntington's disease gene in brain and nonneural tissues. Nat Genet 1993;5:259–65. Epub 1993/11/01.

[107] Schaffar G, Breuer P, Boteva R, Behrends C, Tzvetkov N, Strippel N, Sakahira H, Siegers K, Hayer-Hartl M, Hartl FU. Cellular toxicity of polyglutamine expansion proteins: mechanism of transcription factor deactivation. Mol Cell 2004;15:95–105. Epub 2004/07/01.

[108] Zhai W, Jeong H, Cui L, Krainc D, Tjian R. In vitro analysis of huntingtin-mediated transcriptional repression reveals multiple transcription factor targets. Cell 2005;123:1241–53. Epub 2005/12/27.

[109] Massey TH, Jones L. The central role of DNA damage and repair in CAG repeat diseases. Dis Model Mech 2018:11. Epub 2018/02/09.

[110] Lin MT, Beal MF. Mitochondrial dysfunction and oxidative stress in neurodegenerative diseases. Nature 2006;443:787–95. Epub 2006/10/20.

[111] Costa V, Scorrano L. Shaping the role of mitochondria in the pathogenesis of Huntington's disease. EMBO J 2012;31:1853–64. Epub 2012/03/27.

[112] Orr AL, Li S, Wang CE, Li H, Wang J, Rong J, Xu X, Mastroberardino PG, Greenamyre JT, Li XJ. N-terminal mutant huntingtin associates with mitochondria and impairs mitochondrial trafficking. J Neurosci 2008;28:2783–92. Epub 2008/03/14.

[113] Cui L, Jeong H, Borovecki F, Parkhurst CN, Tanese N, Krainc D. Transcriptional repression of PGC-1alpha by mutant huntingtin leads to mitochondrial dysfunction and neurodegeneration. Cell 2006;127:59–69. Epub 2006/10/05.

[114] Johri A, Chandra A, Beal MF. PGC-1alpha, mitochondrial dysfunction, and Huntington's disease. Free Radic Biol Med 2013;62:37–46. Epub 2013/04/23.

[115] Li JY, Plomann M, Brundin P. Huntington's disease: a synaptopathy?. Trends Mol Med 2003;9:414–20. Epub 2003/10/15.

[116] Okamoto S, Pouladi MA, Talantova M, Yao D, Xia P, Ehrnhoefer DE, Zaidi R, Clemente A, Kaul M, Graham RK, Zhang D, Vincent Chen HS, Tong G, Hayden MR, Lipton SA. Balance between synaptic versus extrasynaptic NMDA receptor activity influences inclusions and neurotoxicity of mutant huntingtin. Nat Med 2009;15:1407–13. Epub 2009/11/17.

[117] Hipp MS, Park SH, Hartl FU. Proteostasis impairment in protein-misfolding and -aggregation diseases. Trends cell Biol 2014;24:506–14. Epub 2014/06/21.

[118] Margulis J, Finkbeiner S. Proteostasis in striatal cells and selective neurodegeneration in Huntington's disease. Front Cell Neurosci 2014;8:218. Epub 2014/08/26.

[119] Gidalevitz T, Ben-Zvi A, Ho KH, Brignull HR, Morimoto RI. Progressive disruption of cellular protein folding in models of polyglutamine diseases. Science 2006;311:1471–4. Epub 2006/02/14.

[120] Olzscha H, Schermann SM, Woerner AC, Pinkert S, Hecht MH, Tartaglia GG, Vendruscolo M, Hayer-Hartl M, Hartl FU, Vabulas RM. Amyloid-like aggregates sequester numerous metastable proteins with essential cellular functions. Cell 2011;144:67–78. Epub 2011/01/11.

[121] Bennett EJ, Shaler TA, Woodman B, Ryu KY, Zaitseva TS, Becker CH, Bates GP, Schulman H, Kopito RR. Global changes to the ubiquitin system in Huntington's disease. Nature 2007;448:704–8. Epub 2007/08/10.

[122] Hipp MS, Patel CN, Bersuker K, Riley BE, Kaiser SE, Shaler TA, Brandeis M, Kopito RR. Indirect inhibition of 26S proteasome activity in a cellular model of Huntington's disease. J Cell Biol 2012;196:573–87. Epub 2012/03/01.

[123] Martinez-Vicente M, Talloczy Z, Wong E, Tang G, Koga H, Kaushik S, de Vries R, Arias E, Harris S, Sulzer D, Cuervo AM. Cargo recognition failure is responsible for inefficient autophagy in Huntington's disease. Nat Neurosci 2010;13:567–76. Epub 2010/04/13.

[124] Morrison BM, Hof PR, Morrison JH. Determinants of neuronal vulnerability in neurodegenerative diseases. Ann Neurol 1998;44:S32–44.

[125] Cowan CM, Raymond LA. Selective neuronal degeneration in Huntington's disease. Curr Top Dev Biol 2006;75:25–71. Epub 2006/09/21.

[126] Shin JY, Fang ZH, Yu ZX, Wang CE, Li SH, Li XJ. Expression of mutant huntingtin in glial cells contributes to neuronal excitotoxicity. J Cell Biol 2005;171:1001–12. Epub 2005/12/21.

[127] Bradford J, Shin JY, Roberts M, Wang CE, Li XJ, Li S. Expression of mutant huntingtin in mouse brain astrocytes causes age-dependent neurological symptoms. Proc Natl Acad Sci USA 2009;106:22480–5. Epub 2009/12/19.

[128] Crotti A, Benner C, Kerman BE, Gosselin D, Lagier-Tourenne C, Zuccato C, Cattaneo E, Gage FH, Cleveland DW, Glass CK. Mutant Huntingtin promotes autonomous microglia activation via myeloid lineage-determining factors. Nat Neurosci 2014;17:513–21. Epub 2014/03/04.

[129] van der Burg JM, Bjorkqvist M, Brundin P. Beyond the brain: widespread pathology in Huntington's disease. Lancet Neurol 2009;8:765–74. Epub 2009/07/18.

[130] Carroll JB, Bates GP, Steffan J, Saft C, Tabrizi SJ. Treating the whole body in Huntington's disease. Lancet Neurol 2015;14:1135–42. Epub 2015/10/16.

[131] Jankovic J, Clarence-Smith K. Tetrabenazine for the treatment of chorea and other hyperkinetic movement disorders. Expert Rev Neurother 2011;11:1509–23. Epub 2011/10/22.

[132] Geraedts JP, De Wert GM. Preimplantation genetic diagnosis. Clin Genet 2009;76:315–25. Epub 2009/10/02.

[133] Braude P, Pickering S, Flinter F, Ogilvie CM. Preimplantation genetic diagnosis. Nat Rev Genet 2002;3:941–53. Epub 2002/12/03.

[134] Harper J, Geraedts J, Borry P, Cornel MC, Dondorp WJ, Gianaroli L, Harton G, Milachich T, Kaariainen H, Liebaers I, Morris M, Sequeiros J, Sermon K, Shenfield F, Skirton H, Soini S, Spits C, Veiga A, Vermeesch JR, Viville S, de Wert G, Macek Jr M, Eshg E. EuroGentest. Current issues in medically assisted reproduction and genetics in Europe: research, clinical practice, ethics, legal issues and policy. Hum Reprod 2014;29:1603–9. Epub 2014/07/10.

[135] Travessa AM, Rodrigues FB, Mestre TA, Ferreira JJ. Fifteen years of clinical trials in Huntington's disease: a very low clinical drug development success rate. J Huntingt Dis 2017;6:157–63. Epub 2017/07/04.

[136] Rodrigues FB, Wild EJ. Clinical trials corner. J Huntingt Dis September 2017;2017(6):255–63. Epub 2017/10/03.

[137] Freeman TB, Cicchetti F, Hauser RA, Deacon TW, Li XJ, Hersch SM, Nauert GM, Sanberg PR, Kordower JH, Saporta S, Isacson O. Transplanted fetal striatum in Huntington's disease: phenotypic development and lack of pathology. Proc Natl Acad Sci USA 2000;97:13877–82. Epub 2000/12/06.
[138] Bachoud-Levi AC, Gaura V, Brugieres P, Lefaucheur JP, Boisse MF, Maison P, Baudic S, Ribeiro MJ, Bourdet C, Remy P, Cesaro P, Hantraye P, Peschanski M. Effect of fetal neural transplants in patients with Huntington's disease 6 years after surgery: a long-term follow-up study. Lancet Neurol 2006;5:303–9. Epub 2006/03/21.
[139] Cicchetti F, Soulet D, Freeman TB. Neuronal degeneration in striatal transplants and Huntington's disease: potential mechanisms and clinical implications. Brain J Neurol 2011;134:641–52. Epub 2011/02/01.
[140] Gonzalez V, Cif L, Biolsi B, Garcia-Ptacek S, Seychelles A, Sanrey E, Descours I, Coubes C, de Moura AM, Corlobe A, James S, Roujeau T, Coubes P. Deep brain stimulation for Huntington's disease: long-term results of a prospective open-label study. J Neurosurg 2014;121:114–22. Epub 2014/04/08.
[141] di Biase L, Munhoz RP. Deep brain stimulation for the treatment of hyperkinetic movement disorders. Expert Rev Neurother 2016;16:1067–78. Epub 2016/06/03.
[142] Sarkar S, Perlstein EO, Imarisio S, Pineau S, Cordenier A, Maglathlin RL, Webster JA, Lewis TA, O'Kane CJ, Schreiber SL, Rubinsztein DC. Small molecules enhance autophagy and reduce toxicity in Huntington's disease models. Nat Chem Biol 2007;3:331–8. Epub 2007/05/09.
[143] Hockly E, Richon VM, Woodman B, Smith DL, Zhou X, Rosa E, Sathasivam K, Ghazi-Noori S, Mahal A, Lowden PA, Steffan JS, Marsh JL, Thompson LM, Lewis CM, Marks PA, Bates GP. Suberoylanilide hydroxamic acid, a histone deacetylase inhibitor, ameliorates motor deficits in a mouse model of Huntington's disease. Proc Natl Acad Sci USA 2003;100:2041–6. Epub 2003/02/11.
[144] Butler R, Bates GP. Histone deacetylase inhibitors as therapeutics for polyglutamine disorders. Nat Rev Neurosci 2006;7:784–96. Epub 2006/09/22.
[145] Reynolds GP, Pearson SJ. Increased brain 3-hydroxykynurenine in Huntington's disease. Lancet 1989;2:979–80. Epub 1989/10/21.
[146] Zwilling D, Huang SY, Sathyasaikumar KV, Notarangelo FM, Guidetti P, Wu HQ, Lee J, Truong J, Andrews-Zwilling Y, Hsieh EW, Louie JY, Wu T, Scearce-Levie K, Patrick C, Adame A, Giorgini F, Moussaoui S, Laue G, Rassoulpour A, Flik G, Huang Y, Muchowski JM, Masliah E, Schwarcz R, Muchowski PJ. Kynurenine 3-monooxygenase inhibition in blood ameliorates neurodegeneration. Cell 2011;145:863–74. Epub 2011/06/07.
[147] Wild EJ, Tabrizi SJ. Therapies targeting DNA and RNA in Huntington's disease. Lancet Neurol 2017;16:837–47. Epub 2017/09/19.
[148] Yamamoto A, Lucas JJ, Hen R. Reversal of neuropathology and motor dysfunction in a conditional model of Huntington's disease. Cell 2000;101:57–66. Epub 2000/04/25.
[149] Yu D, Pendergraff H, Liu J, Kordasiewicz HB, Cleveland DW, Swayze EE, Lima WF, Crooke ST, Prakash TP, Corey DR. Single-stranded RNAs use RNAi to potently and allele-selectively inhibit mutant huntingtin expression. Cell 2012;150:895–908. Epub 2012/09/04.
[150] Hu J, Matsui M, Gagnon KT, Schwartz JC, Gabillet S, Arar K, Wu J, Bezprozvanny I, Corey DR. Allele-specific silencing of mutant huntingtin and ataxin-3 genes by targeting expanded CAG repeats in mRNAs. Nat Biotechnol 2009;27:478–84. Epub 2009/05/05.
[151] Kordasiewicz HB, Stanek LM, Wancewicz EV, Mazur C, McAlonis MM, Pytel KA, Artates JW, Weiss A, Cheng SH, Shihabuddin LS, Hung G, Bennett CF, Cleveland DW. Sustained therapeutic reversal of Huntington's disease by transient repression of huntingtin synthesis. Neuron 2012;74:1031–44. Epub 2012/06/26.
[152] Southwell AL, Skotte NH, Villanueva EB, Ostergaard ME, Gu X, Kordasiewicz HB, Kay C, Cheung D, Xie Y, Waltl S, Dal Cengio L, Findlay-Black H, Doty CN, Petoukhov E, Iworima D, Slama R, Ooi J, Pouladi MA, Yang XW, Swayze EE, Seth PP, Hayden MR. A novel humanized mouse model of Huntington disease for preclinical development of therapeutics targeting mutant huntingtin alleles. Hum Mol Genet 2017;26:1115–32. Epub 2017/01/21.

[153] Lane RM, Smith A, Baumann T, Gleichmann M, Norris D, Bennett CF, Kordasiewicz H. Translating antisense technology into a treatment for Huntington's disease. Methods Mol Biol 2018;1780:497–523. Epub 2018/06/02.

[154] van Roon-Mom WMC, Roos RAC, de Bot ST. Dose-dependent lowering of mutant Huntingtin using antisense oligonucleotides in Huntington disease patients. Nucleic Acid Ther 2018;28:59–62.

[155] Finkel RS, Mercuri E, Darras BT, Connolly AM, Kuntz NL, Kirschner J, Chiriboga CA, Saito K, Servais L, Tizzano E, Topaloglu H, Tulinius M, Montes J, Glanzman AM, Bishop K, Zhong ZJ, Gheuens S, Bennett CF, Schneider E, Farwell W, De Vivo DC, Group ES. Nusinersen versus Sham control in infantile-onset spinal muscular atrophy. N Engl J Med 2017;377:1723–32.

[156] Groen EJN, Talbot K, Gillingwater TH. Advances in therapy for spinal muscular atrophy: promises and challenges. Nat Rev Neurol 2018;14:214–24.

[157] Garriga-Canut M, Agustin-Pavon C, Herrmann F, Sanchez A, Dierssen M, Fillat C, Isalan M. Synthetic zinc finger repressors reduce mutant huntingtin expression in the brain of R6/2 mice. Proc Natl Acad Sci USA 2012;109:E3136–45. Epub 2012/10/12.

[158] Shin JW, Kim KH, Chao MJ, Atwal RS, Gillis T, MacDonald ME, Gusella JF, Lee JM. Permanent inactivation of Huntington's disease mutation by personalized allele-specific CRISPR/Cas9. Hum Mol Genet 2016;25:4566–76. Epub 2017/02/09.

[159] Cox DB, Platt RJ, Zhang F. Therapeutic genome editing: prospects and challenges. Nat Med 2015;21: 121–31. Epub 2015/02/06.

CHAPTER

MULTIPLE SCLEROSIS

7

CHAPTER OUTLINE

7.1 INTRODUCTION

Multiple sclerosis is a neurological disorder characterized by a complex array of symptoms affecting movement and the senses. For many years, it was considered primarily as an autoimmune disorder, driven by neuroinflammation, and so in many ways is distinct from the other neurological diseases covered in this textbook. More recently, however, there has been increasing focus on the neurodegeneration that is a feature of the latter stages of the majority of cases of multiple sclerosis, where the disorder moves from a relapsing/remitting presentation to a progressive decline in function. Conversely, and as covered in detail in previous chapters, there has been an explosion of interest in the role of

The Molecular and Clinical Pathology of Neurodegenerative Disease. https://doi.org/10.1016/B978-0-12-811069-0.00007-0

neuroinflammation in disorders where neurodegeneration is thought to be the primary disease process. As such, there is a substantial amount that can be learned across inflammatory disease and neurodegeneration by comparing and contrasting the events that lead to dysfunction in multiple sclerosis and those that contribute to neuronal loss in disorders such as Alzheimer's and Parkinson's diseases.

Another important distinction is that the past two decades have witnessed a number of successful phase III trials for disease-modifying therapies for some forms of multiple sclerosis. This is in marked contrast to the majority of neurodegenerative diseases, where there is an almost complete absence of any form of disease-modifying therapy. As such, the development of treatments for multiple sclerosis provides an important demonstration that targeting complex central nervous system disorders can be achieved, as well as providing examples of the challenges that must be overcome in developing such therapies.

The history of multiple sclerosis as a disease entity, similar to many of the disorders covered in this textbook, has its origins in the establishment of the field of modern neurology during the 19th century. It is notable, however, that there are putative descriptions of multiple sclerosis going back much further. One of the earliest examples is provided by descriptions of the illness of Saint Lidwina of Schiedam (1380–1433), some of the symptoms of which have been suggested to support a retrospective diagnosis of multiple sclerosis [1]. More recently, the diaries of Augustus d'Esté (1794–1848)—a grandson of the British King George III—have been studied in detail for descriptions of the neurological illness that afflicted him, an illness that displayed many of the characteristics of multiple sclerosis (Box 7.1) [2,3].

A major event in the emergence of the clinical concept of multiple sclerosis occurred during the lifetime of Augustus d'Esté with the work of two pathologists working in parallel in the 1830s—Robert Carswell and Jean Cruveilhier. Working on specimens in Paris, both pathologists noted the formation of plaques in the central nervous system in patients who had exhibited neurological symptoms, although neither Carswell nor Cruveilhier provided a detailed clinical analysis directly correlated with this pathological condition [4,5]. This fell to one of the major figures in the foundation of neurology—Jean-Martin Charcot, based at the Hopital Salpetriere in Paris. During the 1860s and 1870s, he and his students (in particular, Leopold Ordenstein) described and categorized a number of patients displaying neurological dysfunction, including paralysis, a disorder that Charcot described as "sclerosis en plaques" [6–8]. For many years, this was translated into English as disseminated sclerosis, a term that was used in the United States and United Kingdom up until the 1960s; however, the term multiple sclerosis is now the accepted name for the disorder [9].

7.2 CLINICAL PRESENTATION

Multiple sclerosis is an inflammatory disease of the central nervous system and is the most common cause of nontraumatic neurological disability in the young [10]. Acute attacks of inflammation lead to demyelination and cause episodes of neurological dysfunction called relapses. Patients typically recover to varying extents from relapses, but with time, neurological disability accumulates. Although multiple sclerosis was traditionally thought of as an inflammatory disease, it has become clear in recent years that there is a significant neurodegenerative component to it with axonal loss occurring early in the disease [11,12]. The relationship between inflammation and neurodegeneration is unclear, as is the relationship between relapses and the accumulation of neurological disability.

BOX 7.1 AUGUSTUS D'ESTÉ

Augustus d'Esté (1794–1848) was a grandson of King George III of the United Kingdom. He maintained a detailed diary for much of his life, describing a combination of symptoms consistent with a diagnosis of multiple sclerosis. These started in 1822, when 28-year-old Augustus began to experience visual dysfunction, which he recovered from soon after. By charting the improvement and deterioration of Augustus d'Esté's health, it is possible to construct a history of a relapsing remitting disease course culminating in secondary progressive multiple sclerosis [2]. *Image courtesy of the Victoria and Albert museum.*

7.2.1 EPIDEMIOLOGY

The global prevalence of multiple sclerosis is 100–150 per 100,000 population [13]. It is more common in women than men, and it typically presents between the ages of 20 and 40 years although onset in childhood and in later life is recognized. It has traditionally been thought that the prevalence of multiple sclerosis increases with distance from the equator. Geographical variations can be observed even over quite short distances; for example, multiple sclerosis is more prevalent in Scotland than in the south of England [14]. Environmental factors are thought to play a part in this, as migration studies indicate that people who move to countries early in life acquire the risk of the country they move to. Lack of sunlight and lower vitamin D levels have been suggested as a reason why multiple sclerosis is more prevalent in countries farther from the equator, with higher vitamin D levels seeming to be protective for multiple sclerosis (Fig. 7.1) [15].

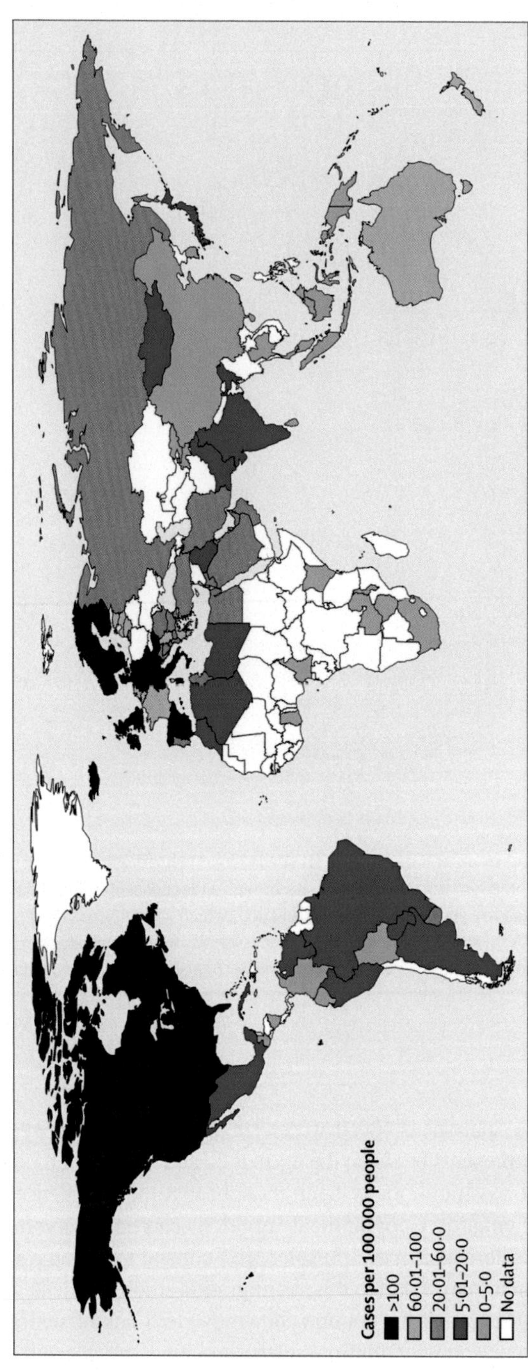

FIGURE 7.1

The global incidence of multiple sclerosis, displaying the marked variation in incidence across the hemispheres and with regard to latitude.

Data are from the atlas of multiple sclerosis (www.atlasofms.org).

However, more recent studies have suggested that this latitudinal gradient in the northern hemisphere has become less apparent with time, and there is an ongoing search for environmental factors that predispose to multiple sclerosis [16]. There are a number of other environmental factors including viral infections that have been implicated in the pathogenesis of multiple sclerosis. Epstein–Barr virus seropositivity in patients with multiple sclerosis is close to 100%, higher than in healthy controls. However, the relevance of this is uncertain given that the background rate of seropositivity is high in the general population [17]. Smoking has also been shown to be a significant but moderate risk factor for the development of multiple sclerosis [18]. The direct impact of any of the above environmental factors on the risk of developing multiple sclerosis is uncertain, and causality has not been proven. There is a certain genetic component to multiple sclerosis with a concordance of 30% among monozygotic twins and an increased risk of the disease in the first-degree relatives. Genome-wide association studies have shown the strongest association to be with major histocompatibility complex genes and the interleukin receptor 7 gene [19,20].

7.2.2 CLINICAL COURSE AND SUBTYPES

Multiple sclerosis can be divided into a number of different subtypes that may overlap. Patients with multiple sclerosis will usually present with a single episode of neurological dysfunction that typically involves the spinal cord, brainstem, or optic nerves. The term "clinically isolated syndrome" is used to describe this first episode. Most patients with a clinically isolated syndrome, who have multiple magnetic resonance imaging lesions, will subsequently go on to develop a subsequent clinical episode, which heralds the onset of clinically definite multiple sclerosis. The most common form of multiple sclerosis is characterized by recurrent relapses—so-called relapsing remitting multiple sclerosis. A relapse is an episode of neurological dysfunction with symptoms typically evolving over days or weeks before plateauing and then remitting with a variable degree of recovery. Secondary progressive multiple sclerosis refers to a group of patients who initially had relapsing remitting multiple sclerosis but then develop progressive accumulation of disability [21]. There may be superimposed relapses during the progressive phase. The proportion of patients developing secondary progressive multiple sclerosis increases with disease duration. Primary progressive is a rarer form of multiple sclerosis, accounting for 10%–15% of cases and characterized by progressive accumulation of disability from the time of disease onset [22]. Primary progressive multiple sclerosis is more likely to affect men and tends to have a later age of onset than relapsing remitting multiple sclerosis [23]. Progressive relapsing multiple sclerosis is the term used to describe people who have progressive disability from the onset but also superimposed relapses [24].

7.2.3 CLINICAL PRESENTATION

The clinical symptoms of multiple sclerosis (summarized in Fig. 7.2) depend on what part of the central nervous system is affected by the demyelinating plaques. The optic nerves, spinal cord, and brainstem are frequently affected. A multitude of different symptoms may occur, including sensory loss and paresthesia, motor weakness, bladder and bowel dysfunction, ataxia, and diplopia. Optic neuritis is a frequent presenting symptom and causes pain that is exacerbated by eye movement and visual impairment. Motor dysfunction and spasticity are seen frequently later in the disease course.

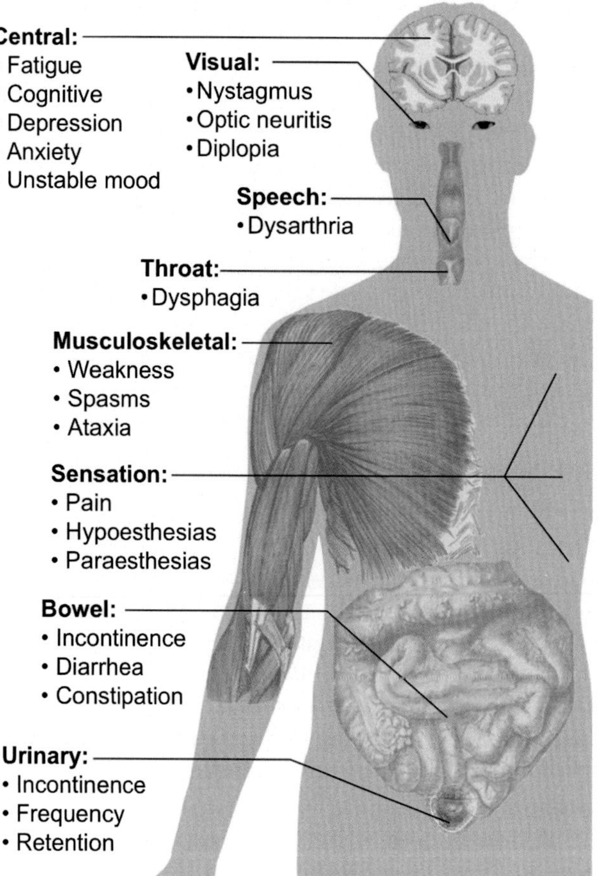

FIGURE 7.2

The symptoms of multiple sclerosis, showing the complex nature of this disorder and the multifaceted impact of the disease across a wide range of systems.

Images derived and modified from the Wikimedia commons collection using a creative commons licence.

Although multiple sclerosis is traditionally thought of as a disease of white matter, it is increasingly recognized that gray matter is affected. Cognitive impairment is frequently seen in both early and late multiple sclerosis. It is important to note that speed of information processing and executive function may be particularly impacted, and cognitive impairment is associated with reduced quality of life [25,26]. Patients may also get brief paroxysmal neurological symptoms such as attacks of neuropathic pain with trigeminal neuralgia or brief episodes of stiffness in the limbs known as tonic spasms. These paroxysmal episodes are thought to be due to electrical instability within lesions. Two eponymous syndromes have been particularly associated with MS. Uthoff's phenomenon is the tendency of multiple sclerosis symptoms to temporarily worsen in heat, and Lhermitte's phenomenon is the term used to describe paresthesia in the limbs on flexing the neck,

BOX 7.2 LHERMITTE AND UHTHOFF

Lhermitte and Uhthoff. Jean Lhermitte (left hand panel) was a French neurologist who noted patients reporting an electrical sensation running up and down the spine, often upon flexion of the neck. Although similar phenomena had previously been reported, Lhermitte was responsible for synthesizing the different aspects of this phenomenon and published a description during the 1920s. Wilhelm Uhthoff (right hand panel) described what became known as Uhthoff's phenomenon in 1890. The phenomenon is a worsening of symptoms upon an increase in body temperature, for example, following exercise or immersion in warm water. He initially noted this in a case of optic neuritis. Taken together, these phenomena are important indicators that an individual may be suffering from multiple sclerosis.

due to dorsal column disease (Box 7.2) [27,28]. Fatigue is one of the most common symptoms observed in multiple sclerosis and can be extremely disabling [29].

7.2.4 DIAGNOSIS

The diagnosis of multiple sclerosis rests on the demonstration of episodes of demyelination that are separated in space and time. Magnetic resonance imaging forms the cornerstone of the investigation of suspected MS. Improvement in magnetic resonance imaging techniques has facilitated a more efficient diagnosis of multiple sclerosis, as both old and new lesions can be seen allowing demonstration of dissemination in time to be seen on a single scan. Plaques of demyelination are seen as high signal lesions on T2-weighted and FLAIR magnetic resonance imaging. Multiple sclerosis lesions are typically located in the preventricular and juxtacortical white matter, adjacent to the corpus callosum, in the brainstem and the cerebellar white matter and the spinal cord. Acute lesions will enhance with gadolinium, whereas new lesions will not.

Although magnetic resonance imaging is very sensitive for the diagnosis of multiple sclerosis and allows the detection of clinically silent plaques of demyelination, it is not specific as other causes of inflammation as well as vascular, infectious, and metabolic disease can cause similar pictures on magnetic resonance imaging. Atrophy of gray matter, as well as white matter inflammatory lesions, is frequently seen in multiple sclerosis, and this correlates to clinical status and disease duration [30].

Examination of the cerebrospinal fluid with a lumbar puncture sometimes forms part of the diagnosis of multiple sclerosis. This was used more frequently before the advent of advanced magnetic resonance imaging techniques. The demonstration of intrathecally synthesized oligoclonal IgG bands in the cerebrospinal fluid is seen in 90% of patients with clinically definite MS [31]. A parallel blood sample is required to demonstrate that the oligoclonal bands are of intrathecal origin. Matched bands, those seen in both the central nervous system and the serum, are not diagnostic for multiple sclerosis and can be seen in a wide variety of infective and inflammatory conditions.

Visual evoked potential, an electrophysiological technique for demonstrating clinically silent optic nerve lesions, is used less frequently in this era of advanced magnetic resonance imaging scanning.

The diagnostic criteria for multiple sclerosis have evolved throughout the years to reflect the advances in magnetic resonance imaging. Previously, multiple sclerosis could be diagnosed only if a patient had two or more clinical attacks, with two or more lesions on magnetic resonance imaging with objective clinical evidence. The 2017 McDonald criteria also allow a diagnosis of multiple sclerosis to be made in a patient with a clinically isolated syndrome if dissemination in space is seen on magnetic resonance imaging and dissemination in time is demonstrated with gadolinium, or if the patient has cerebrospinal fluid–specific oligoclonal bands [32].

7.2.5 OUTCOME

The prognosis of multiple sclerosis is highly variable, and many factors influence disease progression—including gender, age of onset relapse frequency, type of relapse, and the extent of brain lesions on magnetic resonance imaging [33]. Population studies have indicated that approximately 30% of patients progress to using a cane or wheelchair over 10 years. Primary progressive multiple sclerosis is associated with a faster time to progression of disability [34].

7.3 PATHOLOGY

Within the central nervous system, multiple sclerosis is characterized by a number of pathological changes. The first to be described, and a key component of the diagnostic process for this disorder, were demyelinated lesions, or plaques, in the white and gray matter [35]. Lesions, or sclerosis, were observed by two pathologists working in the 1830s—Robert Carswell and Jean Cruveilhier (Fig. 7.3A) [4,5]. Charcot, working in Paris several decades later, went on to catalogue and draw plaques associated with what he called sclerosis en plaques (Fig. 7.3B) [6]. These lesions, which are apparent at a macroscopic level as plaques or scarring visible postmortem, can also be identified in life through the use of magnetic resonance imaging—a key technological aid to diagnosis (Fig. 7.4) [36–38]. The demyelination that underlies these sclerotic plaques is caused by a loss of oligodendrocytes, noted by Joseph Babinski in 1885 [39]. This demyelination is accompanied by significant inflammation and the presence of immune cells from both within the immune system, in the form of activated microglial cells, and without the immune system in the form of macrophages and lymphocytes infiltrating the central nervous system from the periphery [40–42]. The latter cluster around the vasculature, crossing the blood–brain barrier, with the majority of lymphocytes being T cells [43]. A proportion of macrophage cells involved in the inflammatory process in and around active plaques engulf myelin from the oligodendrocytes, adopting a foamy appearance [44,45]. In relapsing/remitting forms of multiple sclerosis, lesions are

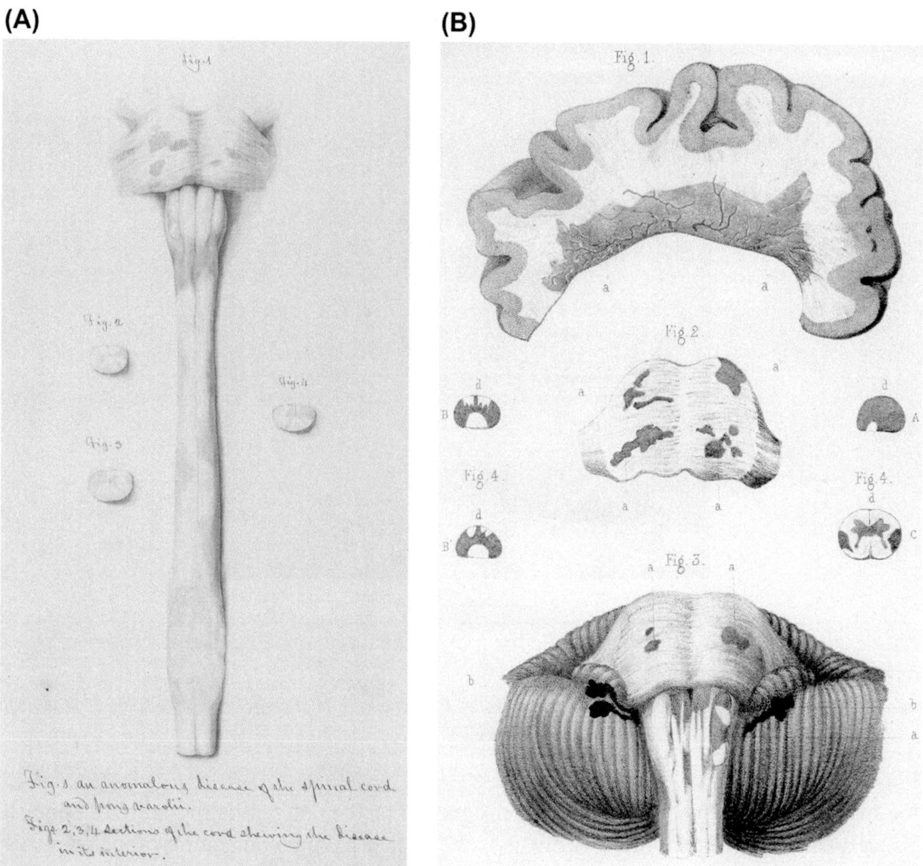

FIGURE 7.3

The pathology of multiple sclerosis as first depicted by Robert Carswell (A) and Jean-Martin Charcot (B). Carswell was the first to report the prototypical lesions found within the central nervous system of individuals with multiple sclerosis—however, he did so before the clinicopathological entity that we now recognize as multiple sclerosis was fully conceived. Charcot, who was responsible for bringing together the symptoms and pathology of multiple sclerosis under a single disease concept, noted some of the microscopic changes that were present in the brain linked to disease.

Image 7.3A courtesy of University College London library.

found predominantly in the white matter [46]. As the disease progresses, however, demyelination is increasingly found in the gray matter, including the cortex [47–49].

A key pathological feature of relapsing remitting multiple sclerosis is a process of remyelination [50], where oligodendrocyte precursor cells act to replace the myelin sheaf [51–53]. This process of remyelination can be observed at a pathological level and is thought to underlie at least part of the regaining of function observed in periods of remission [54].

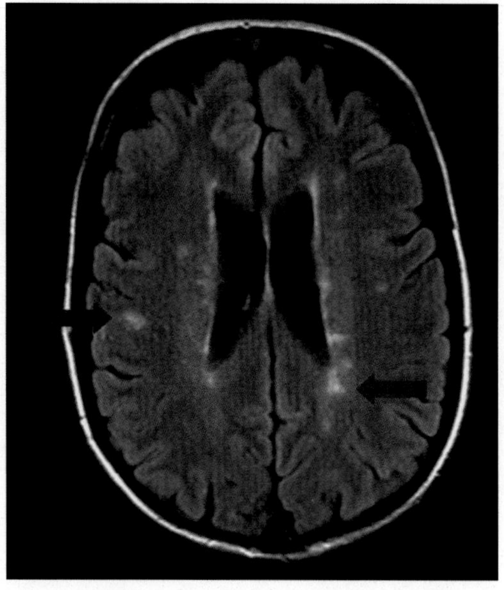

FIGURE 7.4

Sclerotic lesions (indicated by *red arrows*) in the brain of a patient with multiple sclerosis revealed by axial FLAIR magnetic resonance imaging.

Image taken from Raffel J, Wakerley B, Nicholas R. Multiple sclerosis. Medicine 2016;44(9):537–41.

In the majority of patients, the ultimate consequence of the demyelinating and inflammatory processes in relapsing remitting multiple sclerosis is the loss of axonal connections and, eventually, neurodegeneration [21,55]. In addition to initial loss of conductance, the failure of remyelination results in increased metabolic stress and damage to axonal projections [56]. It is important to note, however, that the processes resulting in neuronal loss are likely to be multifactorial and, as yet, incompletely understood (see Section 7.4) [57,58]. The precise pattern of this neurodegeneration, and the impact on cortical function, has important consequences for clinical presentation (in particular, cognition) in progressive multiple sclerosis [59].

7.4 MOLECULAR MECHANISMS OF DEGENERATION

As is clear from Sections 7.2 and 7.3, multiple sclerosis has a complex and heterogeneous disease course. Our understanding of the processes underlying the key pathogenic alterations that drive the clinical presentation of this disorder (in both its relapsing remitting and progressive forms), specifically demyelination, inflammation, and neurodegeneration, has advanced significantly over the past several decades but remains incomplete [13]. The molecular mechanisms that cause these pathological changes are intimately connected to the natural history and chronology of the disease, and so a first step in understanding these mechanisms is to view them in this context.

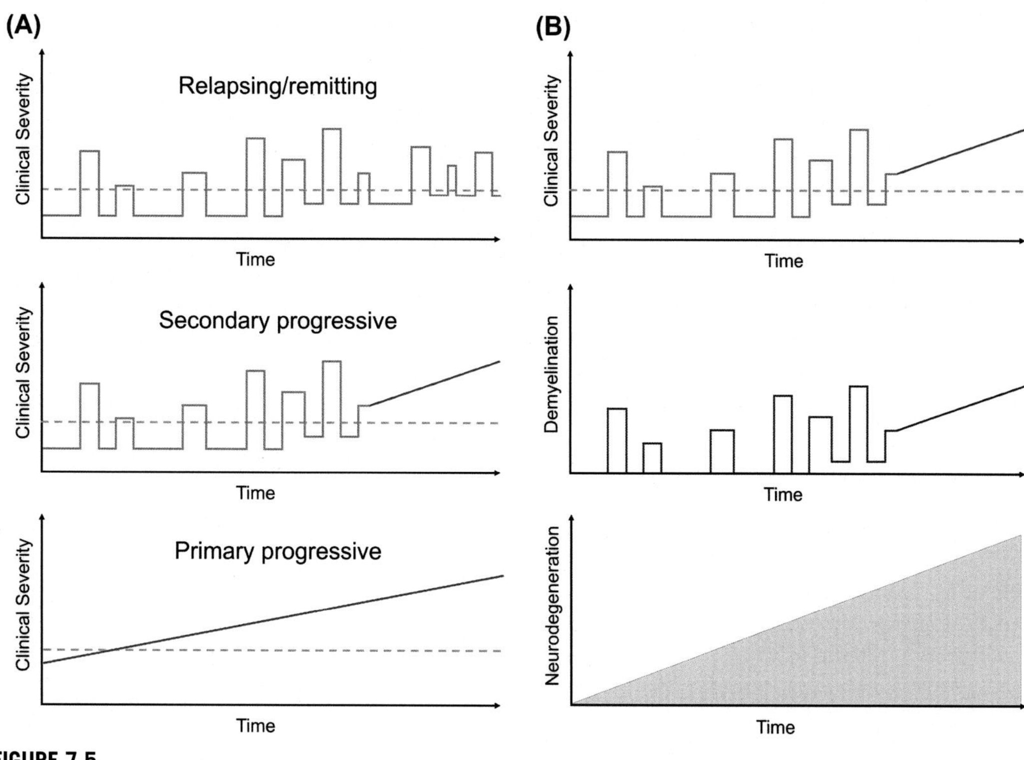

FIGURE 7.5

The different forms and stages of multiple sclerosis. (A) Relapsing remitting disease course (top panel), relapsing remitting disease course leading to secondary progressive disease (middle panel), and primary progressive disease (bottom panel). Clinical threshold indicated by *dotted gray line*. (B) The correlation between clinical status, demyelination, and neurodegeneration in relapsing remitting multiple sclerosis converting to secondary progressive disease.

7.4.1 CHRONOLOGY OF DISEASE

The different forms and clinical stages of multiple sclerosis result from a series of interrelated cellular processes. The initial stage of disease for the majority of patients is of an initial neurological event followed by periods of remission and relapse (that is, relapsing remitting multiple sclerosis). The periods of, and extent of, remission decrease as time progresses. In the majority of patients with relapsing remitting multiple sclerosis, this episodic stage of disease eventually advances to a progressive stage of disease, where remission no longer occurs, and neurological decline is continual. In addition to those cases where progressive disease is secondary to the relapsing remitting stage, a minority of patients undergo progressive decline in the absence of any periods of remission (summarized in Fig. 7.5A). Underlying the clinical manifestation and categorization of multiple sclerosis are three key disease processes: inflammation, driven by autoimmunity; a process of demyelination coupled to remyelination; and axonal damage leading to neurodegeneration. These processes are intertwined with the disease stage—with inflammation and demyelination/remyelination predominating in the relapsing remitting stage of disease. In this stage, demyelination leads to loss of neurological function and relapse, remyelination contributing to a regain of function and remission. In the primary and secondary progressive stages of disease, axonal damage and neurodegeneration predominate (summarized in Fig. 7.5B). Although the chronology of these events has been studied in great detail, it is still not clear how these different stages of disease relate to one another, or how the underlying biological changes are linked. This is exemplified by the ongoing debate as to whether primary progressive multiple sclerosis is a distinct clinical entity, with a potentially distinct molecular etiology [60]. Pathological studies suggest that there may be different roles for inflammation in secondary and primary progressive multiple sclerosis [61]. Equally, the sequential nature of the shift from relapsing remitting disease course to secondary progressive may mask significant differences in disease mechanism, some of which may be shared but some may be independent [62]. The uncertainty as to biological overlap and connections between these stages of disease is of particular importance when considering the molecular etiology that results in neurological dysfunction and neurodegeneration as outlined in the following sections.

7.4.2 GENETICS

A hereditary component to risk of developing multiple sclerosis was recognized soon after the elucidation of multiple sclerosis as a disease entity in the latter half of the 19th century and remained a subject of interest through the first half of the 20th century [63,64]. Similar to the majority of human disease with a genetic component, however, the full nature of this influence on risk of disease took many years to tease apart and remains incompletely understood. In contrast to a number of the neurodegenerative disorders considered in this textbook, there are no examples of monogenic, Mendelian forms of multiple sclerosis [65]. This is partly the result of a critical semantic distinction between inherited disease that exhibits a clinical overlap with multiple sclerosis but quantifiable differences in disease course and prognosis [66]. This is in contrast to the approach taken for disorders such as Parkinson disease, where the genetics of Mendelian forms of Parkinson disease includes a number of disorders that differ markedly from the accepted pathology, clinical course, and symptoms of the idiopathic form of the disease [67]. Although there are no monogenic forms of multiple sclerosis, there is abundant evidence of familial aggregation and shared genetic risk from twin studies [68–70].

As such, the genetics of multiple sclerosis risk are complex and involve the interplay of a large number of genetic variants across the human genome [71]. In common with other human disorders, however, progress over the past several decades has been rapid, driven by technological development in the fields of genetic and then genomic approaches, facilitating ever more speedy and comprehensive analysis of human genetic variation and how this links to clinical disease [72].

The first major breakthrough in the genetics of multiple sclerosis was the identification in 1972 of the major histocompatibility locus on chromosome 6 as the location of risk variants for this disorder, and in particular, the human leukocyte antigen or HLA genes [73,74]. The major histocompatibility locus is complex, incorporating a number of distinct genes, and is linked to a wide range of disorders through its critical role in the production of antigen-presenting proteins and immune response [75]. The initial descriptions of increased risk were associated with variation in the HLA-A and HLA-D genes; however, significant efforts over the past 40 years have significantly refined our understanding of the links between the major histocompatibility locus and risk of multiple sclerosis, with an increased focus on HLA class II risk alleles (including HLA-D) and class I protective alleles (including HLA-A) [76]. These data provide evidence for a critical role of the immune system in modulating risk of developing multiple sclerosis and of underlying changes in immune state resulting in the biological changes that lead to neurological dysfunction and disease [77]. The increasing utility of genome-wide analyses has led to a number of large-scale investigations into common variation in the human genome, leading to increased or decreased risk of disease [19]. Recent iterations of these studies, involving nearly 10,000 cases and over 17,000 controls, have reemphasized the central role of the immune system in risk of multiple sclerosis and highlighted the cellular immune process as being critical to this—in particular, the biology of T-helper cells within the immune system [78]. Similar to other human disorders, there is, therefore, an emerging genetic architecture for multiple sclerosis, one that is heavily weighted towards immune function and common variation [79]. There are, however, a number of outstanding questions relating to how genetic variation relates to risk of developing multiple sclerosis—questions that derive partly from the distinctive staging of this disorder. Specifically, it is not yet clear as to whether there are independent genetic risk factors for the relapsing remitting and progressive forms of multiple sclerosis, or whether the same genetic variation contributes to both presentations of the disease [80,81]. This is of particular importance with regard to understanding the genetic contribution to neurodegenerative aspects of multiple sclerosis but is challenging to address (as is the case for genetic analysis of endophenotypes across human disease) [82]. There are ongoing efforts to achieve this, including analysis of the genetic impact on severity and disability in multiple sclerosis [83].

7.4.3 ENVIRONMENTAL CAUSES

The study of the genetics of multiple sclerosis, and how this links to the pathogenesis of the disease, is complicated by the clear and important role of environmental triggers in the etiology of this disorder. A range of environmental and lifestyle factors have been implicated in increased lifetime risk of developing multiple sclerosis [84]. As for all such associations, a major experimental challenge is to move beyond association and to test whether a causative link exists [85]. Importantly, there is also an increasingly appreciated interplay between genetic variants that predispose to multiple sclerosis and environmental exposures or lifestyle risks [86]. The breadth of potential environmental influences on risk of developing multiple sclerosis is exemplified by a recent review of metaanalyses of environmental risk factors for this disorder, including potential links to vitamin D, smoking and infection [87]. This study also highlights the challenges of replicating studies of this nature, as well as elucidating the mechanism linking a specific environmental exposure to complex disease phenotypes.

One of the most studied aspects of an environmental impact on risk of developing multiple sclerosis is that of a putative role for vitamin D in the etiology of the disease [88]. An association between low serum vitamin D levels and risk of developing multiple sclerosis was first proposed in 1972, linking to a role for vitamin D in myelination and the formation of vitamin D upon exposure to sunlight [89,90]. The latter point has been proposed as a potential explanation for the geographical variation in disease incidence, with increased incidence in northern latitudes with lower sunlight exposure [91]. The timing of the potential role of vitamin D levels in relation to risk of developing multiple sclerosis has been the subject of intense investigation, with some evidence that this may play a role many years before the onset of symptoms [92].

Intriguingly, the use of novel genetic techniques, including a statistical approach called Mendelian randomization, has revealed a potential genetic predisposition to decreased vitamin D levels that may underpin some of the association with increased risk of disease [93]. This supports a causal link, in particular, for early-onset multiple sclerosis [94]. A mechanistic explanation for the association between vitamin D levels could be the role for this vitamin in the immune system, although the precise events in the context of the etiology of multiple sclerosis remain unclear [95–97].

An important implication of a causative role for vitamin D levels is that this may offer a potential therapeutic avenue in disease [98]. This has been the subject of a number of small clinical trials, summarized in a recent Cochrane review [99]. Both the Cochrane review and a more recent metaanalysis of these studies found no clear indication that dietary supplementation with vitamin D had clinical benefits in individuals with multiple sclerosis [100]. Importantly, the timing of these interventions may be critical—the majority of these studies provided vitamin D supplements to individuals who had already developed multiple sclerosis. As studies suggest that the closest association with vitamin D levels and risk of multiple sclerosis occurs potentially many years before the onset of neurological symptoms, it may be that for vitamin D supplementation to be effective, it must occur prior to the development of disease.

Similar to genetic studies of multiple sclerosis, the analysis of vitamin D levels and their association with disease risk has focused on relapsing remitting multiple sclerosis, with relatively few studies examining how vitamin D might link to progressive disease. A small number of studies have investigated an association between vitamin D levels and disability [101] and conversion to the secondary progressive stage of multiple sclerosis [102], but a direct examination of how vitamin D levels relate to the underlying process of neurodegeneration at an experimental level has not been carried out.

Another environmental exposure that has demonstrated a consistent association with increased risk of multiple sclerosis is smoking [103,104]. Large-scale population studies have revealed a dose-dependent increase in risk (as measured by the number of packets of cigarettes smoked), with a decrease in risk upon cessation. Smoking has also been demonstrated to be associated with increased conversion to secondary progressive multiple sclerosis, implicating it in the neurodegenerative process active in this stage of disease [18]. In addition, stratification of patient response to therapeutic intervention (specifically beta interferon treatment) has been shown to be modified by smoking [105]. Whether this reflects a direct mechanistic impact of smoking upon beta interferon activity, or alterations in pharmacokinetics that are associated with smoking, is unclear. Intriguingly, the increased risk of multiple sclerosis associated with smoking status is in marked contrast to the situation in Parkinson disease, where smoking is associated with a reduced risk of disease [106]. This suggests that, if there is indeed a causal link, this is distinct to multiple sclerosis at least with regard to neurodegeneration.

A further key aspect of environmental exposure associated with increased risk of disease and disease course in multiple sclerosis is that of infection [107]. There is widespread epidemiological evidence supporting a connection between a history of infections and increased risk of developing multiple sclerosis [108]. Indeed, this was observed very early in the history of multiple sclerosis as a disease entity, with the neurologist Pierre Marie (a student of Charcot in Paris) reporting this in the 1880s [109]. The association between infections, and in particular viral infections, and risk of multiple sclerosis is a complicated one. An 8-year longitudinal study examining patients with multiple sclerosis, for example, revealed a reduced rate of infections in patients compared with controls but with an exacerbation of symptoms when infections do occur [110]. A more recent study suggested that upper respiratory tract infections are associated with relapse, an association that could be reduced by beta interferon treatment [111]. Exposure to one virus, in particular, has been strongly linked to risk of developing multiple sclerosis—Epstein–Barr virus [112]. Epstein–Barr virus is a human herpesvirus, infection with which causes infectious mononucleosis but is also associated with some forms of cancer [113,114]. Infection with Epstein–Barr virus has also been linked to the incidence of a range of autoimmune disorders, including rheumatoid arthritis and systemic lupus erythematosus [115]. Although the association between Epstein–Barr virus infection and increased risk of multiple sclerosis has been replicated by a number of studies, the nature of this association—whether there is a directly causative link, or that increased rates of infection are a consequence of altered immune biology in patients with multiple sclerosis—remains a matter of debate [116]. Given the strong links to immune function demonstrated by the known biology of Epstein–Barr virus, it is plausible that infection could lead to an alteration in immune function that would favor an autoimmune response within the central nervous system; however, experimental data supporting this are lacking.

As noted above, and in common with genetic analyses, a major challenge in multiple sclerosis research is to separate out the impact of environmental exposure on the etiological events that relate relapsing remitting disease, and potentially different mechanisms underlying primary and secondary progressive disease [117]. This is an underexplored area of multiple sclerosis pathobiology and one that is of great relevance for understanding the processes that result in neurodegeneration and progressive disease.

7.4.4 EXPERIMENTAL ALLERGIC ENCEPHALITIS

The absence of Mendelian forms of multiple sclerosis has limited the use of transgenic models to study this disorder. The autoimmune aspects of this disease, however, have led to the development of a number of animal models where exposure to specific antigens stimulates an acute demyelinating disorder termed experimental allergic encephalitis [118].

This was first developed during the 1930s using nonhuman primates [119]. Initial investigations used injections of tissue from patients following viral infection as an immunogen, resulting in an acute demyelinating episode driven by an autoimmune response. It was subsequently demonstrated that myelin basic protein, a key component of the myelin sheaf, was sufficient to generate an autoimmune response [120,121]. More recently, mice engineered to express a myelin basic protein–specific T cell receptor develop autoimmunity and some of the characteristics of experimental allergic encephalitis, using a genetic approach to reemphasize the role of myelin basic protein in central nervous system autoimmunity [122]. It is notable that a number of proteins specific to the myelin sheaf have been implicated in demyelinating peripheral neuropathies, suggesting a potential overlap in etiology between multiple sclerosis and these disorders [123].

There is controversy as to how closely experimental allergic encephalitis mimics the events that lead to multiple sclerosis and the complex, heterogeneous staging of this disorder [124,125]. It has been suggested that experimental allergic encephalitis is more closely related to the human disorder acute disseminated encephalitis, an acute demyelinating disorder of the central nervous system that shares some clinical aspects with multiple sclerosis [126]. The utility of the experimental allergic encephalitis model with regard to modeling the demyelination aspects of multiple sclerosis has led to its widespread use in investigating the etiology of this disorder and as a platform for drug development [127]. What is clear from this model is that autoimmunity, driven by the identification of myelin basic protein as an antigen, can lead to demyelination similar to that observed in multiple sclerosis. This has provided important insights into the disease process.

7.4.5 INFLAMMATION, AUTOIMMUNITY, AND NEURODEGENERATION

As noted in Section 7.4.1, the nature of the connection between the relapsing remitting stage of multiple sclerosis and progressive disease remains uncertain. It is clear that there is extensive neurodegeneration in progressive multiple sclerosis, with widespread atrophy correlating with disability as a clinical outcome [128]. It is equally clear that inflammation and autoimmunity, involving infiltration of immune cells into the central nervous system, are closely linked to a demyelinating process. A simple model for the disease process and staging of multiple sclerosis is, therefore, that the initial event stimulating disease is autoimmunity, perhaps deriving from the delinquent recognition of myelin basic protein as an antigen, and results in an inflammatory process that leads to demyelination. This causes a loss of function, presenting as an initial neurological event or relapse. Demyelination can be partially addressed through a remyelinating process, leading to a regain of function and periods of remission, or - coupled to loss of support from oligodendrocytes - can result in axonal damage and eventually neuronal loss and neurodegeneration (summarized in Fig. 7.6). As discussed in the preceding sections, there is a substantial body of evidence to support this sequential model for the etiology of multiple sclerosis. This includes genetic evidence strongly supporting inherited changes in immune function as a predisposition to developing disease, as well as data from experimental allergic encephalitis models refining our understanding of the autoimmune process. It is clear, therefore, that inflammation, autoimmunity, and neurodegeneration sit at the heart of the disease process in multiple sclerosis [129,130]. It is also noteworthy that there is an increasing appreciation of neuroinflammation and the immune system as a common denominator in many neurodegenerative disorders [131].

There is, however, an active debate as to how closely this simplified model for the etiology of multiple sclerosis maps on to the complex and heterogeneous nature of the human disorder [55,132]. In particular, there is evidence that atrophy, and so neurodegeneration, is an early event in disease and can run in parallel with a relapsing remitting disease course [133]. There are also imaging data suggesting that there is a decoupling of demyelination and axonal loss [57]. Conversely, the events and changes that cause remyelination to fail in the longer term have not yet been elucidated [56]. Importantly, it is uncertain how primary progressive multiple sclerosis fits into this simplified schema for disease [117].

Because of this, there have been extensive experimental investigations into parallel disease processes that lead to neuronal loss in multiple sclerosis, contributing to neurodegeneration independently of the primary inflammatory events.

Oxidative stress, with increased levels of oxygen-free radicals, has been identified in multiple sclerosis lesions and may be an independent contributor to disease [134]. This is consistent with an extensive literature relating to the role of oxidative stress in neurodegeneration more broadly, although whether raised free radical levels are a cause or consequence of the degenerative process in the diseased brain is unclear [135]. Similar to Alzheimer disease, there is evidence from animal models for multiple sclerosis supporting a role for glutamate excitotoxicity in neuronal death [136,137]. Mitochondrial defects have been identified in a wide range of neurodegenerative disorders and are thought to be a primary disease cause in some forms of Parkinson's disease [138]. Evidence from models for multiple sclerosis, as well as from patients, suggests that mitochondrial function is dysregulated in this disorder [139,140].

A recent study highlighting an important role for oligodendrocytes in supporting neuronal metabolism (where loss of these cells causes neurodegeneration) emphasizes the complex interplay of events that may be occurring in the central nervous systems of individuals with multiple sclerosis [141].

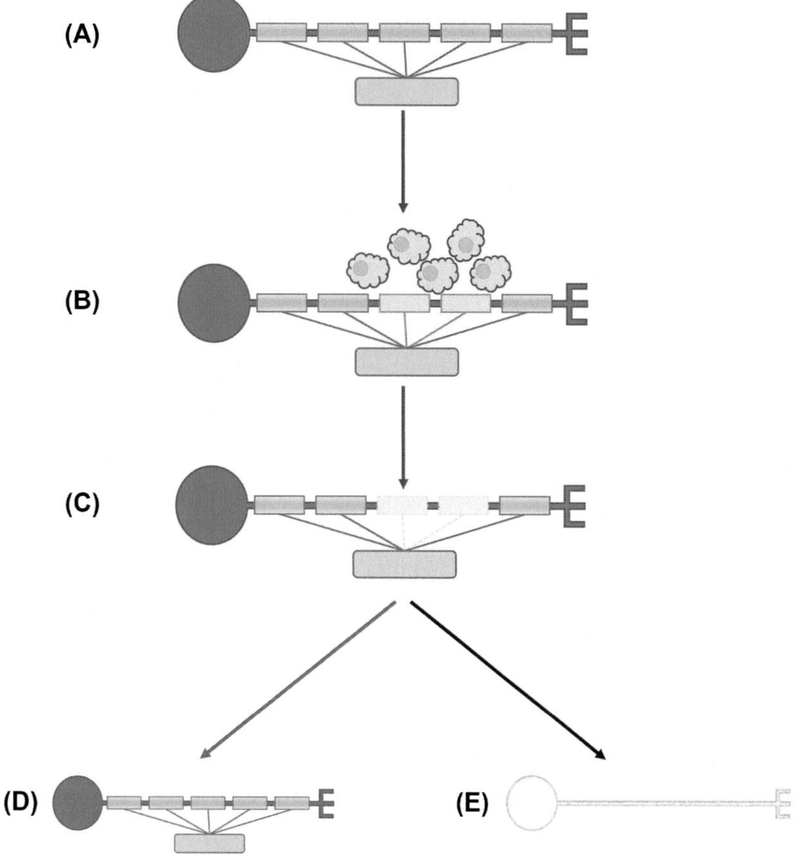

FIGURE 7.6

A simplified model for the disease process in relapsing remitting and progressive multiple sclerosis. Fully myelinated neurons within the central nervous system (A) are subjected to an autoimmune attack from immune cells infiltrating from the periphery (B). This results in demyelination and loss of neurological function (C). If remyelination is successful (D), then the patient enters a period of remission and regains function. If unsuccessful (E), then axonal damage ensues resulting (in combination with additional processes independent of the inflammatory response) in neuronal loss, neurodegeneration, and progressive disease.

The autoimmune and demyelinating process that is central to the etiology of multiple sclerosis contributes to a decrease in oligodendrocytes first noted by Joseph Babinski [39]. As such, and in addition to any deleterious consequences of demyelination on axonal viability, there is an additional outcome where the loss of oligodendrocyte energetic support leads to altered neuronal metabolism, hypoxia, and chronic necrosis [142].

As detailed above, there are a number of common themes across neurodegenerative diseases that are shared with the degenerative/progressive process in multiple sclerosis. It is important to note, however, that there are also marked differences—even if only the progressive disease stage is considered. In particular, the other disorders considered in this textbook (Chapters 2 through to 6) are all characterized by the aggregation and accumulation of misfolded proteins in the central nervous system, a process that is thought to be fundamental to the neuronal cell death observed in these diseases but absent in multiple sclerosis [143].

7.5 THERAPIES

As outlined in Section 7.2, another area where multiple sclerosis diverges from the other forms of neurodegeneration covered in this textbook is in the availability of disease-modifying therapies for the disease. Starting in the 1990s, a number of drugs targeting the process of inflammation and autoimmunity have undergone clinical trials in relapsing remitting multiple sclerosis (Fig. 7.7). Several of these have demonstrated clinical benefits, in particular, with regard to increasing the duration of remission.

7.5.1 SYMPTOMATIC MANAGEMENT

Management of multiple sclerosis can be divided into symptomatic management, the management of acute relapses, and disease-modifying treatment. Symptomatic management centers around managing the complications of multiple sclerosis such as treating pain, spasticity, and bladder dysfunction. 4-Aminopyridine, a potassium channel antagonist, has been shown in clinical trials to improve walking speed in multiple sclerosis [144]. Many of the symptoms in multiple sclerosis are complex and have a significant impact on quality of life and require a multidisciplinary team approach [145].

Steroids are the only drugs recommended for the treatment of a relapse and have been shown to accelerate recovery but do not have any impact on long-term disease outcome [146]. Recent evidence has suggested that oral and intravenous steroids are equally efficacious [147].

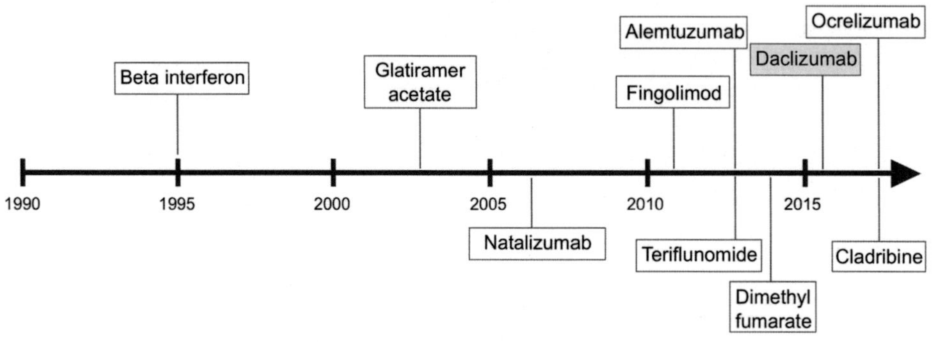

FIGURE 7.7

A timeline for the development of disease-modifying therapies for relapsing remitting multiple sclerosis. Daclizumab, highlighted in red, was licensed for use in 2016 but has since been withdrawn.

7.5.2 DISEASE-MODIFYING THERAPIES FOR RELAPSING REMITTING MULTIPLE SCLEROSIS

There has been a revolution in treatment of multiple sclerosis in recent years with the introduction of new disease-modifying treatments that have been shown to reduce disease activity. In general, treatments target inflammatory processes and all have been shown to decrease the frequency of relapses [148]. Whether disease-modifying therapy has an effect on disability progression has been controversial.

The main focus of disease-modifying treatment in multiple sclerosis to date has been on relapsing remitting multiple sclerosis, and this area of medicine has been revolutionized in the past 10 years with many new drugs now approved for the treatment of multiple sclerosis. All have been shown to decrease relapse rate to varying degrees with the more effective medications having a higher risk of serious adverse events [149].

There are different approaches to disease-modifying therapy depending on disease severity and patient choice. One approach is known as an escalation regime, where patients are initially started on one of the first-line, moderately effective treatments that have a more favorable safety profile with introduction of a more efficacious but potentially less safe agent if there are ongoing relapses.

Alternatively, particularly if a patient has severe disease, with a high lesion load at onset, it may be decided to commence treatment with a highly efficacious agent initially with the aim of obtaining remission—the so-called induction regime. Early treatment has become more commonplace, but the long-term value of this approach with respect to prevention of progressive multiple sclerosis disability remains uncertain. Another change in treatment has been escalating treatment to target no evidence of disease activity (NEDA) as evidenced clinically by lack of relapses and radiologically by the absence of new lesions on magnetic resonance imaging [150].

The oldest disease-modifying treatments for multiple sclerosis have moderate efficacy in reducing relapses, having been shown to reduce the annual relapse rate by approximately 30%. These are injectable treatments and include interferon beta preparations, interferon beta1b and 1a, and glatiramer acetate. They are given by subcutaneous or intramuscular injection. All can cause injection site reactions; the beta interferon treatments can also be associated with flulike symptoms [151,152].

There are three approved oral therapies for multiple sclerosis: dimethylfumarate, fingolimod, and teriflunomide. Dimethylfumarate has been shown to have immunomodulatory properties and has been shown to decrease relapse rate by 50%. The most common side effects are flushing and gastrointestinal symptoms, but there have been reports of serious infections due to depletion of lymphocyte count and also reactivation of the John Cunningham virus causing progressive multifocal leucoencephalopathy has been reported [153,154]. Fingolimod alters lymphocyte migration leading to the sequestration of lymphocytes in lymph nodes. It has been shown to reduce relapse rate by approximately 50%. Significant adverse events include macular edema and bradycardia [155]. Teriflunomide disrupts T cell interaction with antigen-presenting cells. It has moderate efficacy, deceasing relapse rate by 30%. One of the most significant side effects reported has been liver toxicity [156].

Natalizumab is one of the most effective drugs for multiple sclerosis. It is a monoclonal antibody directed against the cell adhesion molecule alpha-4-intergrin and has been shown to reduce relapse rate by 68%. It is given as a monthly intravenous infusion. The most serious adverse effect associated with natalizumab is progressive multifocal leucoencephalopathy—a devastating disease of white matter associated with reactivation of the John Cunningham virus. Patients who are most at risk of developing progressive multifocal leucoencephalopathy can be identified by assessing John Cunningham virus seropositivity and by risk stratification based on duration of treatment with natalizumab and whether any other immunomodulatory drugs were previously used [157].

Alemtuzumab is a humanized monoclonal antibody that causes depletion of T and B lymphocytes that express CD52. It reduces relapse rates by about 50% and is given by intravenous infusion in two courses, 12 months apart. The most serious adverse events have been development of autoimmunity, which can occur in about a third of patients including Goodpasture syndrome (an autoimmune disorder affecting the lungs and kidneys), immune thrombocytopenic purpura, and autoimmune thyroid disease [158].

Ocrelizumab is another monoclonal antibody and binds to CD20 on lymphocytes and has been shown to reduce relapse rate by over 50% in relapsing remitting multiple sclerosis. It can be associated with infusion reactions and hepatotoxicity and is contraindicated in patients with hepatitis B infection [159].

The monoclonal antibody daclizumab, which targeted IL-2 receptors and dampened immune response, was withdrawn in 2017 because of severe liver toxicity and brain inflammation [160].

Cladribine is an efficacious oral treatment, which has been shown to reduce relapse rate by 58%. It is a synthetic purine analogue and depletes T and B cells. It has been associated with lymphopenia, and there have been some reports of progressive multifocal leucoencephalopathy. Concerns about malignancy were raised, and there is a paucity of long-term safety data which so far has limited its use [160].

7.5.3 PROGRESSIVE MULTIPLE SCLEROSIS

In contrast to relapsing/remitting multiple sclerosis, until very recently there were no treatments for primary progressive multiple sclerosis. However, a randomized placebo-controlled trial published in 2017 showed that ocrelizumab, which is licensed for the treatment of relapsing remitting, reduced clinical and magnetic resonance imaging progression of secondary progressive multiple sclerosis [161]. High-dose simvastatin was shown to reduce the rate of brain atrophy in patients with secondary progressive multiple sclerosis, and phase 3 trials are currently underway [162].

One recurring issue with current drugs for relapsing remitting multiple sclerosis is the potential for altered infection rates because of altered immune function [163]. This is a consequence of the majority of licensed drugs for multiple sclerosis directly targeting the immune system as a method for reducing the inflammatory phase of relapse events.

7.5.4 EXPERIMENTAL TREATMENTS

A range of experimental therapies for multiple sclerosis, and in particular for the progressive stage of the disease, are in development [164]. As noted in Section 7.4.3, vitamin D supplementation has for many years been investigated as a potential therapy for multiple sclerosis [88]. Despite some clinical benefits being reported, metaanalyses of the available clinical trials do not reveal a consistent and clear benefit—perhaps partly due to the small size of many of the trials to date [100].

As therapies for progressive disease are the most significant unmet need in multiple sclerosis, a number of groups have explored neuroprotective strategies to improve the survival of neurons within the central nervous systems of individuals with this disorder [165]. An example of this is provided by brain-derived neurotrophic factor, or BDNF, where there have been extensive preclinical investigations of the potential utility of this molecule for neuroprotection in multiple sclerosis [166]. Across neurodegenerative disorders, however, such strategies have, to date, proved disappointing [167].

A similar approach, focusing on glial cells rather than neurons, is the development of strategies to encourage remyelination [168]. The principle underlying this is that remyelination is an active process in the central nervous system of people with multiple sclerosis, and by boosting this process, the ability to recover from relapses and to prolong remission—and potentially stave off conversion to progressive disease—may be enhanced. To date, these studies remain preclinical in nature.

Small-scale studies have suggested that immunoablation followed by autologous stem cell transplantation may have clinical benefits for patients with multiple sclerosis. This has been shown to be effective with sustained remission and improvements in neurological disability in the small number of cases studied to date [169]. Further studies are required to assess the long-term benefit. Given the risks intrinsic to reducing immune system function to this extent, this treatment is likely to be reserved for patients with severe disease who have not responded to other treatments. Trials assessing the efficacy of hematopoietic stem cell transplant in progressive multiple sclerosis are currently underway [170].

7.5.5 THE FUTURE OF DRUG DEVELOPMENT FOR MULTIPLE SCLEROSIS

Despite the successes of the past several decades with regard to developing disease-modifying strategies for relapsing remitting multiple sclerosis, there are a number of major challenges that remain to be overcome [171].

As for the other neurodegenerative disorders considered in this textbook, there is a concerted effort to characterize and stratify patients with multiple sclerosis, and to identify those that are at greatest risk of developing the disease [172,173]. These efforts are of particular relevance to drug discovery, both in terms of targeting the right patients with the most effective drug and in terms of carrying out more efficient and rapid clinical trials [174,175]. Another major area of investigation is what combination of drugs works best for individual patients, and the potential for combination therapy to have clinical outcomes greater than the sum of its parts [176].

7.6 CONCLUSIONS

As noted at the start of this chapter, multiple sclerosis is distinct from the other disorders considered in this textbook in that neurodegeneration is, for the majority of patients, a secondary presentation in the disease. It is very clear, however, that neurodegeneration is a critical part of the overall disease process—and notably, this aspect of multiple sclerosis remains poorly understood.

Multiple sclerosis stands as a template for the development of disease-modifying therapies for neurological disease, demonstrating that it is possible to target a disease process that is active within the central nervous system. It is instructive that, just as research into multiple sclerosis moves to focus on the progressive form of the disease, research in a number of disorders where the primary presentation is neurodegenerative in nature—Parkinson, Alzheimer, and motor neuron disease—is moving toward understanding immune function and inflammation as a central part of the disease process. As discussed in Section 7.5, there are substantial ongoing efforts to develop novel therapies for progressive multiple sclerosis, and it is to be hoped that these efforts will bear fruit over the coming years.

FURTHER READING

Haghikia A, Hohlfeld R, Gold R, Fugger L. Therapies for multiple sclerosis: translational achievements and outstanding needs. Trends Mol Med 2013;19:309–19.

Thompson AJ, Baranzini SE, Geurts J, Hemmer B, Ciccarelli O. Multiple sclerosis. Lancet 2018;391:1622–36.

Wingerchuk DM, Weinshenker BG. Disease modifying therapies for relapsing multiple sclerosis. BMJ 2016;354:i3518. Epub 2016/08/24.

REFERENCES

[1] Medaer R. Does the history of multiple sclerosis go back as far as the 14th century? Acta Neurol Scand 1979;60:189–92. Epub 1979/09/01.

[2] Landtblom AM, Fazio P, Fredrikson S, Granieri E. The first case history of multiple sclerosis: Augustus d'Este (1794-1848). Neurol Sci 2010;31:29–33. Epub 2009/10/20.

[3] Firth D. The case of Augustus d'Este (1794-1848): the first account of disseminated sclerosis: (section of the history of medicine). Proc R Soc Med 1941;34:381–4. Epub 1941/05/01.

[4] Carswell R. Pathological anatomy: illustrations of the elementary forms of disease. Longman, Orme, Brown, Green and Longman; 1838.

[5] Cruveilhier J. Anatomie pathologique du corps humain, ou descriptions, avec figures lithographiées et coloriées, des diverses altérations morbides dont le corps humain est susceptible. Chez JB Baillière; 1835.

[6] Charcot M. Histologie de la sclerose en plaque. Gaz Hôsp 1868;41:554–6.

[7] Lehmann HC, Compston A, Hartung HP. 150th anniversary of clinical description of multiple sclerosis: Leopold Ordenstein's legacy. Neurology 2018;90:1011–6. Epub 2018/05/29.

[8] Ordenstein L. In: Delahaye A, editor. Sur la paralysie agitante et la sclérose en plaques généralisée. 1868.

[9] Kurtzke JF. Multiple sclerosis: what's in a name? Neurology 1988;38:309–14. Epub 1988/02/01.

[10] Browne P, Chandraratna D, Angood C, Tremlett H, Baker C, Taylor BV, Thompson AJ. Atlas of multiple sclerosis 2013: a growing global problem with widespread inequity. Neurology 2014;83:1022–4. Epub 2014/09/10.

[11] Trapp BD, Peterson J, Ransohoff RM, Rudick R, Mork S, Bo L. Axonal transection in the lesions of multiple sclerosis. N Engl J Med 1998;338:278–85. Epub 1998/01/29.

[12] Filippi M, Bozzali M, Rovaris M, Gonen O, Kesavadas C, Ghezzi A, Martinelli V, Grossman RI, Scotti G, Comi G, Falini A. Evidence for widespread axonal damage at the earliest clinical stage of multiple sclerosis. Brain J Neurol 2003;126:433–7. Epub 2003/01/23.

[13] Thompson AJ, Baranzini SE, Geurts J, Hemmer B, Ciccarelli O. Multiple sclerosis. Lancet 2018;391:1622–36. Epub 2018/03/27.

[14] Mackenzie IS, Morant SV, Bloomfield GA, MacDonald TM, O'Riordan J. Incidence and prevalence of multiple sclerosis in the UK 1990-2010: a descriptive study in the general practice research database. J Neurol Neurosurg Psychiatry 2014;85:76–84. Epub 2013/09/21.

[15] Munger KL, Zhang SM, O'Reilly E, Hernan MA, Olek MJ, Willett WC, Ascherio A. Vitamin D intake and incidence of multiple sclerosis. Neurology 2004;62:60–5. Epub 2004/01/14.

[16] Koch-Henriksen N, Sorensen PS. The changing demographic pattern of multiple sclerosis epidemiology. Lancet Neurol 2010;9:520–32. Epub 2010/04/20.

[17] Ascherio A, Munch M. Epstein-Barr virus and multiple sclerosis. Epidemiology 2000;11:220–4. Epub 2000/10/06.

[18] Hernan MA, Jick SS, Logroscino G, Olek MJ, Ascherio A, Jick H. Cigarette smoking and the progression of multiple sclerosis. Brain J Neurol 2005;128:1461–5. Epub 2005/03/11.

[19] International Multiple Sclerosis Genetics C, Hafler DA, Compston A, Sawcer S, Lander ES, Daly MJ, De Jager PL, de Bakker PI, Gabriel SB, Mirel DB, Ivinson AJ, Pericak-Vance MA, Gregory SG, Rioux JD, McCauley JL, Haines JL, Barcellos LF, Cree B, Oksenberg JR, Hauser SL. Risk alleles for multiple sclerosis identified by a genomewide study. N Engl J Med 2007;357:851–62. Epub 2007/07/31.
[20] Sawcer S, Franklin RJ, Ban M. Multiple sclerosis genetics. Lancet Neurol 2014;13:700–9. Epub 2014/05/24.
[21] Scalfari A, Neuhaus A, Daumer M, Muraro PA, Ebers GC. Onset of secondary progressive phase and long-term evolution of multiple sclerosis. J Neurol Neurosurg Psychiatry 2014;85:67–75. Epub 2013/03/15.
[22] Koch M, Kingwell E, Rieckmann P, Tremlett H. The natural history of primary progressive multiple sclerosis. Neurology 2009;73:1996–2002. Epub 2009/12/10.
[23] Tremlett H, Zhao Y, Devonshire V, Neurologists UBC. Natural history comparisons of primary and secondary progressive multiple sclerosis reveals differences and similarities. J Neurol 2009;256:374–81. Epub 2009/03/25.
[24] Tullman MJ, Oshinsky RJ, Lublin FD, Cutter GR. Clinical characteristics of progressive relapsing multiple sclerosis. Mult Scler 2004;10:451–4. Epub 2004/08/26.
[25] Piras MR, Magnano I, Canu ED, Paulus KS, Satta WM, Soddu A, Conti M, Achene A, Solinas G, Aiello I. Longitudinal study of cognitive dysfunction in multiple sclerosis: neuropsychological, neuroradiological, and neurophysiological findings. J Neurol Neurosurg Psychiatry 2003;74:878–85. Epub 2003/06/18.
[26] Chiaravalloti ND, DeLuca J. Cognitive impairment in multiple sclerosis. Lancet Neurol 2008;7:1139–51. Epub 2008/11/15.
[27] Uhthoff W. Untersuchungen über die bei der multiplen Herdsklerose vorkommenden Angenstörungen. Arch für Psychiatr Nervenkrankh 1890;21:55–116.
[28] Lhermitte J, Bollak NM, Nicolas M. Les douleurs a type de decharge electrique consecutives a la flexion cephalique dans la sclerose en plaques. Rev Neurol 1924;2:56–62.
[29] Induruwa I, Constantinescu CS, Gran B. Fatigue in multiple sclerosis - a brief review. J Neurol Sci 2012;323:9–15. Epub 2012/09/01.
[30] Sanfilipo MP, Benedict RH, Sharma J, Weinstock-Guttman B, Bakshi R. The relationship between whole brain volume and disability in multiple sclerosis: a comparison of normalized gray vs. white matter with misclassification correction. Neuroimage 2005;26:1068–77. Epub 2005/06/18.
[31] Link H, Huang YM. Oligoclonal bands in multiple sclerosis cerebrospinal fluid: an update on methodology and clinical usefulness. J Neuroimmunol 2006;180:17–28. Epub 2006/09/02.
[32] Thompson AJ, Banwell BL, Barkhof F, Carroll WM, Coetzee T, Comi G, Correale J, Fazekas F, Filippi M, Freedman MS, Fujihara K, Galetta SL, Hartung HP, Kappos L, Lublin FD, Marrie RA, Miller AE, Miller DH, Montalban X, Mowry EM, Sorensen PS, Tintore M, Traboulsee AL, Trojano M, Uitdehaag BMJ, Vukusic S, Waubant E, Weinshenker BG, Reingold SC, Cohen JA. Diagnosis of multiple sclerosis: 2017 revisions of the McDonald criteria. Lancet Neurol 2018;17:162–73. Epub 2017/12/26.
[33] Runmarker B, Andersen O. Prognostic factors in a multiple sclerosis incidence cohort with twenty-five years of follow-up. Brain J Neurol 1993;116(Pt 1):117–34. Epub 1993/02/01.
[34] Pittock SJ, Mayr WT, McClelland RL, Jorgensen NW, Weigand SD, Noseworthy JH, Weinshenker BG, Rodriguez M. Change in MS-related disability in a population-based cohort: a 10-year follow-up study. Neurology 2004;62:51–9. Epub 2004/01/14.
[35] Frohman EM, Racke MK, Raine CS. Multiple sclerosis–the plaque and its pathogenesis. N Engl J Med 2006;354:942–55. Epub 2006/03/03.
[36] Simon JH. MRI outcomes in the diagnosis and disease course of multiple sclerosis. Handb Clin Neurol 2014;122:405–25. Epub 2014/02/11.
[37] Young IR, Hall AS, Pallis CA, Legg NJ, Bydder GM, Steiner RE. Nuclear magnetic resonance imaging of the brain in multiple sclerosis. Lancet 1981;2:1063–6. Epub 1981/11/14.
[38] Filippi M, Rocca MA, Ciccarelli O, De Stefano N, Evangelou N, Kappos L, Rovira A, Sastre-Garriga J, Tintore M, Frederiksen JL, Gasperini C, Palace J, Reich DS, Banwell B, Montalban X, Barkhof F, Group MS. MRI criteria for the diagnosis of multiple sclerosis: MAGNIMS consensus guidelines. Lancet Neurol 2016;15:292–303. Epub 2016/01/30.

[39] Babinski J. Recherches sur l'anatomie pathologique de la sclerose en plaque et étude comparative des diverses variétés de la scleroses de la moelle. Arch Physiol Paris 1885;5:186–207.
[40] Bogie JF, Stinissen P, Hendriks JJ. Macrophage subsets and microglia in multiple sclerosis. Acta Neuropathol 2014;128:191–213. Epub 2014/06/24.
[41] Bruck W, Sommermeier N, Bergmann M, Zettl U, Goebel HH, Kretzschmar HA, Lassmann H. Macrophages in multiple sclerosis. Immunobiology 1996;195:588–600. Epub 1996/10/01.
[42] Sospedra M, Martin R. Immunology of multiple sclerosis. Annu Rev Immunol 2005;23:683–747. Epub 2005/03/18.
[43] Esiri MM. Multiple sclerosis: a quantitative and qualitative study of immunoglobulin-containing cells in the central nervous system. Neuropathol Appl Neurobiol 1980;6:9–21. Epub 1980/01/01.
[44] Li H, Cuzner ML, Newcombe J. Microglia-derived macrophages in early multiple sclerosis plaques. Neuropathol Appl Neurobiol 1996;22:207–15. Epub 1996/06/01.
[45] Boven LA, Van Meurs M, Van Zwam M, Wierenga-Wolf A, Hintzen RQ, Boot RG, Aerts JM, Amor S, Nieuwenhuis EE, Laman JD. Myelin-laden macrophages are anti-inflammatory, consistent with foam cells in multiple sclerosis. Brain J Neurol 2006;129:517–26. Epub 2005/12/21.
[46] Frischer JM, Weigand SD, Guo Y, Kale N, Parisi JE, Pirko I, Mandrekar J, Bramow S, Metz I, Bruck W, Lassmann H, Lucchinetti CF. Clinical and pathological insights into the dynamic nature of the white matter multiple sclerosis plaque. Ann Neurol 2015;78:710–21. Epub 2015/08/05.
[47] Schob F. Ein Beitrag zur pathologischen Anatomie der multiplen Sklerose. Eur Neurol 1907;22:62–87.
[48] Kutzelnigg A, Lucchinetti CF, Stadelmann C, Brück W, Rauschka H, Bergmann M, Schmidbauer M, Parisi JE, Lassmann H. Cortical demyelination and diffuse white matter injury in multiple sclerosis. Brain J Neurol 2005;128:2705–12.
[49] Brownell B, Hughes JT. The distribution of plaques in the cerebrum in multiple sclerosis. J Neurol Neurosurg Psychiatry 1962;25:315 20. Epub 1962/11/01.
[50] Prineas JW, Connell F. Remyelination in multiple sclerosis. Ann Neurol 1979;5:22–31. Epub 1979/01/01.
[51] Prineas JW, Barnard RO, Revesz T, Kwon EE, Sharer L, Cho ES. Multiple sclerosis. Pathology of recurrent lesions. Brain J Neurol 1993;116(Pt 3):681–93. Epub 1993/06/01.
[52] Prineas JW, Barnard RO, Kwon EE, Sharer LR, Cho ES. Multiple sclerosis: remyelination of nascent lesions. Ann Neurol 1993;33:137–51. Epub 1993/02/01.
[53] Scolding N, Franklin R, Stevens S, Heldin CH, Compston A, Newcombe J. Oligodendrocyte progenitors are present in the normal adult human CNS and in the lesions of multiple sclerosis. Brain J Neurol 1998;121(Pt 12): 2221–8. Epub 1999/01/05.
[54] Smith KJ, Blakemore WF, McDonald WI. Central remyelination restores secure conduction. Nature 1979;280:395–6. Epub 1979/08/02.
[55] Friese MA, Schattling B, Fugger L. Mechanisms of neurodegeneration and axonal dysfunction in multiple sclerosis. Nat Rev Neurol 2014;10:225–38. Epub 2014/03/19.
[56] Franklin RJ. Why does remyelination fail in multiple sclerosis? Nat Rev Neurosci 2002;3:705–14. Epub 2002/09/05.
[57] DeLuca GC, Williams K, Evangelou N, Ebers GC, Esiri MM. The contribution of demyelination to axonal loss in multiple sclerosis. Brain J Neurol 2006;129:1507–16. Epub 2006/04/07.
[58] Bitsch A, Schuchardt J, Bunkowski S, Kuhlmann T, Bruck W. Acute axonal injury in multiple sclerosis. Correlation with demyelination and inflammation. Brain J Neurol 2000;123(Pt 6):1174–83. Epub 2000/05/29.
[59] Steenwijk MD, Geurts JJ, Daams M, Tijms BM, Wink AM, Balk LJ, Tewarie PK, Uitdehaag BM, Barkhof F, Vrenken H, Pouwels PJ. Cortical atrophy patterns in multiple sclerosis are non-random and clinically relevant. Brain J Neurol 2016;139:115–26. Epub 2015/12/08.
[60] Antel J, Antel S, Caramanos Z, Arnold DL, Kuhlmann T. Primary progressive multiple sclerosis: part of the MS disease spectrum or separate disease entity? Acta Neuropathol 2012;123:627–38. Epub 2012/02/14.

[61] Revesz T, Kidd D, Thompson AJ, Barnard RO, McDonald WI. A comparison of the pathology of primary and secondary progressive multiple sclerosis. Brain J Neurol 1994;117(Pt 4):759–65. Epub 1994/08/01.
[62] Larochelle C, Uphaus T, Prat A, Zipp F. Secondary progression in multiple sclerosis: neuronal exhaustion or distinct pathology? Trends Neurosci 2016;39:325–39. Epub 2016/03/19.
[63] Eichhorst H. Über infantile und hereditäre multiple Sklerose. Arch für Pathol Anat Physiol für Klin Med 1896;146:173–92.
[64] Mackay RP. The familial occurrence of multiple sclerosis and its implications. Ann Intern Med 1950;33: 298–320. Epub 1950/08/01.
[65] Weisfeld-Adams JD, Katz Sand IB, Honce JM, Lublin FD. Differential diagnosis of Mendelian and mitochondrial disorders in patients with suspected multiple sclerosis. Brain J Neurol 2015;138:517–39. Epub 2015/02/01.
[66] Natowicz MR, Bejjani B. Genetic disorders that masquerade as multiple sclerosis. Am J Med Genet 1994;49:149–69. Epub 1994/01/15.
[67] Jenner P, Morris HR, Robbins TW, Goedert M, Hardy J, Ben-Shlomo Y, Bolam P, Burn D, Hindle JV, Brooks D. Parkinson's disease–the debate on the clinical phenomenology, aetiology, pathology and pathogenesis. J Park Dis 2013;3:1–11. Epub 2013/08/14.
[68] Ebers GC, Sadovnick AD, Risch NJ. A genetic basis for familial aggregation in multiple sclerosis. Canadian Collaborative Study Group. Nature 1995;377:150–1. Epub 1995/09/14.
[69] O'Gorman C, Lin R, Stankovich J, Broadley SA. Modelling genetic susceptibility to multiple sclerosis with family data. Neuroepidemiology 2013;40:1–12. Epub 2012/10/19.
[70] Mumford CJ, Wood NW, Kellar-Wood H, Thorpe JW, Miller DH, Compston DA. The British Isles survey of multiple sclerosis in twins. Neurology 1994;44:11–5. Epub 1994/01/01.
[71] Sawcer S. The complex genetics of multiple sclerosis: pitfalls and prospects. Brain J Neurol 2008;131:3118–31. Epub 2008/05/21.
[72] Baranzini SE, Oksenberg JR. The genetics of multiple sclerosis: from 0 to 200 in 50 years. Trends Genet 2017;33:960–70. Epub 2017/10/11.
[73] Jersild C, Svejgaard A, Fog T. HL-A antigens and multiple sclerosis. Lancet 1972;1:1240–1. Epub 1972/06/03.
[74] Naito S, Namerow N, Mickey MR, Terasaki PI. Multiple sclerosis: association with HL-A3. Tissue Antigens 1972;2:1–4. Epub 1972/01/01.
[75] Shiina T, Hosomichi K, Inoko H, Kulski JK. The HLA genomic loci map: expression, interaction, diversity and disease. J Hum Genet 2009;54:15–39. Epub 2009/01/23.
[76] Moutsianas L, Jostins L, Beecham AH, Dilthey AT, Xifara DK, Ban M, Shah TS, Patsopoulos NA, Alfredsson L, Anderson CA, Attfield KE, Baranzini SE, Barrett J, Binder TMC, Booth D, Buck D, Celius EG, Cotsapas C, D'Alfonso S, Dendrou CA, Donnelly P, Dubois B, Fontaine B, Fugger L, Goris A, Gourraud PA, Graetz C, Hemmer B, Hillert J, International IBDGC, Kockum I, Leslie S, Lill CM, Martinelli-Boneschi F, Oksenberg JR, Olsson T, Oturai A, Saarela J, Sondergaard HB, Spurkland A, Taylor B, Winkelmann J, Zipp F, Haines JL, Pericak-Vance MA, Spencer CCA, Stewart G, Hafler DA, Ivinson AJ, Harbo HF, Hauser SL, De Jager PL, Compston A, McCauley JL, Sawcer S, McVean G. Class II HLA interactions modulate genetic risk for multiple sclerosis. Nat Genet 2015;47:1107–13. Epub 2015/09/08.
[77] Hollenbach JA, Oksenberg JR. The immunogenetics of multiple sclerosis: a comprehensive review. J Autoimmun 2015;64:13–25. Epub 2015/07/05.
[78] International Multiple Sclerosis Genetics C, Wellcome Trust Case Control C, Sawcer S, Hellenthal G, Pirinen M, Spencer CC, Patsopoulos NA, Moutsianas L, Dilthey A, Su Z, Freeman C, Hunt SE, Edkins S, Gray E, Booth DR, Potter SC, Goris A, Band G, Oturai AB, Strange A, Saarela J, Bellenguez C, Fontaine B, Gillman M, Hemmer B, Gwilliam R, Zipp F, Jayakumar A, Martin R, Leslie S, Hawkins S, Giannoulatou E, D'Alfonso S, Blackburn H, Martinelli Boneschi F, Liddle J, Harbo HF, Perez ML, Spurkland A, Waller MJ, Mycko MP, Ricketts M, Comabella M, Hammond N, Kockum I, McCann OT, Ban M, Whittaker P,

Kemppinen A, Weston P, Hawkins C, Widaa S, Zajicek J, Dronov S, Robertson N, Bumpstead SJ, Barcellos LF, Ravindrarajah R, Abraham R, Alfredsson L, Ardlie K, Aubin C, Baker A, Baker K, Baranzini SE, Bergamaschi L, Bergamaschi R, Bernstein A, Berthele A, Boggild M, Bradfield JP, Brassat D, Broadley SA, Buck D, Butzkueven H, Capra R, Carroll WM, Cavalla P, Celius EG, Cepok S, Chiavacci R, Clerget-Darpoux F, Clysters K, Comi G, Cossburn M, Cournu-Rebeix I, Cox MB, Cozen W, Cree BA, Cross AH, Cusi D, Daly MJ, Davis E, de Bakker PI, Debouverie M, D'Hooghe MB, Dixon K, Dobosi R, Dubois B, Ellinghaus D, Elovaara I, Esposito F, Fontenille C, Foote S, Franke A, Galimberti D, Ghezzi A, Glessner J, Gomez R, Gout O, Graham C, Grant SF, Guerini FR, Hakonarson H, Hall P, Hamsten A, Hartung HP, Heard RN, Heath S, Hobart J, Hoshi M, Infante-Duarte C, Ingram G, Ingram W, Islam T, Jagodic M, Kabesch M, Kermode AG, Kilpatrick TJ, Kim C, Klopp N, Koivisto K, Larsson M, Lathrop M, Lechner-Scott JS, Leone MA, Leppa V, Liljedahl U, Bomfim IL, Lincoln RR, Link J, Liu J, Lorentzen AR, Lupoli S, Macciardi F, Mack T, Marriott M, Martinelli V, Mason D, McCauley JL, Mentch F, Mero IL, Mihalova T, Montalban X, Mottershead J, Myhr KM, Naldi P, Ollier W, Page A, Palotie A, Pelletier J, Piccio L, Pickersgill T, Piehl F, Pobywajlo S, Quach HL, Ramsay PP, Reunanen M, Reynolds R, Rioux JD, Rodegher M, Roesner S, Rubio JP, Ruckert IM, Salvetti M, Salvi E, Santaniello A, Schaefer CA, Schreiber S, Schulze C, Scott RJ, Sellebjerg F, Selmaj KW, Sexton D, Shen L, Simms-Acuna B, Skidmore S, Sleiman PM, Smestad C, Sorensen PS, Sondergaard HB, Stankovich J, Strange RC, Sulonen AM, Sundqvist E, Syvanen AC, Taddeo F, Taylor B, Blackwell JM, Tienari P, Bramon E, Tourbah A, Brown MA, Tronczynska E, Casas JP, Tubridy N, Corvin A, Vickery J, Jankowski J, Villoslada P, Markus HS, Wang K, Mathew CG, Wason J, Palmer CN, Wichmann HE, Plomin R, Willoughby E, Rautanen A, Winkelmann J, Wittig M, Trembath RC, Yaouanq J, Viswanathan AC, Zhang H, Wood NW, Zuvich R, Deloukas P, Langford C, Duncanson A, Oksenberg JR, Pericak-Vance MA, Haines JL, Olsson T, Hillert J, Ivinson AJ, De Jager PL, Peltonen L, Stewart GJ, Hafler DA, Hauser SL, McVean G, Donnelly P, Compston A. Genetic risk and a primary role for cell-mediated immune mechanisms in multiple sclerosis. Nature 2011;476:214–9. Epub 2011/08/13.

[79] Kingwell K. Multiple sclerosis: surveying the genetic architecture of MS. Nat Rev Neurol 2011;7:535. Epub 2011/09/21.

[80] Cree BA. Genetics of primary progressive multiple sclerosis. Handb Clin Neurol 2014;122:211–30. Epub 2014/02/11.

[81] Bush WS, McCauley JL, DeJager PL, Dudek SM, Hafler DA, Gibson RA, Matthews PM, Kappos L, Naegelin Y, Polman CH, Hauser SL, Oksenberg J, Haines JL, Ritchie MD. International Multiple Sclerosis Genetics C. A knowledge-driven interaction analysis reveals potential neurodegenerative mechanism of multiple sclerosis susceptibility. Genes Immun 2011;12:335–40. Epub 2011/02/25.

[82] Kendler KS, Neale MC. Endophenotype: a conceptual analysis. Mol Psychiatry 2010;15:789–97. Epub 2010/02/10.

[83] International Multiple Sclerosis Genetics C. Genome-wide association study of severity in multiple sclerosis. Genes Immun 2011;12:615–25. Epub 2011/06/10.

[84] Ebers GC. Environmental factors and multiple sclerosis. Lancet Neurol 2008;7:268–77. Epub 2008/02/16.

[85] Hill AB. The environment and disease: association or causation? Proc R Soc Med 1965;58:295–300. Epub 1965/05/01.

[86] Olsson T, Barcellos LF, Alfredsson L. Interactions between genetic, lifestyle and environmental risk factors for multiple sclerosis. Nat Rev Neurol 2017;13:25–36. Epub 2016/12/10.

[87] Belbasis L, Bellou V, Evangelou E, Ioannidis JP, Tzoulaki I. Environmental risk factors and multiple sclerosis: an umbrella review of systematic reviews and meta-analyses. Lancet Neurol 2015;14:263–73. Epub 2015/02/11.

[88] Ascherio A, Munger KL, Simon KC. Vitamin D and multiple sclerosis. Lancet Neurol 2010;9:599–612. Epub 2010/05/25.

[89] Goldberg P. Multiple sclerosis: vitamin D and calcium as environmental determinants of prevalence: (A viewpoint) part 1: sunlight, dietary factors and epidemiology. Int J Environ Stud 1974;6:19–27.

[90] Jelinek GA, Marck CH, Weiland TJ, Pereira N, van der Meer DM, Hadgkiss EJ. Latitude, sun exposure and vitamin D supplementation: associations with quality of life and disease outcomes in a large international cohort of people with multiple sclerosis. BMC Neurol 2015;15:132. Epub 2015/08/06.

[91] Tremlett H, Zhu F, Ascherio A, Munger KL. Sun exposure over the life course and associations with multiple sclerosis. Neurology 2018;90:e1191–9. Epub 2018/03/09.

[92] Handel AE, Giovannoni G, Ebers GC, Ramagopalan SV. Environmental factors and their timing in adult-onset multiple sclerosis. Nat Rev Neurol 2010;6:156–66. Epub 2010/02/17.

[93] Mokry LE, Ross S, Ahmad OS, Forgetta V, Smith GD, Goltzman D, Leong A, Greenwood CM, Thanassoulis G, Richards JB. Vitamin D and risk of multiple sclerosis: a Mendelian randomization study. PLoS Med 2015;12:e1001866. Epub 2015/08/26.

[94] Gianfrancesco MA, Stridh P, Rhead B, Shao X, Xu E, Graves JS, Chitnis T, Waldman A, Lotze T, Schreiner T, Belman A, Greenberg B, Weinstock-Guttman B, Aaen G, Tillema JM, Hart J, Caillier S, Ness J, Harris Y, Rubin J, Candee M, Krupp L, Gorman M, Benson L, Rodriguez M, Mar S, Kahn I, Rose J, Roalstad S, Casper TC, Shen L, Quach H, Quach D, Hillert J, Baarnhielm M, Hedstrom A, Olsson T, Kockum I, Alfredsson L, Metayer C, Schaefer C, Barcellos LF, Waubant E. Network of Pediatric Multiple Sclerosis C. Evidence for a causal relationship between low vitamin D, high BMI, and pediatric-onset MS. Neurology 2017;88:1623–9. Epub 2017/03/31.

[95] Kamen DL, Tangpricha V. Vitamin D and molecular actions on the immune system: modulation of innate and autoimmunity. J Mol Med Berl 2010;88:441–50. Epub 2010/02/02.

[96] Correale J, Ysrraelit MC, Gaitan MI. Vitamin D-mediated immune regulation in multiple sclerosis. J Neurol Sci 2011;311:23–31. Epub 2011/07/05.

[97] Correale J, Ysrraelit MC, Gaitan MI. Immunomodulatory effects of Vitamin D in multiple sclerosis. Brain J Neurol 2009;132:1146–60. Epub 2009/03/27.

[98] Ganesh A, Apel S, Metz L, Patten S. The case for vitamin D supplementation in multiple sclerosis. Mult Scler Relat Disord 2013;2:281–306. Epub 2013/10/01.

[99] Jagannath VA, Fedorowicz Z, Asokan GV, Robak EW, Whamond L. Vitamin D for the management of multiple sclerosis. Cochrane Database Syst Rev 2010:CD008422. Epub 2010/12/15.

[100] Zheng C, He L, Liu L, Zhu J, Jin T. The efficacy of vitamin D in multiple sclerosis: a meta-analysis. Mult Scler Relat Disord 2018;23:56–61. Epub 2018/05/20.

[101] Smolders J, Menheere P, Kessels A, Damoiseaux J, Hupperts R. Association of vitamin D metabolite levels with relapse rate and disability in multiple sclerosis. Mult Scler 2008;14:1220–4. Epub 2008/07/26.

[102] Muris AH, Rolf L, Broen K, Hupperts R, Damoiseaux J, Smolders J. A low vitamin D status at diagnosis is associated with an early conversion to secondary progressive multiple sclerosis. J Steroid Biochem Mol Biol 2016;164:254–7. Epub 2015/11/26.

[103] Hernan MA, Olek MJ, Ascherio A. Cigarette smoking and incidence of multiple sclerosis. Am J Epidemiol 2001;154:69–74. Epub 2001/06/28.

[104] Riise T, Nortvedt MW, Ascherio A. Smoking is a risk factor for multiple sclerosis. Neurology 2003;61:1122–4. Epub 2003/10/29.

[105] Petersen ER, Oturai AB, Koch-Henriksen N, Magyari M, Sorensen PS, Sellebjerg F, Sondergaard HB. Smoking affects the interferon beta treatment response in multiple sclerosis. Neurology 2018;90:e593–600. Epub 2018/01/19.

[106] Hernan MA, Takkouche B, Caamano-Isorna F, Gestal-Otero JJ. A meta-analysis of coffee drinking, cigarette smoking, and the risk of Parkinson's disease. Ann Neurol 2002;52:276–84.

[107] Ascherio A, Munger KL. Environmental risk factors for multiple sclerosis. Part I: the role of infection. Ann Neurol 2007;61:288–99. Epub 2007/04/21.

[108] Kurtzke JF. Epidemiologic evidence for multiple sclerosis as an infection. Clin Microbiol Rev 1993;6:382–427. Epub 1993/10/01.

[109] Marie DP. In: Pierre Marie M, Delahaye A, Lecrosnier E, editors. Sclérose en plaques et maladies infectieuses. 1884.

[110] Sibley WA, Bamford CR, Clark K. Clinical viral infections and multiple sclerosis. Lancet 1985;1:1313–5. Epub 1985/06/08.

[111] Panitch HS. Influence of infection on exacerbations of multiple sclerosis. Ann Neurol 1994;36(Suppl.): S25–8. Epub 1994/01/01.

[112] Lucas RM, Hughes AM, Lay ML, Ponsonby AL, Dwyer DE, Taylor BV, Pender MP. Epstein-Barr virus and multiple sclerosis. J Neurol Neurosurg Psychiatry 2011;82:1142–8. Epub 2011/08/13.

[113] Thorley-Lawson DA, Hawkins JB, Tracy SI, Shapiro M. The pathogenesis of Epstein-Barr virus persistent infection. Curr Opin Virol 2013;3:227–32. Epub 2013/05/21.

[114] Taylor GS, Long HM, Brooks JM, Rickinson AB, Hislop AD. The immunology of Epstein-Barr virus-induced disease. Annu Rev Immunol 2015;33:787–821. Epub 2015/02/24.

[115] Draborg AH, Duus K, Houen G. Epstein-Barr virus in systemic autoimmune diseases. Clin Dev Immunol 2013;2013:535738. Epub 2013/09/26.

[116] Pakpoor J, Giovannoni G, Ramagopalan SV. Epstein-Barr virus and multiple sclerosis: association or causation? Expert Rev Neurother 2013;13:287–97. Epub 2013/03/02.

[117] Mahad DH, Trapp BD, Lassmann H. Pathological mechanisms in progressive multiple sclerosis. Lancet Neurol 2015;14:183–93. Epub 2015/03/17.

[118] Constantinescu CS, Farooqi N, O'Brien K, Gran B. Experimental autoimmune encephalomyelitis (EAE) as a model for multiple sclerosis (MS). Br J Pharmacol 2011;164:1079–106. Epub 2011/03/05.

[119] Rivers TM, Schwentker FF. Encephalomyelitis accompanied by myelin destruction experimentally produced in monkeys. J Exp Med 1935;61:689–702. Epub 1935/04/30.

[120] Han H, Myllykoski M, Ruskamo S, Wang C, Kursula P. Myelin-specific proteins: a structurally diverse group of membrane-interacting molecules. Biofactors 2013;39:233–41. Epub 2013/06/20.

[121] Wucherpfennig KW, Hafler DA. A review of T-cell receptors in multiple sclerosis: clonal expansion and persistence of human T-cells specific for an immunodominant myelin basic protein peptide. Ann NY Acad Sci 1995;756:241–58. Epub 1995/07/07.

[122] Goverman J, Woods A, Larson L, Weiner LP, Hood L, Zaller DM. Transgenic mice that express a myelin basic protein-specific T cell receptor develop spontaneous autoimmunity. Cell 1993;72:551–60. Epub 1993/02/26.

[123] Keller MP, Chance PF. Inherited neuropathies: from gene to disease. Brain Pathol 1999;9:327–41. Epub 1999/04/29.

[124] Sriram S, Steiner I. Experimental allergic encephalomyelitis: a misleading model of multiple sclerosis. Ann Neurol 2005;58:939–45. Epub 2005/11/30.

[125] Behan PO, Chaudhuri A. EAE is not a useful model for demyelinating disease. Mult Scler Relat Disord 2014;3:565–74. Epub 2015/08/13.

[126] Lassmann H. Acute disseminated encephalomyelitis and multiple sclerosis. Brain J Neurol 2010;133:317–9. Epub 2010/02/05.

[127] Steinman L, Zamvil SS. Virtues and pitfalls of EAE for the development of therapies for multiple sclerosis. Trends Immunol 2005;26:565–71. Epub 2005/09/13.

[128] Miller DH, Barkhof F, Frank JA, Parker GJ, Thompson AJ. Measurement of atrophy in multiple sclerosis: pathological basis, methodological aspects and clinical relevance. Brain J Neurol 2002;125:1676–95. Epub 2002/07/24.

[129] Frischer JM, Bramow S, Dal-Bianco A, Lucchinetti CF, Rauschka H, Schmidbauer M, Laursen H, Sorensen PS, Lassmann H. The relation between inflammation and neurodegeneration in multiple sclerosis brains. Brain J Neurol 2009;132:1175–89. Epub 2009/04/03.

[130] Ellwardt E, Zipp F. Molecular mechanisms linking neuroinflammation and neurodegeneration in MS. Exp Neurol 2014;262(Pt A):8–17. Epub 2014/02/18.

[131] Glass CK, Saijo K, Winner B, Marchetto MC, Gage FH. Mechanisms underlying inflammation in neurodegeneration. Cell 2010;140:918–34. Epub 2010/03/23.

[132] Lassmann H, van Horssen J. The molecular basis of neurodegeneration in multiple sclerosis. FEBS Lett 2011;585:3715–23. Epub 2011/08/23.

[133] Chard DT, Griffin CM, Parker GJ, Kapoor R, Thompson AJ, Miller DH. Brain atrophy in clinically early relapsing-remitting multiple sclerosis. Brain J Neurol 2002;125:327–37. Epub 2002/02/15.

[134] Haider L, Fischer MT, Frischer JM, Bauer J, Hoftberger R, Botond G, Esterbauer H, Binder CJ, Witztum JL, Lassmann H. Oxidative damage in multiple sclerosis lesions. Brain J Neurol 2011;134:1914–24. Epub 2011/06/10.

[135] Andersen JK. Oxidative stress in neurodegeneration: cause or consequence? Nat Med 2004;10(Suppl.):S18–25. Epub 2004/08/10.

[136] Pitt D, Werner P, Raine CS. Glutamate excitotoxicity in a model of multiple sclerosis. Nat Med 2000;6:67–70. Epub 1999/12/29.

[137] Hynd MR, Scott HL, Dodd PR. Glutamate-mediated excitotoxicity and neurodegeneration in Alzheimer's disease. Neurochem Int 2004;45:583–95. Epub 2004/07/06.

[138] Schapira AH. Mitochondria in the aetiology and pathogenesis of Parkinson's disease. Lancet Neurol 2008;7:97–109.

[139] Dutta R, McDonough J, Yin X, Peterson J, Chang A, Torres T, Gudz T, Macklin WB, Lewis DA, Fox RJ, Rudick R, Mirnics K, Trapp BD. Mitochondrial dysfunction as a cause of axonal degeneration in multiple sclerosis patients. Ann Neurol 2006;59:478–89. Epub 2006/01/05.

[140] Mahad D, Ziabreva I, Lassmann H, Turnbull D. Mitochondrial defects in acute multiple sclerosis lesions. Brain J Neurol 2008;131:1722–35. Epub 2008/06/03.

[141] Lee Y, Morrison BM, Li Y, Lengacher S, Farah MH, Hoffman PN, Liu Y, Tsingalia A, Jin L, Zhang PW, Pellerin L, Magistretti PJ, Rothstein JD. Oligodendroglia metabolically support axons and contribute to neurodegeneration. Nature 2012;487:443–8. Epub 2012/07/18.

[142] Trapp BD, Stys PK. Virtual hypoxia and chronic necrosis of demyelinated axons in multiple sclerosis. Lancet Neurol 2009;8:280–91. Epub 2009/02/24.

[143] Irvine GB, El-Agnaf OM, Shankar GM, Walsh DM. Protein aggregation in the brain: the molecular basis for Alzheimer's and Parkinson's diseases. Mol Med 2008;14:451–64. Epub 2008/03/28.

[144] Jensen HB, Ravnborg M, Dalgas U, Stenager E. 4-Aminopyridine for symptomatic treatment of multiple sclerosis: a systematic review. Ther Adv Neurol Disord 2014;7:97–113. Epub 2014/03/04.

[145] Toosy A, Ciccarelli O, Thompson A. Symptomatic treatment and management of multiple sclerosis. Handb Clin Neurol 2014;122:513–62. Epub 2014/02/11.

[146] Filippini G, Brusaferri F, Sibley WA, Citterio A, Ciucci G, Midgard R, Candelise L. Corticosteroids or ACTH for acute exacerbations in multiple sclerosis. Cochrane Database Syst Rev 2000:CD001331. Epub 2000/10/18.

[147] Burton JM, O'Connor PW, Hohol M, Beyene J. Oral versus intravenous steroids for treatment of relapses in multiple sclerosis. Cochrane Database Syst Rev 2012;12:CD006921. Epub 2012/12/14.

[148] Wingerchuk DM, Weinshenker BG. Disease modifying therapies for relapsing multiple sclerosis. BMJ 2016;354:i3518. Epub 2016/08/24.

[149] Rae-Grant A, Day GS, Marrie RA, Rabinstein A, Cree BAC, Gronseth GS, Haboubi M, Halper J, Hosey JP, Jones DE, Lisak R, Pelletier D, Potrebic S, Sitcov C, Sommers R, Stachowiak J, Getchius TSD, Merillat SA, Pringsheim T. Comprehensive systematic review summary: disease-modifying therapies for adults with multiple sclerosis: report of the guideline development, dissemination, and implementation subcommittee of the American academy of neurology. Neurology 2018;90:789–800. Epub 2018/04/25.

[150] Rotstein DL, Healy BC, Malik MT, Chitnis T, Weiner HL. Evaluation of no evidence of disease activity in a 7-year longitudinal multiple sclerosis cohort. JAMA Neurol 2015;72:152–8. Epub 2014/12/23.

[151] Rudick RA, Goelz SE. Beta-interferon for multiple sclerosis. Exp Cell Res 2011;317:1301–11. Epub 2011/03/15.

[152] Scott LJ. Glatiramer acetate: a review of its use in patients with relapsing-remitting multiple sclerosis and in delaying the onset of clinically definite multiple sclerosis. CNS Drugs 2013;27:971–88. Epub 2013/10/17.
[153] Bomprezzi R. Dimethyl fumarate in the treatment of relapsing-remitting multiple sclerosis: an overview. Ther Adv Neurol Disord 2015;8:20–30. Epub 2015/01/15.
[154] Misbah SA. Progressive multi-focal leucoencephalopathy - driven from rarity to clinical mainstream by iatrogenic immunodeficiency. Clin Exp Immunol 2017;188:342–52. Epub 2017/03/01.
[155] Derfuss T, Ontaneda D, Nicholas J, Meng X, Hawker K. Relapse rates in patients with multiple sclerosis treated with fingolimod: subgroup analyses of pooled data from three phase 3 trials. Mult Scler Relat Disord 2016;8:124–30. Epub 2016/07/28.
[156] Confavreux C, O'Connor P, Comi G, Freedman MS, Miller AE, Olsson TP, Wolinsky JS, Bagulho T, Delhay JL, Dukovic D, Truffinet P, Kappos L, Group TT. Oral teriflunomide for patients with relapsing multiple sclerosis (TOWER): a randomised, double-blind, placebo-controlled, phase 3 trial. Lancet Neurol 2014;13:247–56. Epub 2014/01/28.
[157] Derfuss T, Kuhle J, Lindberg R, Kappos L. Natalizumab therapy for multiple sclerosis. Semin Neurol 2013;33:26–36. Epub 2013/05/28.
[158] Havrdova E, Horakova D, Kovarova I. Alemtuzumab in the treatment of multiple sclerosis: key clinical trial results and considerations for use. Ther Adv Neurol Disord 2015;8:31–45. Epub 2015/01/15.
[159] Stahnke AM, Holt KM. Ocrelizumab: a new B-cell therapy for relapsing remitting and primary progressive multiple sclerosis. Ann Pharmacother 2018;52:473–83. Epub 2017/12/14.
[160] Holmoy T, Torkildsen O, Myhr KM. An update on cladribine for relapsing-remitting multiple sclerosis. Expert Opin Pharmacother 2017;18:1627–35. Epub 2017/09/01.
[161] Montalban X, Hauser SL, Kappos L, Arnold DL, Bar-Or A, Comi G, de Seze J, Giovannoni G, Hartung HP, Hemmer B, Lublin F, Rammohan KW, Selmaj K, Traboulsee A, Sauter A, Masterman D, Fontoura P, Belachew S, Garren H, Mairon N, Chin P, Wolinsky JS, Investigators OC. Ocrelizumab versus placebo in primary progressive multiple sclerosis. N Engl J Med 2017;376:209–20. Epub 2016/12/22.
[162] Chataway J, Schuerer N, Alsanousi A, Chan D, MacManus D, Hunter K, Anderson V, Bangham CR, Clegg S, Nielsen C, Fox NC, Wilkie D, Nicholas JM, Calder VL, Greenwood J, Frost C, Nicholas R. Effect of high-dose simvastatin on brain atrophy and disability in secondary progressive multiple sclerosis (MS-STAT): a randomised, placebo-controlled, phase 2 trial. Lancet 2014;383:2213–21. Epub 2014/03/25.
[163] Winkelmann A, Loebermann M, Reisinger EC, Hartung HP, Zettl UK. Disease-modifying therapies and infectious risks in multiple sclerosis. Nat Rev Neurol 2016;12:217–33. Epub 2016/03/05.
[164] Narayan RN, Forsthuber T, Stuve O. Emerging drugs for primary progressive multiple sclerosis. Expert Opin Emerg Drugs 2018;23:97–110. Epub 2018/04/12.
[165] Franklin RJ, ffrench-Constant C, Edgar JM, Smith KJ. Neuroprotection and repair in multiple sclerosis. Nat Rev Neurol 2012;8:624–34. Epub 2012/10/03.
[166] De Santi L, Annunziata P, Sessa E, Bramanti P. Brain-derived neurotrophic factor and TrkB receptor in experimental autoimmune encephalomyelitis and multiple sclerosis. J Neurol Sci 2009;287:17–26. Epub 2009/09/18.
[167] Allen SJ, Watson JJ, Shoemark DK, Barua NU, Patel NK. GDNF, NGF and BDNF as therapeutic options for neurodegeneration. Pharmacol Ther 2013;138:155–75. Epub 2013/01/26.
[168] Stangel M. Remyelinating and neuroprotective treatments in multiple sclerosis. Expert Opin Investig Drugs 2004;13:331–47. Epub 2004/04/23.
[169] Sormani MP, Muraro PA, Schiavetti I, Signori A, Laroni A, Saccardi R, Mancardi GL. Autologous hematopoietic stem cell transplantation in multiple sclerosis: a meta-analysis. Neurology 2017;88:2115–22. Epub 2017/04/30.
[170] Rush CA, Atkins HL, Freedman MS. Autologous hematopoietic stem cell transplantation in the treatment of multiple sclerosis. Cold Spring Harb Perspect Med 2018. Epub 2018/04/04.

[171] Haghikia A, Hohlfeld R, Gold R, Fugger L. Therapies for multiple sclerosis: translational achievements and outstanding needs. Trends Mol Med 2013;19:309–19. Epub 2013/04/16.

[172] Ramagopalan SV, Dobson R, Meier UC, Giovannoni G. Multiple sclerosis: risk factors, prodromes, and potential causal pathways. Lancet Neurol 2010;9:727–39. Epub 2010/07/09.

[173] Comabella M, Sastre-Garriga J, Montalban X. Precision medicine in multiple sclerosis: biomarkers for diagnosis, prognosis, and treatment response. Curr Opin Neurol 2016;29:254–62. Epub 2016/04/15.

[174] Kalincik T, Manouchehrinia A, Sobisek L, Jokubaitis V, Spelman T, Horakova D, Havrdova E, Trojano M, Izquierdo G, Lugaresi A, Girard M, Prat A, Duquette P, Grammond P, Sola P, Hupperts R, Grand'Maison F, Pucci E, Boz C, Alroughani R, Van Pesch V, Lechner-Scott J, Terzi M, Bergamaschi R, Iuliano G, Granella F, Spitaleri D, Shaygannejad V, Oreja-Guevara C, Slee M, Ampapa R, Verheul F, McCombe P, Olascoaga J, Amato MP, Vucic S, Hodgkinson S, Ramo-Tello C, Flechter S, Cristiano E, Rozsa C, Moore F, Luis Sanchez-Menoyo J, Laura Saladino M, Barnett M, Hillert J, Butzkueven H, Group MSS. Towards personalized therapy for multiple sclerosis: prediction of individual treatment response. Brain J Neurol 2017;140:2426–43. Epub 2017/10/21.

[175] Ontaneda D, Fox RJ, Chataway J. Clinical trials in progressive multiple sclerosis: lessons learned and future perspectives. Lancet Neurol 2015;14:208–23. Epub 2015/03/17.

[176] Conway D, Cohen JA. Combination therapy in multiple sclerosis. Lancet Neurol 2010;9:299–308. Epub 2010/02/23.

Index

Note: 'Page numbers followed by "f" indicate figures, "t" indicate tables, "b" indicate boxes.'

O

P

R

S

Printed in the United States
By Bookmasters